# Advances in Asian Human-Environmental Research



***Aims and Scope***

The series aims at fostering the discussion on the complex relationships between physical landscapes, natural resources, and their modification by human land use in various environments of Asia. It is widely acknowledged that human-environment-interactions become increasingly important in area studies and development research, taking into account regional differences as well as bio-physical, socio-economic and cultural particularities.

The book series seeks to explore theoretic and conceptual reflection on dynamic human-environment systems applying advanced methodology and innovative research perspectives. The main themes of the series cover urban and rural landscapes in Asia. Examples include topics such as land and forest degradation, glaciers in Asia, mountain environments, dams in Asia, medical geography, vulnerability and mitigation strategies, natural hazards and risk management concepts, environmental change, impacts studies and consequences for local communities. The relevant themes of the series are mainly focused on geographical research perspectives of area studies, however there is scope for interdisciplinary contributions.

For further volumes:
http://www.springer.com/series/8560

Bualtar Glacier descends northwards from Minapin Peak (7,266 m) in the Rakaposhi Range, a fall of some 4,965 m in 22 km to its snout. The extreme steepness and elevation ranges of larger Karakoram glacier basins are indicated, and a sense of the great vertical changes in conditions from the debris covered ice of lower tongue to precipitous, avalanched walls at the head

Kenneth Hewitt

# Glaciers of the Karakoram Himalaya

## Glacial Environments, Processes, Hazards and Resources

Springer

Kenneth Hewitt
Department of Geography and Environmental Studies
Wilfrid Laurier University
Waterloo, ON, Canada

ISSN 1879-7180 ISSN 1879-7199 (electronic)
ISBN 978-94-007-6310-4 ISBN 978-94-007-6311-1 (eBook)
DOI 10.1007/978-94-007-6311-1
Springer Dordrecht Heidelberg New York London

Library of Congress Control Number: 2013942813



*Cover image*: Nomads near Nanga Parbat, 1995. Copyright © Marcus Nüsser (used with permission)

Printed on acid-free paper

Springer is part of Springer Science+Business Media (www.springer.com)

*To Farida*

# Preface

This book mainly concerns the present-day glaciers of the Karakoram Himalaya: the conditions that sustain them, the landscapes they have helped to shape and broader environmental, water resources and hazards issues associated with them. The core region involves about 16,500 $km^2$ of perennial snow and ice in the headwaters of the Indus and about 4,000 $km^2$ in the Yarkand drainage. In an otherwise extreme continental, arid region, the glaciers comprise large stores of freshwater. Meltwaters from glacier basins make up between 30 % and 40 % of the average annual flows of these rivers where they leave the Karakoram and dominate their discharges for 2–3 months of summer. Seasonal snowmelt makes up most of the balance. Yields from snowmelt seem, on an average, to be somewhat greater than those from glaciers but may be less in some years. By contrast, glaciers are relatively minor contributors and snowmelt much more important for tributaries of the indus draining the southern flank of the Greater Himalaya. Rainfall is the largest contributor, as it is in most years for the Indus basin as a whole. The upper Indus and Yarkand are among the few large river basins in South Asia where rainfall is not the dominant source of stream flows. It would be difficult to overemphasise the unique ecological and human significance of the glaciers and snowfields in the two basins. They will become more critical with anticipated climate change and economic developments.

Water demands and shortages in surrounding, populous lowlands are placing ever-greater pressures on mountain resources or the desire to exploit them. A sense of urgency arises from national and trans-boundary development plans for water and power. There are existing conflicts in the high mountain areas that could be aggravated, and new ones may arise. More hopefully, rational approaches based on mutual benefits could bring improved relations and greater security in the high mountains.

At the time of writing, a commonly expressed concern is how climate change may greatly reduce the glaciers and water supplies from them. The concern is justified, although some misleading or exaggerated accounts and the talk of 'disappearing glaciers' have confused the issue. In many parts of High Asia, glaciers have undergone large retreats and loss of mass in recent decades. However, changes turn out to have been less rapid than widely suggested, quite variable in different regions and, to date, substantially different in the Karakoram (Hewitt 2005; Raina 2009; Scherler et al. 2011). There were appreciable losses of glacier mass through much of the twentieth century. Since the 1960s, they have been relatively small and inconsistent. In some cases, glacier advances are causing problems. The most likely explanation is that increased warming is being compensated in these high mountains by increased snowfall and summer cloudiness. As global warming intensifies, that may, or may not, continue. For the moment, the more worrying responses concern the timing or greater unpredictability of river flows, and environmental extremes.

From a scientific perspective, a balanced assessment of snow, ice and glacial environments in the region, and just how changes have occurred, is constrained by patchy and limited research (Kaul 1999; Hagg et al. 2009; Raina 2009; Armstrong 2010; Shroder and Bishop 2010). Although it is widely stated that the Karakoram glaciers are vital for water supply and at great risk from climate change, there is almost no continuous or widespread monitoring of them. Indeed, no adequate account is available of what is already known about them. It is over 75 years since the last book-length treatment was published. Moreover, this last major study has never been translated into English (Visser and Visser-Hooft 1935–38). The most extensive overview from a decade earlier is still only available in Italian (Dainelli 1924–35); the exceptional study of High Asian snow and ice by von Wissmann (1959) is available only in German. There have certainly been some valuable, more narrowly focused papers and overview chapters in more recent publications (Mercer 1975; Goudie et al. 1984; Haserodt 1984; Kuhle 2004; Hewitt 2006, 2011; Smiraglia et al. 2008; Shroder and Bishop 2010; Shroder 2011). Constraints of space, interests and language limit their coverage.

The Karakoram has a fairly long history of modern investigations, more so perhaps than newcomers to the topic might imagine. Scientific studies go back almost 200 years (LIGC 1984; Allen 1995). A number of large expeditions and scientists from several countries brought great advances in knowledge (see Chap. 1). Their work is an essential resource, although its limitations must be noted. As a whole, coverage is patchy – discontinuous in time and space. Much of the work pre-dates the development of some basic glaciological concepts and, as noted, may not be available in English or any other but the original language. Rugged terrain and harsh climates have limited scientific work in much of the region, and security issues often prevent it. Few investigations have gone beyond the lower parts of the glaciers and rarely above 4,000 m.[1] Yet, 80 % of glacier basin areas and all the sources of glacier ice occur above this elevation. It makes the few studies at higher elevations that much more important, although these too are biased towards a few valleys and their larger glaciers. Most of the region has received little or no attention, including some of the largest glaciers, but especially the lesser ranges where smaller glaciers prevail. 'Small' in this context is taken to mean less than 15 km in length. Collectively,

[1]Measurements of elevations above sea level will be written with a comma separating 100s from 1000s (e.g. 3,000 rather than 3000). This will distinguish elevations, from elevation spans, length or relative height measurements which, unless over 10,000 units, will not include commas.

however, these glaciers alone comprise a greater area than, say, the 2,896 $km^2$ of glaciers in the European Alps (Haeberli 1998).

Until recently, information about the glaciers was almost entirely a product of expeditions coming from outside the region and, usually, a secondary concern compared to mountaineering or military and commercial intelligence. Most were more or less short term, logistically and spatially limited, usually for a few weeks or months of summer. Those making observations in winter can be counted on the fingers of one hand and for most of the region are non-existent. Members have tended to be new to the region and visited only once, rarely more than two or three times, with some important exceptions (Dainelli 1959; Hewitt 1989, 2007). A few major expeditions by Italian, German, Austrian and Japanese teams were extensive in space and time and made detailed investigations of the glaciers. Many of their results, however, are only available in brief summaries and short papers in English. The more numerous and oft-quoted English sources are mainly from before 1947, when the region was part of the British Empire.

Nevertheless, there are some great advantages of writing this study today: the richness of the literature on glacial environments and processes worldwide and ever-improving coverage of the Karakoram region by satellite imagery. Invaluable background is provided by a number of substantial texts and works of reference on glaciers (Drewry 1986; Hambrey 1994; Paterson 1994; Menzies 1995; Benn and Evans 1998; Singh et al. 2011). The present work benefits hugely from these sources. It should be added, however, that Karakoram glaciers have only rare and minor appearances in any of them, usually in terms of certain exceptional or hazardous phenomena such as glacial lake outburst floods and surge-type glaciers.

In the past decade or so, more frequent and higher-definition satellite imagery has transformed spatial and temporal coverage of the region (Williams and Ferrigno 2010; Bishop and Colby 2011). New monitoring and analytical possibilities are opening up all the time. They offer opportunities for more representative, region-wide and all-season characterisations. An attempt is made here to take advantage of this, especially in relation to the higher, more rugged parts of glacier basins and the least-visited areas. Nevertheless, the dangers of limited or no ground control, or lack of experience in areas of interest, are constant problems. Not only has most information about Karakoram glaciers themselves come from ground-based observation and phenomena, so too has the development of the basic concepts of mountain environments and glaciology. Fully effective translation between them and remotely sensed information and analyses is a work in progress.

Problems can also arise because the most intensively studied mountain glaciers occur in other, very different environments – typically, midlatitude and subpolar regions. They tend to be relatively small ice masses, generally at lower elevations and of much less elevation range than Karakoram glaciers. This applies to most reference or 'benchmark' glaciers so fundamental for tracking and comparing mountain ice globally (Oerlemanns 2001; Haeberli et al. 1998; Bolch 2011). The Karakoram is in an extreme continental, subtropical location. More than half of all its glacier ice occurs in 15 basins, in glaciers more than 40 km long and with the highest areas well over 7,000 m. Many glaciers have elevation ranges of more than 4,000 m, some over

5,000 m. As a result, sets of distinctive conditions must be addressed that are of minor significance elsewhere and absent from most of the glaciers literature and research. It is important to have some of these in mind from the beginning.

In the Karakoram, it will be shown that the greater fraction of all glacier ice is input to the glaciers by avalanches rather than direct snowfall. Yet, while there has been a great deal of research into avalanches, it has largely ignored avalanche-fed glaciers, let alone valley glaciers that can be more than 50 km in length. Rock walls too steep to support seasonal or perennial snow build-up account for over 60 % of most glacier basins, more than 80 % in many. These are the main source areas of Karakoram glaciers. In addition, wind action and redistribution of snow have major influences. They probably affected the larger part of all the snow mass that is eventually incorporated into glaciers. The greater part of the vertical descent of most of the glacier ice itself is in icefalls. Much or all of the ice in the most widely observed and studied main and lower ice streams has passed through one or more icefalls. This will be shown to open up a range of situations that can affect or modify the properties and behaviour of the ice. Such uniquely 'Himalayan' conditions are frequently remarked upon but only rarely investigated, even in research on Himalayan glaciers (Yafeng and Wenying 1980; Benn et al. 2003; Cogley 2011; Hewitt 2011). They have to be addressed here, despite the lack of well-developed empirical and conceptual work.

This book can hardly escape the constraints of past work, whether in the studies on which discussion must be based or the author's experience and limitations. The strategy adopted is, however, partly chosen so as to deal with this: a regional approach paying particular attention to the terrain and environmental conditions in and around glacier basins and, to the extent possible, types, forms and features of the ice masses. Although essentially descriptive, terrain classification and distributions offer a more representative sense of the region's glacierised areas and environments and a balanced way to explore where further research is most needed.

The approach is described in more detail at the end of Chap. 1. Chapters 2, 3, 4, 5, 6 and 7 focus on present-day glacier basins, ice masses and their maintenance. Glacial landforms and earth surface processes related to them are considered in Chap. 8. In Chaps. 9 and 10, some more extreme, short-lived glacial phenomena and hazards to human inhabitants are looked at, in particular surge-type glaciers, glacial impoundments and outburst floods. Chapter 11 draws attention to the huge numbers and great diversity of rock glaciers in the region, a part of the Karakoram cryosphere largely neglected in the past. Questions of recent glacier change are elaborated in Chap. 12. A concluding chapter looks at issues relating to people, glaciers and, especially, regional water supply.

## References

Allen N Jr (1995) Karakorum Himalaya: Sourcebook for a protected area. IUCN The World Conservation Union, Pakistan

Armstrong RL (2010) The Glaciers of the Himalayan-Hindu-Kush region. ICIMOD, Technical paper, Kathmandu

Benn DI, Evans DJA (1998) Glaciers and glaciation. Arnold, London
Benn DI, Kirkbride MP, Owen LA, Brazier V (2003) Glaciated valley landsystems. In: Evans DJA (ed) Glacial landsystems. Hodder Arnold, London, pp 370–406
Bishop MP, Colby JD (2011) Topographic normalization of multispectral satellite imagery, In: Singh et al. (eds), pp 1187–1197
Bolch T (2011) Benchmark glacier. In: Singh et al. (eds), pp 95–98
Cogley JG (2011) Himalayan glaciers in 2010 and 2035, In: Singh et al. (eds), pp 520–525
Dainelli G (1924–35) Relazioni Scientifiche della Spedizione Italiana De Filippi nell'Himalaia, Caracorum e Turchestan Chinese (1913–1914). Series II, 10 vols. Zanichelli, Bologna
Dainelli G (1959) Esploratori e alpinisti nel Caracorum. Unione Tipografico-editrice Torinese, Turin
Drewry D (1986) Glacial geologic processes, Edward Arnold, London
Goudie AS, Jones DK, Brunsden D (1984) Recent fluctuations in some glaciers of the western Karakoram Mountains, Pakistan. In: Miller KJ (ed) The International Karakoram Project. Cambridge University Press, Cambridge, pp 411–455
Haeberli W (1998) Historical evolution and operational aspects of worldwide glacier monitoring. In: Haeberli W, Hoelzle M, Suter S (eds), pp 35–51
Haeberli W, Hoelzle M, Suter S (1998) Into the second century of worldwide glacier monitoring: Prospects and strategies, UNESCO Publishing, Paris
Hagg W, Severskiy IV, Young G (2009) Assessment of snow, glacier and water resources in Asia. IHP/HWRP, Koblenz
Hambrey MJ (1994) Glacial environments. University College London Press, London
Haserodt K (1984) Abflussverhalten der Flüsse mit Bezügen zur Sonnenscheindauer und zum Niederschlag zwischen Hindukusch (Chitral) und Hunza-Karakorum (Gilgit, Nordpakistan). Mitteilungen der Geographischen Gesellschaft, München, 96:29–36
Hewitt K (1989) European science in High Asia: Geomorphology in the Karakoram Himalaya to 1939. In: Tinkler KJ (ed) History of geomorphology; Hutton to Hack. Unwin Hyman, Boston, pp 165–203
Hewitt K (2005) The Karakoram anomaly? Glacier expansion and the elevation effect, Karakoram Himalaya. Mt Res Dev 25(4):332–340
Hewitt K (2006) Glaciers of the Hunza Basin and related features. In: Kreutzmann H (ed) Karakoram in transition: Culture, development and ecology in the Hunza Valley. Oxford University Press, Oxford, Chap 5, pp 49–72
Hewitt K (2007) Rediscovering colonized landscapes: the first Europeans at the Mustagh Pass, Karakoram Himalaya, Inner Asia. In: Gervers M, Bulag U, Long G (eds) The exploitation of the landscape of central and inner Asia, Toronto Studies in Central and Inner Asia, 9. Asian Institute, University of Toronto, Toronto, pp 41–67
Hewitt K (2011) Glacier change, concentration and elevation effects in the Karakoram Himalaya, upper Indus Basin. Mt Res Dev 31(3):188–200
Kaul MK (1999) Inventory of the Himalaya glaciers: a contribution to the international hydrological programme GSI Special Publication. No. 34, Indian Geological Survey, Calcutta
Kuhle M (2004) The Pleistocene Glaciation in the Karakoram Mountains: reconstruction of past glacier extensions and ice thicknesses. J Mt Sci 1(3):17–298
LIGC (Lanzhou Institute of Glaciology and Cryopedology) (1984) A Bibliography of the glaciology and cryopedology in China and Its Adjacent Districts (1820–192), Academia Sinica, Gansu People's Publishing House (in Chinese and English)
Menzies J (1995) Modern glacial environments: Processes, dynamics and sediments. Butterworth-Heinemann, Oxford
Mercer JH (1975) Glaciers of the Karakoram. In: Field WO (ed) Mountain glaciers of the northern hemisphere, vol 1, Cold Regions Research and Engineering Laboratory, Hanover NH, pp 371–409
Oerlemans J (2001) Glaciers and climate change. AA Balkema, Lisse
Paterson WSB (1994) The physics of glaciers, 3rd edn, Pergamon, New York
Raina VK (2009) Himalayan glaciers: A state-of-art review of glacial studies, glacial retreat and climate change. Ministry of Environment and Forests, Government of India, New Delhi

Scherler D, Bookhagen B, Strecker MR (2011) Spatially variable response of Himalayan glaciers to climate change affected by debris cover. Nat Geosci 4:156–159. doi:10.1038/NGEO1068

Shroder JF Jr (2011) Himalaya. In: Singh et al. (eds), pp 510–20

Shroder JF Jr, Bishop MP (2010) Glaciers of Pakistan. In: Williams RS, Ferrigno JG (eds), F201–F257

Singh VP, Singh P, Haritashya UK (2011) Encyclopedia of snow, ice and glaciers, Encyclopedia of Earth science series, Springer, Dordrecht

Smiraglia C, Mayer C, Mihalcea C, Diolaiuti G, Belo M, Vassena G (2008) Himalayan-Karakoram glaciers: results and problems in the study of recent variations of major non-polar glaciers. In: Bonardi L (ed) Terra Glacialis, Special issue, Mountain glaciers and climate changes in the last century. Servizio Glaciologico Lombardo, Milan, pp 149–164

Visser PhC, Visser-Hooft J (1935–1938) Wissenschaftliche Ergebnisse der niederländischen Expeditionen in den Karakorum und die angrenzenden Gebiete in den Jahren 1922, 1925 und 1929–30. E.J. Brill, Leiden

Williams RS, Ferrigno JG (2010) Satellite image atlas of glaciers of the world: Asia. Professional paper 1386-F, U.S. Geological Survey, U.S. Government Printing Office, Washington DC

Wissmann, H von (1959) Die heutige Vergletscherung und Schneegrenze in Hochasien mit Hinweisen auf die Vergletscherung der letzten Eiszeit. Akademie der Wissenschaften und der Literatur in Mainz. Abhandlungen der mathematisch-naturwissenschaftlichen Klasse 14:1103–1431

Yafeng S, Wenying W (1980) Research on snow cover in China and the avalanche phenomena of Batura Glacier in Pakistan. J Glaciol 26(94):25–30

# Acknowledgements

Materials and work reported on in the book emerged over some decades. There are a great many persons, colleagues and institutions have offered advice and critical discussion as well as assistance in working safely in a challenging environment.

First I would thank the many guides, helpers and hosts from towns and villages in the Karakoram mountains. Without them, none of this would have been possible. In particular, I owe thanks to my first guide and companion, Haji Mahdi of Askole, and many others from that village who accompanied me over several decades. From Nagar, I particularly thank Shaffi Ahmed, his family and uncle, Captain Salim, who offered me much help in the 1980s. Since then, my main guide has been Ghulam Muhammad of Halde, whose skills and those of his family and companions from that village have made much of the work in the past 20 years possible and safe. I hope that in some way what is written here will be of benefit to the people of the region.

I need to mention the help and advice of many better versed in these subjects than myself. This began at King's College London with Clifford Embleton, Denis Brunsden and John Thornes. During the Snow and Ice Hydrology Project, between 1985 and 1990, I learned much from various colleagues, in particular, James Gardner, Michael Quick and Gordon Young. The project and participants benefitted from the work of a group of quite remarkable graduate students, most of whom went on to professional, research or teaching careers in related matters: in particular, Ghazanfar Ali, David Butz, Fes de Scally, Inamullah Khan, Erik Mattson, Ken MacDonald, Jeff Schmok and Cameron Wake. Their work remains valuable and, in most cases, not repeated, to the present time. Hence, it is basic for what is presented in this volume. Over the years, from both direct contact and reading their publications, I have benefitted especially from the work of David Archer, John Clague, Garry Clarke, Monique Fort, Lasafam Iturrizaga, Jack Ices, Uli Kamp, Matthias Kuhle, Paul Mayewski, Bruno Messerli, Jack Shroder Jr., Lewis Owen, Mike Searle, John Menzies, Claudio Smiraglia and Shi Yafeng. Similarly, I was fortunate to meet with members of the German 'Culture Area Karakoram' Project, notably Eckart Ehlers, Hermann Kreutzmann and Matthias Winiger and, more recently, Jingshi Liu of the Tibetan Plateau Research Institute, Academia Sinica, in Beijing.

I have been greatly helped by meetings and exchanges with colleagues at the Cold Regions Research Centre, Wilfrid Laurier University (WLU), and thank its successive directors, Drs. Mike English, Scott (SP.) Slocombe and Richard Petrone. Thanks also to colleagues at ICIMOD, Kathmandu; The Italian Geographical Society, Rome; and the Central and Inner Asian Seminars, University of Toronto. At WLU, I have been fortunate to always find Houston Saunderson, an invaluable colleague ready to discuss these matters and offer wise comments.

In preparing the book, special thanks go to Nina Hewitt who carried out final manuscript preparations. I am indebted to Dr. Marcus Nüsser, an anonymous reviewer for technical comments on the manuscript. I thank Pam Schaus of WLU and Nils Harm of Heidelberg University for preparing the figures, roughly half each. I thank Robert Doe, Robert van Gameren and Naomi Portnoy at Springer Science for their timely help in getting the manuscript to completion.

The following publishers and individuals have kindly given permission for the reproduction of illustrations and other materials from the sources cited (specific citations in text):

Canadian Association of Geographers for Fig. 1.3 (Modified after Canadian Geographer. 1968, 12/2: 85–98)
Gesellschaft für Erdkunde zu Berlin (Berlin Geographical Society) for Figure 71. a–d. from Schneider, HJ, *Die Erde* 1969, 100, pp. 266–286
International Glaciological Society for Figs. 6.4, 8.1 and 9.1 from *Journal of Glaciology*, 2011 53, No. 181, 181–188
The International Mountain Society (IMS) and the United Nations University (UNU), c/o MRD Editorial Office, Bern, Switzerland, for Fig. 12.1, *Mountain Research and Development*, 2005, 25(4):332–340
*Mountain Research and Development* for Figs. 3.2., 4.1., 5.3., and 12.2 (modified after) MRD 2011, 31(3):188–200)
Springer Science+Business Media B.V. 2011 for Fig. 5.3 from Encyclopedia of Snow, Ice and Glaciers (eds. Singh VP, Singh P, Haritashya UK)
Taylor and Francis, Figs 10.1, 10.2 and 10.3 and Tables 10.1, 10.2, 10.3 and 10.4 from *Physical Geography,* 31/6, pp. 528–551

Personal permissions were given by:
Inamullah Khan for original Figs. 5.2. and 5.3
Cameron Wake for data for compiling Fig. 4.1
Jingshi Liu for data in Table 10.1
Marcus Nüsser for Fig. 2.1 (Modified after Umwelt und Entwicklung im Himalaya:> Forschungsgeschichte und aktuelle Themen. In: Geographische Rundschau 64 (4): 4–9 & supplement (2012))

Toronto, Ontario, Canada Kenneth Hewitt

# Contents

# Chapter 1
# The Regional Context

**Abstract** The Karakoram Himalaya comprises the highest, most heavily glacierised watersheds of the upper Indus and Yarkand River basins. It is set within other vast mountain systems of High Asia that support more than 100,000 $km^2$ of perennial snow and ice cover. The chapter situates the Karakoram in this Central Asian context and outlines key aspects of the regional environment. A variety of other cold region or cryosphere features are of interest including seasonal snow covers and perennially frozen ground or permafrost. These affect much greater areas than the glaciers, as do periglacial conditions and snow avalanches. Rock glaciers are concentrated in some parts, periglacial systems closely related to glaciation in distribution, climate responses and genesis. Two main aspects of the Karakoram environment are reviewed: the region's geology and geotectonic evolution, and its climate. The glaciers would not exist at all, or be so extensive, without the great elevations created by mountain-building forces. Significant aspects of their morphology and behaviour relate to rugged, steep landscapes developed through deep dissection by rivers and past glaciations. The Karakoram climate is influenced by three seasonally varying weather systems: the predominantly winter Westerlies, the summer monsoon and Inner Asian high-pressure systems. The glaciers are affected by each system, their relations to one another, by their strong year-to-year variations and how, in turn, they are influenced by the high mountain terrain. In the lower Karakoram valleys and surrounding high plateaux, dry conditions are found, including where most weather stations are located. However, the glacierised area is, in fact, largely humid. Measured snowfall and water yields from glacier basins challenge a long-held view of the Karakoram as part of what has been termed 'the semiarid Himalaya'. Snowfalls in glacier source areas are in the range 1,000–2,000 mm water equivalent. Also, summer snowfall at high elevations will be shown to be a major factor in sustaining the glaciers, whereas winter precipitation dominates in weather station records. Estimates of the much-studied snowlines are shown to be problematic since they occur where freeze–thaw cycles, wind and avalanche redistribution of snow, are concentrated. The importance of Karakoram glaciers for the flows of the Yarkand River and main stem of the Indus is outlined.

K. Hewitt, *Glaciers of the Karakoram Himalaya: Glacial Environments, Processes, Hazards and Resources*, Advances in Asian Human-Environmental Research,
DOI 10.1007/978-94-007-6311-1_1, © Springer Science+Business Media Dordrecht 2014

**Keywords** High Asia • Cryosphere • Greater Karakoram Range • Indus River • Yarkand River • Regional tectonics and geology • Regional climate • Snowlines

## 1.1 High Asia

The Karakoram is distinguished, primarily, by the extent and development of high mountain topography and by its location in the southwest central part of High Asia. These two geographical elements combine to affect the occurrence, maintenance and forms of glaciers in the region. The broad geographical setting is variously referred to as the 'Greater Himalayan Region', 'the Tibetan Plateau and Adjoining Regions' or, for some, 'The Third Pole' (Tandong 2007; Qui 2008; Hilton 2012). It includes the Hindu Kush, Greater and Lesser Himalaya and Tibetan Plateau (Fig. 1.1). Here are the headwaters of rivers that drain to the vast surrounding lowlands: the Amudar'ya, Indus, Ganges, Tsangpo-Brahmaputra, Irrawaddy and Sichuan basins. The mountains also form rugged fringes to the basins of interior drainage north and west of the Tibetan Plateau, including the Yarkand system.

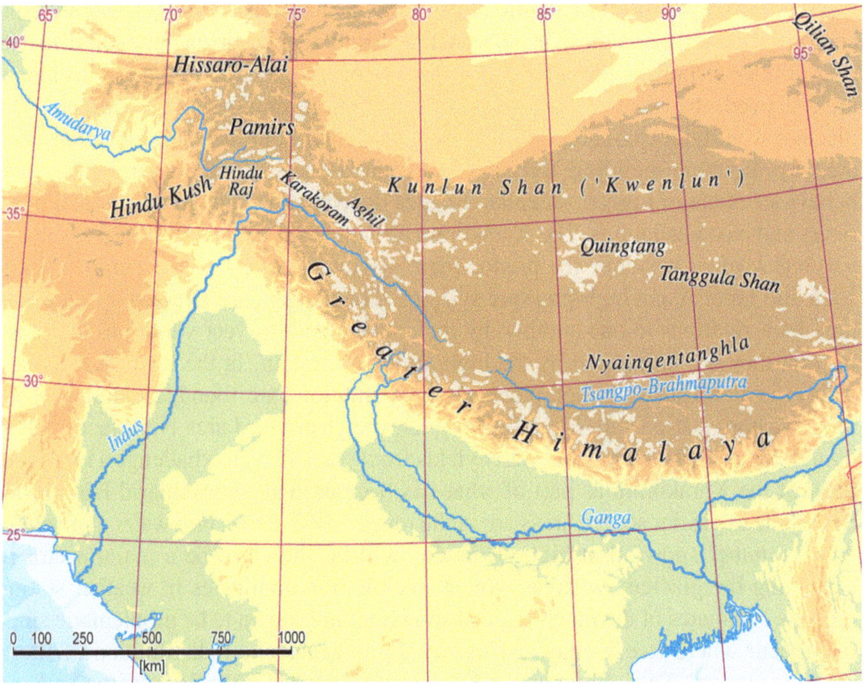


**Fig 1.1** The Greater Himalayan Region showing major river systems, the Karakoram and other main mountain ranges with concentrations of glaciers

**Table 1.1** Summary of four inventories of major glaciated areas in the Greater Himalayan Region

| | Mountain range (countries) | Glacier area (km²) | | | |
|---|---|---|---|---|---|
| # E-W | Source | von Wissmann (1959) | UNESCO 1998 | Dyurgerov and Meier (2005) | USGS 2010 |
| 1 | Hissaro-Alai | 1,830 | 2,328 | 1,750 | 2,336 |
| 2 | Pamirs | – | 7,116 | – | 10,200 |
| | CIS | 11,738 | 4,493 | 12,260 | 7,493 |
| | China | – | 2,623 | – | 2,696 |
| 3 | Tien Shan | 16,500 | – | 15,417 | – |
| | CIS[a] | 9,000 | 7,311 | – | 7,251 |
| | China | 7,500 | 9,257 | – | 9,225 |
| 4 | Hindu Kush | 3,900 | – | 3,200 | – |
| | Afghan | – | – | – | 2,700 |
| 5 | Hindu Raj | 2,300 | – | 2,700 | – |
| 6 | Karakoram | | | | |
| | Pakistan/India | 15,145 | – | 16,600 | 15,000 |
| | China | – | 4,769 | – | 6,262 |
| 7 | Aghil | 1,700 | – | – | – |
| 8 | Kunlun Shan ('Kwenlun') | 11,500 | 12,263 | 12,260 | 12,267 |
| 9 | Greater Him. | 31,530 | 27,960 | – | 35,100 |
| | India | – | 21,590 | – | 8,500 |
| | China | – | 11,000 | 33,050 | 8,418 |
| | Nepal | – | 2,620 | – | 5,324 |
| | Bhutan | 3,300 | 2,730 | – | 1,317 |
| 10 | C. and E. Tibet[b] | | | | |
| | Quingtang | | 3,355 | 1,802 | – |
| | Tanggula Shan | 1,195 | 2,206 | 2,206 | 2,213 |
| | Nyainqentanglha | | 11,000 | 7,536 | 10,700 |
| 11 | Quilian Shan | – | 1,930 | 1,970 | 2,206 |
| | Totals | | | | |
| | High Asia | 120,091 | – | – | – |
| | CIS | – | 18,501 | 116,180 | 22,044 |
| | China | – | 58,000 | – | 59,425 |

[a]*CIS* Commonwealth of Independent States (Former Inner Asian members of USSR)
[b]Central and Eastern Tibet

In the Greater Himalayan Region, there are about a dozen major concentrations of glaciers at high elevations (Table 1.1; Plate 1.1). The total perennial snow and ice cover is thought to exceed 100,000 km² of which the Karakoram share is about 20 %.[1] This refers to the areas covered by snow that remains through the summer

[1]There are considerable differences in the glacial areas quoted in the literature, evidently due to differing nomenclatures, sources of information and classifications. Over the years, various students, including myself, have come up with different numbers as measurement opportunities and concepts have changed (see Annex 1).

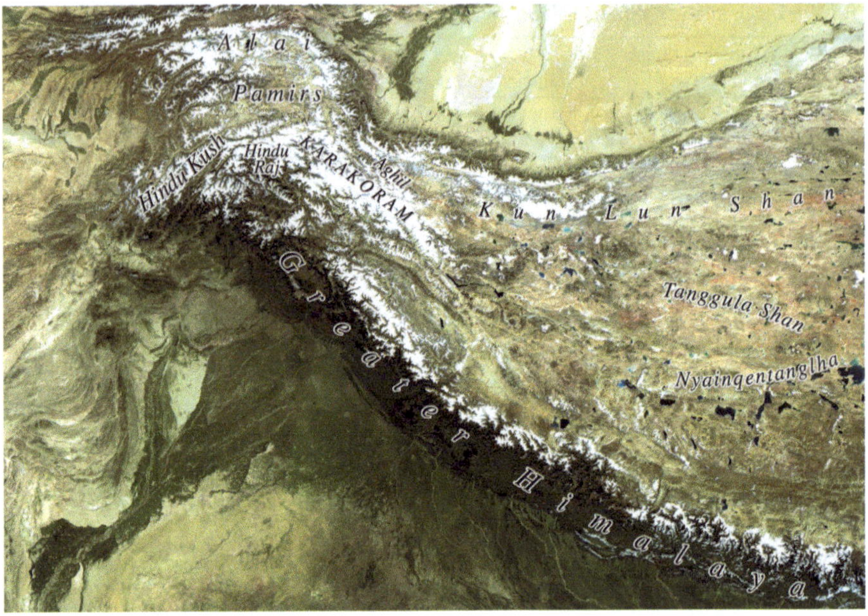


**Plate 1.1** Satellite image showing main concentrations of perennial snow and ice in the Greater Himalayan Region

and ice above glacier termini. In much of High Asia the true glacierised area, meaning that which is actually covered by glacier ice, is rarely more than half the perennial snow and ice cover (Mayer et al. 2006, p. 126; Hewitt 2011).[2] Higher-resolution and more frequent satellite coverage has made clearer the extent of the areas that are perennially frozen but too steep to support glacier ice or parts of active ice streams. There may be a permanent snow cover or thin veneers of ice or avalanche flutings. They are certainly part of glacier basins and will be shown as critical for glacier nourishment in the Karakoram, but they are not underlain by glacier ice (Plate 1.2). Most work on the region has, however, treated glacier area as synonymous with, or close to, the perennial snow and ice cover. Most estimates including my own have tended, therefore, to exaggerate the true *glacierised* area in the more rugged parts of High Asia while underestimating the scale and roles of off-ice areas within glacier basins (Chaps. 3 and 4).

The southern arc of high mountains, sometimes referred to as the Hindu Kush–Karakoram–Greater Himalaya (HKH), is of singular interest. Rarely more than 200 km wide, it nevertheless supports substantial and diverse glacier systems. The belt spans almost 30° of longitude and 13° of latitude, that is, more than

[2]For citations, page numbers are only given to refer to a specific statement, measure or finding within a longer work. They are not given where the whole work is involved, or studies deal in several places or more broadly with topics of interest, or are important, useful or representative background reading.

**Plate 1.2** High elevation source zones of Baltoro Glacier in the Karakoram illustrate how large areas comprise off-glacier rock walls. View from Concordia (4,691 m) of the Broad Peak (8,051 m) watershed (Hewitt July 2005)

2,500 km from west to east and 800 km from the most southerly to the most northerly glaciers. The strongest gradients tend to be north–south across the grain of the mountains. A predominantly southeast to northwest trend also introduces considerable climatic and ecological diversity. Rugged terrain and extreme elevations occur throughout and reflect a more than 50 million-year geological history of intense mountain building as described below.

The numbers of individual glaciers in the region have been estimated at around 50,000, over 7,000 of them in the Karakoram (Williams and Ferrigno 2010). The real numbers may be an order of magnitude greater if all minor ice masses are included. The glacier cover is, however, dominated by valley glaciers of exceptional size. Much of the greatest concentrations of glacier ice occur in a few dozen valley glaciers, tens of kilometres in length. Most of them are in the Karakoram (Plate 1.3).

North of the HKH are some small ice caps on peaks rising from the Tibetan Plateau. On the lesser ranges are many small valley glaciers, cirque glaciers and countless minor, avalanche-fed or 'fall' glaciers. Individual masses are tiny compared to the major valley glaciers but as a whole comprise substantial parts of the regional and global mountain ice. Those of the Karakoram exceed in total area the ice cover of the European Alps. Moreover, the smaller masses are commonly more critical for local water supply and ecosystems in the HKH than the larger glaciers. Equally numerous at high elevations are small and minor, disconnected glacier masses within the basins of larger valley glaciers. Mostly they form along the lee flank of interfluves, at breaks of slope and other irregularities. Repeated ice avalanches from their termini or icefall sections link them to the main glaciers.

**Plate 1.3** Biafo Glacier mid-ablation zone (ice surface approx. 4,000 m), an example of the large valley glaciers that dominate the Karakoram ice cover and comprise a large fraction of all HKH ice (Hewitt October 2010)

Differences among the various HKH glacierised mountains are related partly to elevation, the mass, ruggedness and alignment of the ranges, partly to latitude and the degree of continentality. Of special significance is location with respect to atmospheric circulation as it controls prevailing or seasonal air masses and how these, in turn, are affected by the high mountains. The more extensive ice covers occur on high ranges facing west, south or east, because they are the first encountered by moisture-bearing winds, respectively, from the western to eastern HKH.

These glaciers are all in more or less extreme continental locations, yet some even have 'maritime' qualities due to heavy snowfalls. Chinese inventories and classifications, for example, call glaciers 'maritime' if they receive at least 1,000 mm water equivalent of snowfall annually (Tandong 2007). Glaciers in much of the high Karakoram are 'maritime' in this sense. Although hemmed in by massifs of the Hindu Kush, Pamir and Northwestern Greater Himalaya, the Karakoram has sufficiently high elevation terrain to generate heavy winter and summer snowfalls (Chap. 4).

## 1.2 The Greater Karakoram Region

The Karakoram is the most heavily glacierised core of a mountain complex in the northwestern part of the HKH. It has a total area of about 60,000 km$^2$ and, as noted, about one-third is covered by perennial snow and ice. Other nearby mountains with lesser but substantial glacier covers are the Pamirs, Hindu Kush, Aghil, Kunlun and Kailas Ranges and the Zanskar and Nanga Parbat Himalayas. Together they comprise the 'South West Central Asian Mountain System' of the Vissers (Visser-Hooft and Visser-Hooft 1935–1938).[3] It serves to define the Greater Karakoram Region within the High Asian context.

The Karakoram glaciers extend over almost 5° of longitude and 3° of latitude.[4] The westernmost parts are 2° or, roughly, 220 km north of the eastern parts. Its glaciers are in the subtropics and some 10° south of the best-known ones of the European Alps. They are 6–7° north of the high Himalayan glaciers of Nepal, Bhutan and Sikkim, which are 660–770 km further into the tropics and much further east. Those of the Indian Himalaya, nearer to the Karakoram, are some 3–4° to the south. The geographical coordinates are associated with various ways in which the Karakoram glaciers differ from the others. However, more crucial are location relative to the circulation of the atmosphere and how the mountain masses affect that. The Everest Massif, for example, has a smaller perennial snow and ice cover and much smaller glaciers than, say, the Batura Mustagh or Haramosh Massif (7,409 m) in the Karakoram, which culminate almost 2,000 m lower.

Elevation is a recurrent theme in identifying and assessing features of high mountain ice. In the Greater Karakoram Region, the mountains that rise above 6,500 m are generally picked out by more extensive glacier covers. The large glaciers of the Karakoram are all associated with watersheds that rise above 7,000 m elevation, and the glacier cover increases in proportion to the height and extent of high elevations up to the highest, the K2 Massif (Chap. 2).

## 1.3 The High Asian Cryosphere

Glaciers are the main focus here but other cold climate or cryosphere phenomena interact with them and help to put glacial environments in context. Some occupy a much larger part of the Karakoram and High Asia generally.

Snowfall is a key concern in relation to glacier nourishment, water supply, ecosystems and climate change. Temporary or seasonal snowfall blankets most of the HKH and all of the Karakoram, varying in duration from days to over 11 months, and in depth with elevation and exposure to moisture-bearing winds. Snowfall is a larger source of streamflows than glacier meltwaters in the upper Indus and Yarkand Basins, on the Tibetan Plateau and High Asia river basins as a whole. The snowfall and glacier contributions exceed that of rainfall in HKH Basins except south of the Greater Himalayan crest where monsoon rainfall dominates.

Permafrost or perennially frozen ground comprises a vast area of Central Asia, the world's largest outside higher latitudes (Brown et al. 1997; Zhang et al. 2003).

[3] With due respect to the Vissers, this is a mouthful and SWCAMS not much handier. 'The Greater Karakoram Region' is more manageable.

[4] That is from Long. 73° 50′ to 78° 10′ East and Lat. 34° and 36°50′ North.

The Tibetan Plateau contains the greatest area of continuous permafrost. The lower limits decrease from south to north over the plateau and increase in depth with elevation above the lower limit. The picture is more complex in the more rugged mountain terrain where permafrost tends to be discontinuous, interrupted by deeply incised valleys and limited by the dryness of rain-shadowed valleys and, on south-facing slopes, by the greater and more intense sunshine. Permafrost may occur beneath and beside glaciers, especially those with ice temperatures below freezing and frozen to their beds. In the Karakoram, permafrost extent exceeds the glacier cover and occurs throughout intermediate elevations. It can reach down as low as 3,500 m on north-facing slopes and down to 4,000 m or so on south-facing slopes.

It is thought that permafrost is diminishing over most of Central Asia, mainly due to climate warming (Cheng and Wu 2007; Zhao et al. 2010). Mountain permafrost in the HKH is likely to be equally or more sensitive to climate change (Haeberli and Gruber 2009). To date, there are no data that track what changes are happening in the Karakoram.

At intermediate elevations, permafrost is associated with what are termed 'periglacial conditions'. Unlike the perennially frozen high elevations, surface processes involve seasonal freezing and thawing, including a near-surface 'active layer' in permafrost (French 2011). In unconsolidated surface materials, patterned ground is observed, generated by needle ice growth and frost heaving. More generally, freeze–thaw processes are of interest in relation to water locked up in, or released by, melting of snow, glaciers, permafrost and other forms of ice (Harris and Murton 2005). Frequent freeze–thaw cycles affect most of the Karakoram below the perennial snow zone and, in summer, can involve south-facing slopes as high as 6,000 m (Hewitt 1968a). In midwinter, freeze–thaw is confined to the lowest valley areas. These cycles are associated with frost weathering, which can dislodge debris from surrounding slopes onto the glaciers. Permafrost is also present in the vast areas of rock wall in the glacier zone. At lower elevations, the countless talus cones, fans and aprons of debris below rock slopes and chutes indicate activity in which freeze–thaw has a major role (Hewitt 1968a). Steep rock walls will be shown to be the dominant landform in the Karakoram, playing a key role in what happens to snowfall in and around glacier basins (Chap. 8). Everywhere, on and below them, the work and deposits of snow avalanches are observed.

At elevations near and somewhat below where glaciers form, the combination of snow avalanches and debris falling from rock walls is a key to the widespread occurrence of rock glaciers. These are debris-covered lobes that resemble glaciers, typically moving very slowly and with steep rims (Chap. 11). They are closely associated with permafrost areas and periglacial processes. In the Karakoram, most occur in valleys below existing glaciers, sometimes issuing from them or seeming to have replaced former glaciers (Owen and England 1998; Shroder and Bishop 2010). They share some common features of location and genesis with the glaciers but are among the most neglected cryogenic features in the region.

In general, these other cryosphere phenomena have received much less attention than the glaciers, especially in the Karakoram, and there are only a few studies to draw upon. An effort will be made to sketch their main characteristics, but the lack of research limits definitive conclusions. It does seem important to identify questions for future research and how they are involved in environmental change (Chap. 12).

The Karakoram glacial environment is also singled out by the widespread occurrence of surge-type glaciers – ice masses that advance suddenly and quickly over exceptional distances in a few months or years. They may amount to as much as a third of Karakoram valley glaciers (Kotlyakov 1997; Hewitt 1998b, 2007). The Pamirs form the only other High Asian mountains with many surge-type glaciers, and there are few or none in the Greater Himalaya. This seems related to distinctive conditions in the Karakoram and raises major questions about the stability and dynamics of its glaciers and in tracking glacier change (Chap. 9).

Glacial lakes of small and large size are associated with glaciers and glacier change. A wide variety of impoundments is found in the HKH region, and there is a history of catastrophic floods (Chap. 10). In recent years, those posing unusual dangers of outburst floods have received particular attention (Richardson and Reynolds 2000; Mool et al. 2001; Kattelmann 2003). Reports suggest that climate warming and glacier retreat have increased the incidence of these lakes and dangers of outburst floods. How far, and in what way, this has affected the Karakoram will be shown to involve special problems, and concerns in relation to a history of very large ice dams (Hewitt and Liu 2010).

Another development concerning high mountain glaciers arises from increased evidence of catastrophic landslides in and near glacier basins (Hewitt 1988). They arise from the prevalence of steep rock walls around the glaciers and interact with the ice in ways that modify its behaviour and the fate of landslide materials. A sense of their importance has only emerged in the past couple of decades (McSaveney 2002; Hewitt et al. 2010). It now seems that many hundreds if not thousands could have occurred in the Greater Karakoram Region in the Holocene. They have had significant impacts on the advance and retreat of some glaciers (Hewitt 2009a).

The glacial legacy is not confined to today's perennial snow and ice areas. Much of High Asia was glaciated in the Quaternary and most if not all of the Karakoram one or more times. Throughout the upper Indus and Yarkand valleys, and far beyond today's glaciers, are ice-sculpted features and glacial deposits. In lesser and surrounding ranges of the Greater Karakoram Region are hundreds of glacial cirques and innumerable small lakes derived from glacial activity. Neoglaciation, glacier fluctuations since the last major glaciation, has also left distinctive marks. Of special interest is the 'Little Ice Age' (Kick 1989). It was the last major event of cooling and ice advance, which extended from roughly the fifteenth to the early twentieth centuries (Chap. 12).

## 1.4 Indus and Yarkand River Basins

The broad ecological, hydrological and human significance of the Karakoram glaciers relate to location in the Indus and Yarkand Basins and contributions to their rivers. Most of the meltwaters drain to the Arabian Sea through the Indus Plains, some 12–15 % to the Yarkand River and Tarim Basin. On the western fringes of the Karakoram, glaciers of the Hindu Raj and Hindu Kush drain to the Indus through the Kabul River. In the more distant past, Karakoram ice also flowed into the adjacent Wakhan Valley, on the north slope of the Hindu Kush in Afghanistan, and contributed to the Aral Sea drainage.

The Karakoram tributaries of the Yarkand River drain northwards to the southern territories of Xinjiang Uygur Autonomous Region, northwestern China, and on into the interior Tarim Basin where they sink into desert sands. The Yarkand is the largest tributary of the Tarim River. Above 3,800 m a total of 2,811 glaciers have been counted with an area of 5,725 $km^2$, about 11 % of the catchment (Yang and An 1991). Those in the Karakoram Range cover about 4,200 $km^2$. Given the aridity of so much of the Yarkand Basin and its lesser mountains, the glaciers are of overwhelming importance for freshwater yields.

The larger part of the Karakoram lies in the trans-Himalayan upper Indus Basin. In total, the basin is almost 200,000 $km^2$ in area. It includes extensive high desert tracts on the margins of the Tibetan Plateau and a series of rivers draining the north slopes of the Greater Himalaya. The latter include other important glacier areas such as those of the Zanskar and Nanga Parbat Ranges.

Areas of perennial snow and ice cover roughly 10 % of the upper Indus Basin. The actual glacier cover is about half of this. However, by far the heaviest snowfalls occur in the glacierised areas. Yields of water from ice ablation are much higher again and their contributions much greater than their areas might suggest. They amount to half or more of the flows of rivers that drain the main or Mustagh Karakoram, and as much as 35–40 % of the average flow of the main stem Indus where it leaves the high mountains.

The state of the glaciers is an important factor in, as well as indicator of, ecological and societal sustainability (Alford 1985; Bandyopadhyay and Gyawali 1994; WWF for Nature-Pakistan 2008; Kaser et al. 2010). It is thought these may be especially vulnerable to climate change through less dependable water supplies and possible increases in glacier hazards such as glacial lake outburst floods (Alam 2002; Bajracharya et al. 2008; Cogley 2010; Kreutzmann 2011). Climate change is expected to magnify rather than diminish the importance of the glaciers, mainly in relation to increasing demand, development plans and a predicted reduction in the ice cover. Exactly how it is affecting the high Karakoram glaciers is not well understood and has been subject to some exaggerated reports, as discussed in Chap. 12.

In the lowlands, as many as 200 million folk depend to some extent on the glacier meltwaters, and some millions are at risk from certain glacial hazards. The glacier contribution became of special importance for Pakistan as a result of the Indus Waters Treaty which made the country uniquely dependent on the main stem of the Indus. It has had to provide water for replacement works, as control of the predominantly rain- and snow-fed eastern Indus tributaries goes to India (Chap. 13).

## 1.5 The Karakoram Geological Environment

The major environmental factors in the distribution and behaviour of Karakoram glaciers relate to the regional geology and history of mountain building and to the regional climate as modified by the influence of rugged terrain and high elevations.

### *1.5.1 Geology and Tectonics*

The extreme high elevations of the Karakoram and their extent, necessary conditions for the existence of large glaciers in these subtropical latitudes, reflect a history of mountain building or orogeny. Extreme ruggedness, deep dissection and distinctive drainage patterns arise from erosion in response to long-term, relatively rapid uplift and compression of the mountain belt. The latest and ongoing episode, the Himalayan orogeny, is identified primarily with the Greater Himalaya Range but affects all of the Karakoram.

The region's geology and geophysics have received intense scrutiny, far more so than the glaciers. They are described in many papers, important monographs and collections (Wadia 1931; Gansser 1964; Tahirkheli 1982; Malinconico and Lillie 1989; Treloar and Searle 1993; MacFarlane et al. 1999; Khan et al. 2000). These identify the origins of the major geological formations and multiple phases of uplift, crustal scale faulting, batholithic igneous intrusions and metamorphism. They provide the background to ongoing developments due to tectonics and erosion.

Recent geological interpretations and terminology depend upon the theory of plate tectonics and how it interprets mountain building (Fig. 1.2). The key events, leading to and following from the Himalayan orogeny, involve the northward movement of a crustal plate represented by the Indian subcontinent and its collision with the vast Asian Plate. This is referred to as a continent-to-continent collisional event. Intense tectonic activity is involved along a front of more than 2,500 km. The Greater Himalaya comprises the northern, uplifted and compressed part of the Indian Plate while the trans-Himalayan regions, including the Karakoram, are the highly disturbed parts of the Asian Plate. Prior to collision, an ocean separated the two. Each carries the legacies of older episodes of sedimentation, igneous and metamorphic activity and more ancient orogenies (Searle 1991). Collision is thought to have begun roughly 50 million years ago. Since then, the Indian Plate has moved northwards some 2,000 km, indenting the Asian continent and giving rise to the mountain belts of the HKH.

The main rock formations of the Karakoram trend in a southeast to northwest direction. At the western boundaries, they begin to turn towards and merge into the west and southwest trend of the Hindu Kush. At the eastern margins, in Ladakh, the typical Karakoram high mountain terrain gives way to a series of lesser ranges and more open, sediment-filled basins, which, in turn, merge into the great Tibetan

**Fig 1.2** Main geological formations and structures of the Karakoram and Nanga Parbat Ranges (After Searle 1991)

Plateau. The Karakoram formations generally get younger from north to south, but the picture is complicated in the highest mountains by exposures of rocks of diverse ages, mainly intrusives unroofed by erosion but also associated with complex tectonic thrust and fault structures.

The Karakoram Mountains broadly coincide with a small crustal plate that was welded to the Asian Plate more than 350 million years ago. Most rocks of the mountain range predate the Himalayan orogeny but have been modified by it and uplifted to form the most consistently high terrain in High Asia. Volcanic activity is absent today except for hot springs along many of the valleys. The Earth's crust is at its thickest in the Greater Karakoram Region, more than 70 km in depth. Plate tectonics explains this by crustal shortening and thickening and the downward plunge or subduction of Indian Plate rocks beneath the Asian Plate (Searle 1991, pp. 334–337).

The Nanga Parbat–Haramosh Massif (NPHM) complicates the separation of Karakoram and Greater Himalayan rocks, since both are involved in a common, post-collisional evolution. The massif comprises the northwestern culmination of the Greater Himalaya around Nanga Parbat (8,125 m) and Haramosh (7,409 m) in the south central part of the Karakoram. To emphasise its distinctive character and role, it is sometimes referred to the 'Northwestern Syntaxis' (Wadia 1931). It is a region of exceptionally rapid uplift in an antiform or domal structure, bounded by near-vertical fault systems and exhumed by almost equally rapid erosion. It is singled out as having both the region's oldest and its youngest rocks exposed at the surface. Indian Plate basement rocks of Proterozoic and Palaeozoic origin have been forced to the surface in what is called a 'pop-up' or 'window'. In the process they have become host to very young intrusive rocks and metamorphism of Pliocene to Pleistocene ages (Khan et al. 2000). Remarkably, the latter, though evidently formed some kilometres within the crust, are already exposed near the summit of Nanga Parbat.

Present-day rates of uplift are estimated at between 3 and 10 mm/year, highest in the NPHM and tending to fall off away from it, but sufficient to sustain high mountain conditions throughout the Karakoram. Present-day sediment yields for the whole trans-Himalayan Indus suggest net denudation of 1.3 mm/year, but values reported for tributaries of the NPHM are from 6 mm/year up to 22 mm/year (Shroder and Bishop 2000; Cornwell et al. 2003). In the Indus gorge, through parts of the NPHM, rates of bedrock incision from 3 to 12 mm/year have been reported, suggesting fluvial downcutting quite able to match rates of uplift (Burbank et al. 1996). The amounts of debris carried by the glaciers, especially on their surfaces, suggest that high, if not the highest, rates of primary erosion apply to their basins. Most of it comes more or less directly from surface erosion by avalanches, rockfalls and other slope processes in glacier basins. In general, the high rates of uplift and erosion are aspects of geological evolution bearing most directly on glaciation.

Age determinations for Karakoram and NPHM rocks and implied 'cooling histories' – the time taken for a given rock or mineral to transit from the higher temperatures at depths of origin in the Earth's crust to exposure at the surface – indicate 10–20 km of rock removed in the last 20 million years or so (Zeitler 1985; Searle 1991). Most or all of the denudation has been by earth surface processes: glaciers, rivers and mass movements. This argues against major changes in available relief and steepness in the Quaternary and against suggestions of a subdued landscape in the early Pleistocene or late Tertiary (Gansser 1983). It seems more likely that high mountain terrain similar to the present has existed here since the Miocene, at least (Hewitt 2009b).

The geological formations are associated with distinctive terrain surrounding the ice in glacier basins (Plate 1.4). Bedrock occurs at the surface over most of the Karakoram and forms more than half the area of glacier basins, a significant factor in the morphology and development of these landscapes. Rock at the surface is generally of hard, often massive formations, formerly buried at great depths in the crust. In much of the region, it has been transformed by intense igneous and metamorphic activity in successive mountain-building episodes. The resistant formations support huge areas of steep to near-vertical rock walls that can be continuous over elevation spans of thousands of metres.

**Plate 1.4** The north or right flank of the lower Baltoro Glacier, with landforms in monzogranites and leucogranites of the Baltoro Plutonic Unit in the Karakoram Batholith of Searle (see Fig. 1.3). This illustrates how rock types influence the terrain in glacier basins (Hewitt July 2005)

Patterns of uplift and rock resistance to erosion are expressed in the elevation of different sub-ranges and massifs, and they influence the local extent and forms of the glaciers. The trend and morphologies of glacier basins are associated with complex structural elements or geological lineaments (Cronin 1989). The geometry and texture of valley side slopes commonly reflect rock composition or lithology and structures related to intense folding and faulting. The volumes and mix of debris transported by each glacier reflect the rock formations in their basins, their composition and resistance to erosion. This can impart some distinctive properties to moraines on and beside the ice. However, the diversity of lithologies and structures exposed in the larger basins means that their glacial sediments are of heterogeneous composition. There can be as much variety within as between basins.

### *1.5.2 Quaternary Events and Intermontane Sediments*

> In ages past the Himalaya contained glaciers of enormous length and thickness, the only types of which are seen in those now filling the valleys of the Karakoram… huge masses of alluvial deposits seen in the valleys of all the great rivers of the western Himalaya from the Chang Chingmo and Leh to Skardu… I think are clearly glacial…. (Godwin-Austen 1867, p. 350)

Present-day glacial conditions are the focus here. Nevertheless, interest in Karakoram glaciers and their landscapes has focused largely on the Quaternary Period and glaciations of the last 3 million years or so. There is debate about the

**Plate 1.5** The extensive valley fill in Braldu Valley lies under, up- and downstream of Biafo Glacier terminus in middle ground. The sediments, including those of large, former lakes, have been backed up behind large landslides blocking the Braldu down valley. The nearer mountains are in rocks of the Karakoram Metamorphic Complex of Searle (1991), those in the highest background in the Karakoram Batholith (Hewitt 1996)

exact number of glaciations, their extent and relations to tectonics, landscape development and climate variations. One or more are believed to have covered much or all of the Karakoram. To date, more work has been devoted to Quaternary reconstruction than to the present-day glaciers (Paffen et al. 1956; Kalvoda 1992; Derbyshire et al. 1984; Shroder 1993; Kuhle 1998, 2006; Kamp and Haserodt 2004; Owen 2006). Views of past developments and glaciations do strongly influence how today's glacier cover is interpreted and, in turn, it has influenced notions of past activity. They cannot be ignored here.

The vast intermontane sediments in the trans-Himalayan valleys, around and below the glaciers, are a key concern. The main Indus, Yarkand and their Karakoram tributaries flow mostly in valley fill tens to hundreds of metres thick (Plate 1.5). One of the few major gorges not floored by deep valley fill is where the Indus enters the NPHM, and high rates of bedrock incision are found (Plate 1.6; Leland et al. 1998). Channels in bedrock are otherwise rare. Intermontane river valleys repeatedly open into extensive braided reaches, and during high summer flows, streams engulf river flats to seem like lakes. Even greater quantities and areas of valley sediments comprise river terraces, fans, abandoned river flats and old lakebeds. Extensive aeolian deposits, including dune fields and loess, occur along the valley

**Plate 1.6** The full bedrock gorge where the main Indus enters the Nanga Parbat–Haramosh Massif or 'Northwest Himalayan Syntaxis' (see text). This important feature is rare for the Indus streams in the Karakoram (Hewitt 1987)

floors, and dust storms are common. The valley fill materials clearly belong with the existing drainage systems and channels and record episodes of deposition and aggradation within them. All of this is, perhaps, unexpected in such high, steep and rapidly uplifted mountains.

The earliest scientific explorers remarked upon these sediments and offered interpretations of them (Strachey 1853; Cunningham 1854; Hewitt 1989). From the 1860s, most studies have treated those in Ladakh, Baltistan and Gilgit–Hunza as evidence of Quaternary glaciation(s), following Godwin-Austen (1867) as quoted above. Others developed their ideas to establish the glacial sequences that are still generally accepted (Drew 1875; Dainelli 1924–1935; Norin 1925; Kalvoda 1992; Derbyshire and Owen 1990; Owen 1996).

However, newly emerging discoveries could alter these interpretations of Karakoram glaciers (Hewitt 1999). Much of the valley fill formerly attributed to glaciers now appears to be due to postglacial developments, in particular temporary blockages of the rivers by massive rock slope failures (Hewitt 1988, 1998a, 1999; Shroder et al. 1989; Shroder 1998). Most of the sediments along the Karakoram Indus streams are derived from episodes of rapid aggradation behind natural dams and subsequent trenching of the fill. Many have occurred during the last few thousand years, making the fill much younger than hitherto supposed (Korup et al. 2006; Hewitt et al. 2011). The valley fill has some important links with recent and present-day events around glacier termini, as discussed in Chaps. 8 and 12.

## 1.6 Karakoram Climatic Environments

> The weather in this region, from the middle of June to the last week in August is extremely uncertain.... The monsoon makes its influence felt with prevailing winds from the south and west and a large number of stormy days. During the summer of 1902 there were rarely more than two clear days in succession.... In 1903 with one interval of three fine days early in July, and five from August 8 to 12, there were only a few days when it was not stormy or the sky covered with heavy clouds.... (Workman and Workman 1905, p. 264)

Karakoram glaciers are affected by two distinct sets of climatic controls. The first derives from location with respect to the circulation of the atmosphere and the air masses entering the region. The second set involves the high mountain terrain and such effects as the typical orographic lifting, obstructing and funnelling of air movement. Differences also arise between north- and south-facing slopes and between summit, slope and valley floor conditions (Barry 1992). The exceptional height over large areas and abrupt edges of the mountain masses create thermal and other effects that isolate much of the region. This is especially obvious in the deeply incised valleys, separated from the weather layer and conditions observed in surrounding lands even at similar elevations. The regional climatic situation and seasonally varying air masses will be considered first, then how the mountains impose their influence.

The location relative to the global circulation in the subtropics and to the layout of the High Asian land mass governs the regional climatology. It means that the Karakoram is singled out by the influence and relative strengths of three very different weather systems. Two bring the snowfall that nourishes the glaciers: the winter Westerlies and summer monsoon-related storms. A third set of conditions derives from the continental anticyclones over and to the west of the Tibetan Plateau. At the same time, the Karakoram shares in the climatic divide created by the Himalayan ranges between South and Central Asia and divisions within this great divide, gradients and seasonal differences across the mountains whose significance will emerge in later chapters.

The prevailing circulation throughout the year is westerly, but it exercises greatest influence on the glaciers in the winter half. Between October and early June, the Westerly Jet steers cyclonic storms from the Atlantic and Mediterranean towards the Hindu Kush, immediately west and southwest of the Karakoram (Mooley 1957; Moazzam 1966). The storms are referred to as 'Westerly disturbances' or depressions. They represent the easternmost extension of the winter precipitation regime of the Mediterranean climates. The disturbances arrive as typical, well-developed cyclonic storms with cold and warm fronts and sectors. In the high mountain areas, the frontal structure tends to break down and movement becomes more complicated, or cyclonic activity can be slowed and blocked. At times, it is also renewed and intensified to draw in additional moisture from the Arabian Sea. Disturbances that track directly into the mountains have the most influence, their numbers and intensity being decisive for winter snowfall in and around the glacier basins. Their tracks may, however, be centred well to the south over the Indus Plains or Himalayan foothills. This varies considerably year to year and, potentially, could be affected by global climatic change.

The westerly disturbances are most common and influential between October and June. In winter, there can be seven or eight of them in a month. In summer, they tend to pass to the north of the Karakoram although, over the decades, I have observed snowfall from such storms in every month of the year.

In the first half of June, the Jet normally shifts northwards from its route south of the Himalaya, and more or less simultaneously, the summer monsoon begins to move over the subcontinent. A thermal anticyclone develops over the Tibetan Plateau which tends to keep the monsoon out of the mountains, apparently a more significant factor in this regard than the mountain barrier. The pattern of events in the Karakoram is less clear through the summer.

While the regional airstream continues to be predominantly westerly in summer, occasional, powerful storms from the south can bring heavy snowfall to the Karakoram. Some derive from depressions from the southwest, drawing in moisture from the Arabian Sea, but strong northward thrusts of the monsoon are observed in most years. One explanation, offered by Flohn (1959, p. 303), involves the movement of horizontal waves in the Westerlies over Siberia, generating troughs that penetrate far to the south. These weaken or destroy the Tibetan Plateau anticyclones, allowing monsoonal depressions to enter the Greater Himalaya and cross into the Karakoram.

It used to be thought that the influence of the monsoon is weak or absent from the Karakoram in most years. Significant 'bursts' were considered very rare, a return period of 50 years being widely cited (Finsterwalder 1960). Although many studies perpetuate this belief, there now seems little doubt that, in most summers, monsoonal air and storms bring high-altitude snowfall to much of the Karakoram (Mayewski and Jeschke 1979; Wake 1989; Barros et al. 2000; Böhner 2006).

At the same time, the influence of the monsoon creates differences between glaciers in the western, central and eastern parts of the Karakoram. Those to the east are closer to the summer precipitation regime of, say, the glaciers of the Indian Himalaya and Nepal. There is a lack of high elevation data, but precipitation in the western Karakoram seems likely to be more winter dominated. Work in the high Central Karakoram, detailed below, indicates almost equal amounts of snowfall in winter and summer and a monsoonal influence in most years.

Anticyclonic conditions are associated with clear weather, high levels of daytime solar radiation and nocturnal heat loss. The development and duration of these conditions are important in two main ways: they affect or relate to the tracking and strength of the other two weather systems and to the incidence of clear weather. The potential for blocking moisture-bearing winds is important. The balance of clear-versus-cloudy weather in summer is critical since, as will be shown, direct solar radiation is the overwhelming cause of melting in glacier basins.

The three different weather systems vary in strength from year to year. How they do so is one way in which global climate change may be felt in the region (see Chap. 12). Nevertheless, the influence of the different air masses is greatly modified by their interactions with rugged terrain and extreme elevations. Neglect of these impacts or failure to fully appreciate them has led to some misleading impressions.

### *1.6.1 Conflicting Impressions of Climate*

From the earliest modern reports, the Karakoram Mountains have been described as part of the so-called semiarid Himalaya. Until quite recently, and sometimes even now, maps based on station records show very low precipitation in the whole region, including the glacierised areas of the Karakoram (GP 1985; Scherler et al. 2011). When the large glaciers became known, some early observers thought that they must be anomalous remnants of former times, since the contemporary climate appeared unable to sustain them (Strachey 1853; Visser-Hooft and Visser-Hooft1935–1938). Such ideas were being put forward at, or near, the culmination of the Little Ice Age, that is, possibly during the coldest period and largest glacier advances in several thousand of years (Grove 1988). An alternative suggestion is that the heavy debris mantles on most glaciers preserve them out of phase with today's climate (Kick 1989). These impressions are reinforced for visitors who mostly travel along arid valleys, the occasional views of the glaciers being distant and foreshortened or heavily debris covered.

Large parts of the upper Indus Basin are indeed arid, especially its lower, inhabited areas and towards the margins of the Tibetan Plateau. In addition to extreme rain shadowing by surrounding mountains, valley wind systems intensify the dryness, specifically the desiccating katabatic winds from the cooler slopes and cold glacier valleys above. Weather station records can seem to support the same notions, but that is because nearly all are sited on the floors of dry intermontane basins (Table 1.2).

However, conditions are substantially different in the glacier basins of the Karakoram, most of which are more accurately called 'humid'. Water yields in tributaries draining the main range represent more than five times the precipitation measured at the weather stations of Chitral, Gilgit, Skardu and Leh. Water equivalents of snowfall in the upper basins of the glaciers can be ten times greater (see Chap. 4).

Another former view, rarely challenged, is that winter westerly disturbances dominate precipitation in the region (Burrard and Hayden 1933, p. 147; Mooley 1957). This does apply to weather station records through the twentieth century, which record lowest monthly precipitation in summer (Archer and Fowler 2004). A different picture is found in work on snowfall in the high Karakoram, where summer snowfall is almost equal to winter and exceeds it in some years. Reports from mountaineering expeditions also suggest frequent and heavy summer contributions. From a dozen expeditions of my own, field notes echo conditions described in the quote above (Workmans 1905; Ding 1992). This appears crucial for glacier maintenance and the implications of climate change. Summer storms in glacier basins not only bring additional snowfall, they dramatically reduce sunshine hours and ablation of ice (Chap. 5).

**Table 1.2** Monthly weather data for Gilgit meteorological station, Lat. 35° 55′ N; Long. 74° 17′ E; elevation 1,500 m, 1959–2010 (After World Meteorological Organisation)

| Month | Temp. °C (Mean daily min.) | Temp. °C (Mean daily max.) | Precip. (mm)[a] |
|---|---|---|---|
| Jan | −2.7 | 9.3 | 4.0 |
| Feb | 0.2 | 12.0 | 6.0 |
| Mar | 5.4 | 17.9 | 12.6 |
| Apr | 9.4 | 23.9 | 23.0 |
| May | 11.7 | 28.4 | 25.3 |
| Jun | 15.2 | 34.3 | 6.1 |
| Jul | 18.8 | 36.1 | 15.6 |
| Aug | 18.1 | 35.3 | 15.5 |
| Sep | 12.7 | 31.6 | 6.5 |
| Oct | 6.6 | 25.3 | 8.4 |
| Nov | 0.6 | 17.8 | 1.8 |
| Dec | −2.4 | 11.0 | 4.1 |

[a]The total annual precipitation is 129 mm

In fairness, the semiarid notion also arose from the patchiness of information from higher elevations. Conditions in glacier basins are still among the least known aspects of the region's climate and hydrology. Systematic observations in or near glacier basins have been rare. Reports are mostly by expeditions of limited duration, usually in summer and confined to one valley and glacier. Barely a handful of observation sets come from above 4,000 m elevation where over 80 % of the ice cover occurs, fewer still above 5,000 m where maximum snowfalls occur and the glaciers are nourished. Fewest of all are observations for winter conditions on the glaciers. Satellite imagery does have promise in this regard but still suffers from the shortness of the record, extreme year-to-year variability and inadequate ground control (Forsythe et al. 2012).

Thus, it is understandable why records from permanent stations in the valleys have been relied upon (Haserodt 1984; WWF-Pakistan 2008; Bocchiola and Diolaiuti 2012). They extend back many decades and are practically the only basis for estimating parameters for two-thirds of the year. They allow all-year and longer-term reconstructions and comparisons of weather conditions across the region (Archer and Fowler 2004). Nevertheless, extrapolating from valley weather stations to the glacier zone continues to raise problems, along with the 'semiarid' fallacy. It seems preferable, therefore, to begin by characterising Karakoram climate with such observations as are available for glacier basins. These and a glacier zone perspective are given priority here.

## 1.7 Glacier Climates and Orographic Effects

Cryosphere phenomena in the Karakoram are closely related to elevation and elevation range (Hewitt 1993). Within given watersheds, these influence the total amounts of snowfall, the intensity and duration of sub-zero temperatures, freeze–thaw cycles, cloudiness and solar radiation. All have vertical gradients sufficient to create huge differences in conditions between higher and lower parts of glacier basins.

The gradients in precipitation and thermal conditions, which can both exceed and be less than those for the free atmosphere, assume special importance because of the great total elevation ranges. Low down, valleys have as little as 170 mm annual precipitation so arid conditions prevail. Yet, just a few kilometres away, the higher parts of glacier basins, generally above 4,800 m, can receive snowfall of 1,500–2,000 mm water equivalent (Hewitt 2011).

At and below 2,300 m in the Central Karakoram, where glaciers have their lowest reach, there are 1–2 months with an average temperature below 0 °C. Winter snowfalls are generally small and a few days of snow on the ground or frost. At 2,500 m, average daily temperatures above freezing continue for 8–10 months, for about 10 weeks at 3,800 m and around 4,700, less than 4 weeks. They are rare or absent above 5,000 m. Mainly because of aridity, cryogenic processes have limited presence or impact up to about 3,500 m in the central and western Karakoram except where snow, ice and meltwater are carried down from above by avalanches and glaciers. Between about 4,000 and 5,000 m, seasonal snow cover and freeze–thaw are dominant in surface conditions. The highest elevations, above about 5,000 m, are snowbound all year round and subject to sub-zero temperatures, avalanches and wind drift of snow in all months. This altitudinal zoning of precipitation and the seasonal migration of temperatures help to govern where the glaciers are, their extent, hydrology and activity.

### *1.7.1 Area–Altitude and Seasonal Relations of Glacier Climates*

The scope of vertical atmospheric gradients is critical for the area-altitude distribution of the ice cover. The entire glacial zone extends from the highest observed ice at over 8,600 m to the lowest around 2,300 m. However, more than 80 % occurs between about 4,000 and 6,000 m (Chap. 2). Given the latitude, the way seasonal conditions migrate vertically is also critical, determining just how long particular thermal and moisture conditions occur at given heights.

Since the higher parts of the Karakoram rise into the mid-troposphere, there is increased windiness and much higher wind speeds compared to lower areas. This assumes an overwhelming role in the redistribution and packing of snow, especially around the higher interfluves. Drifting snow is encountered at all elevations and high up almost all the snow is affected by the wind. Some areas are stripped, some wind-packed and others, usually lee slopes, are buried in wind-deposited snow. The patterns relate to the interactions of wind fields and topography and are a good indicator of them.

Wind speeds, fluctuations and directions are strongly affected by topography, including the interaction of weather systems with mountain terrain. Valley wind systems are set up by differential heating and cooling, and in many valley locations,

**Plate 1.7** Dust storms driven by afternoon katabatic winds along the Shyok valley near Khapalu, Baltistan (Hewitt June 2012)

powerful down valley or katabatic winds are common. They are usually set up by daytime heating of bare, arid valley floors, while high-altitude snow and ice maintains low temperatures leading to cold air drainage. By the time these winds reach valley floors, they have expanded, heated up and developed extremely low relative humidity. They often generate severe dust storms (Middleton 1986). These are a common occurrence in Skardu or Khapalu Basin, for example (Plate 1.7).

The rain–snow interface is important in summer precipitation, how and how far it moves up the glaciers. In midsummer, rain may occur as high as 4,500 m, although most summer storms bring snow hundreds of metres lower down (Plate 1.8). Rainfall plays a comparatively minor role because of the shortness of the season higher up and reduced precipitation lower down. When fresh snow falls, it shuts down melting on the ice (Chap. 5).

Seasonal patterns also change with elevation. Around 75 % of precipitation measured at valley weather stations of Gilgit or Skardu occurs in the winter half of the year, predominantly between February and April. However, as noted, the only detailed observations available on glaciers between 4,800 and 5,500 m elevation show summer and winter precipitation about equal (see Chap. 4).

In general, the air masses entering the region or given weather systems are affected in distinct ways by high elevation conditions, including much more cloud and duration of cloudiness, enhanced moisture inflow and higher humidity and precipitation. Conversely, when a run of clear weather does occur, the intense solar radiation at higher elevations assumes a major role, notably through rapid melting of snow and ice.

**Plate 1.8** A rare example of the rain–snow interface reaching exceptional elevations on Karakoram glaciers, almost 5,000 m during an August, monsoonal storm, Baltoro Glacier, Vigne tributary (Hewitt 2005)

### *1.7.2 Freeze–Thaw Cycles*

Thermal conditions not only affect glaciers according to whether temperatures are above or below zero but also through fluctuations across the freezing point, the frequency, depth and duration of freeze–thaw cycles. At all times of the year, there is a zone where frequent freeze–thaw cycles occur on a daily basis. The zone shifts vertically through the year, from low down in winter to around the climatic snowlines in mid- to late summer (Hewitt 1968b). In the absence of continuous records in glacier basins, assembling as many temperature readings as possible from expeditions offers a way to track the seasonal pattern (Fig. 1.3). This is important for all aspects of the cryosphere, as much on the glaciers as the off-ice periglacial zone where freeze–thaw has received the greatest attention (French 2011). It will be returned to in most subsequent chapters.

### *1.7.3 Topoclimatic Effects*

Topoclimate refers to conditions modified in particular elevation bands by slope steepness, ruggedness and orientation. It involves differences at intermediate scales

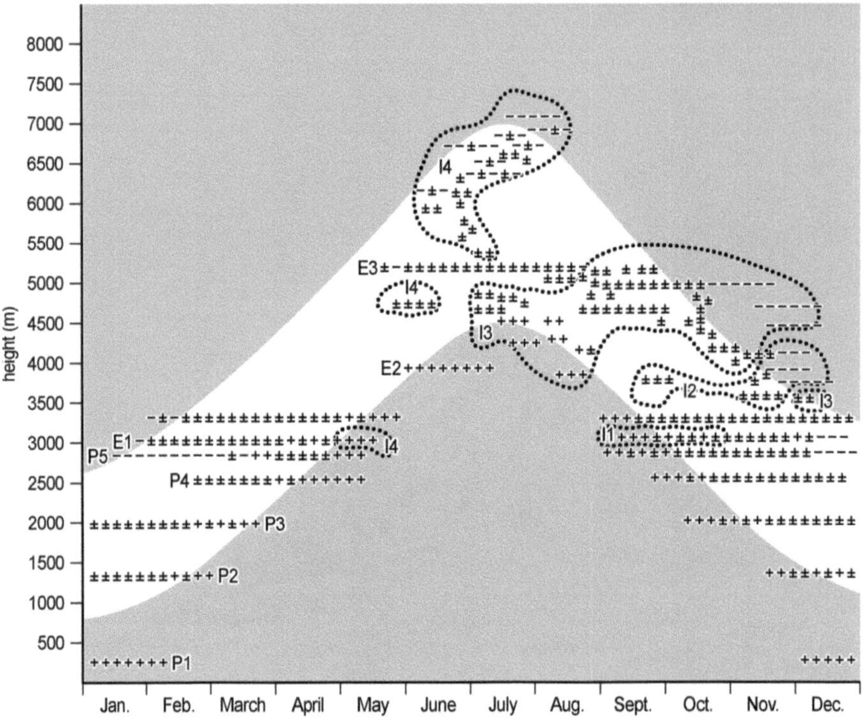


**Fig. 1.3** Freeze-Thaw cycles in the Karakoram reconstructed from temperature readings at high elevations by expeditions (P1 etc) from the late 19th to the mid-20th centuries (After Hewitt 1968a)

between the larger regional and mountain mass or orographic variations already described and those of the most localised or microclimate conditions. In Barry's (1992, p. 83) terms, it is '...a mosaic due to the effects of topography... over distances of 100 m to 1–10 km...'

Topoclimate is involved in the distribution and depth of snow cover and, below the perennial snow, of vegetation patterns. Slope orientation is important with respect to sun and prevailing winds. Slope angle and its variability have to be considered and shading by surrounding topography. Topoclimatic conditions involve important differences between on-glacier and off-glacier locations. Temperatures, humidity and wind speeds can change markedly with height above the glacier and with distance from the ice margins and surrounding terrain. Lateral moraines have sufficient scale to create topoclimatic variations between the ice and valley walls. On-ice topography and debris covers influence periods and rates of melting in summer and the, usually very uneven, winter distribution of snow cover. Such effects are detailed in connection with mass balance of glaciers discussed in Chaps. 4, 5, and 6 and for landforms in Chap. 8.

## 1.8 Snowlines and Limits

Where snow is found, its quantities and attributes vary seasonally, with weather conditions, and from year to year. Two main zones are of interest: a zone of perennial snow at higher elevations and, below it, areas with seasonal snow cover and a snow-free period. At times in midwinter, snow may blanket the entire Karakoram region including valley floors far below the glacier tongues. Snow covers migrate up and down with season and, to a lesser extent with weather conditions, creating what are called 'transient snowlines'. The 'climatic' snowline, sometimes called the 'regional' snowline, is the lowest elevation at which snow persists at summer's end – or the highest level reached by the transient snowline. More exactly, the climatic snowline is supposed to refer to the average position over a period of years. It has been widely invoked as an indicator of climatic conditions and to track changes in them. It is commonly treated as equivalent to the highest level to which glacier ice is exposed by melting and the lowest to which on-ice snow persists, also referred to as the 'firn limit'.[5]

More attention has been given to transient and climatic snowlines than any other cryosphere phenomenon in the region. They are among the very few measurements likely to appear in global glacial and mountain studies (Price 1981, p. 130; Messerli and Ives 1997, pp. 6–7). Estimates began in the early nineteenth century (Schlagintweit and Schlagintweit 1860–1866). Von Wissmann (1959) made an exhaustive study, and his estimates are still commonly accepted. He placed the climatic snowlines for the western Karakoram and Hindu Kush at around 4,600–4,800 m elevation on north-facing slopes and 4,800–5,000 m on south-facing slopes (his Fig. 1.4). They are shown to rise to between 5,200 and 5,800 m in the far eastern upper Shyok River basin and as high as 6,100 m on the north flank in the Shaksgam–Yarkand drainage. Von Wissmann's (1959) work also suggests that snowlines can be lower in the highest, most heavily glacierised areas while rising above the more deeply dissected basins, as well as for mountains with lesser glacier covers and lower watersheds.

Von Wissmann (1959) based his estimates on reports for 100 years prior to the mid-1950s. There are no substantial differences between his estimates, those of the Schlagintweits' from the mid-nineteenth century, Dainelli and Marinelli's (1928), based on observations in 1913–1914 or earlier, or those of the Vissers (1935–1938). There are also few discernable differences from estimates in the 1970s (BIG 1979), the 1980s (Goudie et al. 1984, p. 391; Haserodt 1989, p. 55) and more recently (Kamp and Haserodt 2004, p. 301). These are estimates that were found to apply during or near the culmination of the Little Ice Age in the Karakoram, possibly the coldest episode in the Holocene. Yet, the same or very similar snowline elevations are still widely cited after what is generally perceived as a period of exceptional or

[5]The term 'firn' refers to snow that persists from year to year on the glacier surface, but there are problems with it in the Karakoram (see Chap. 4).

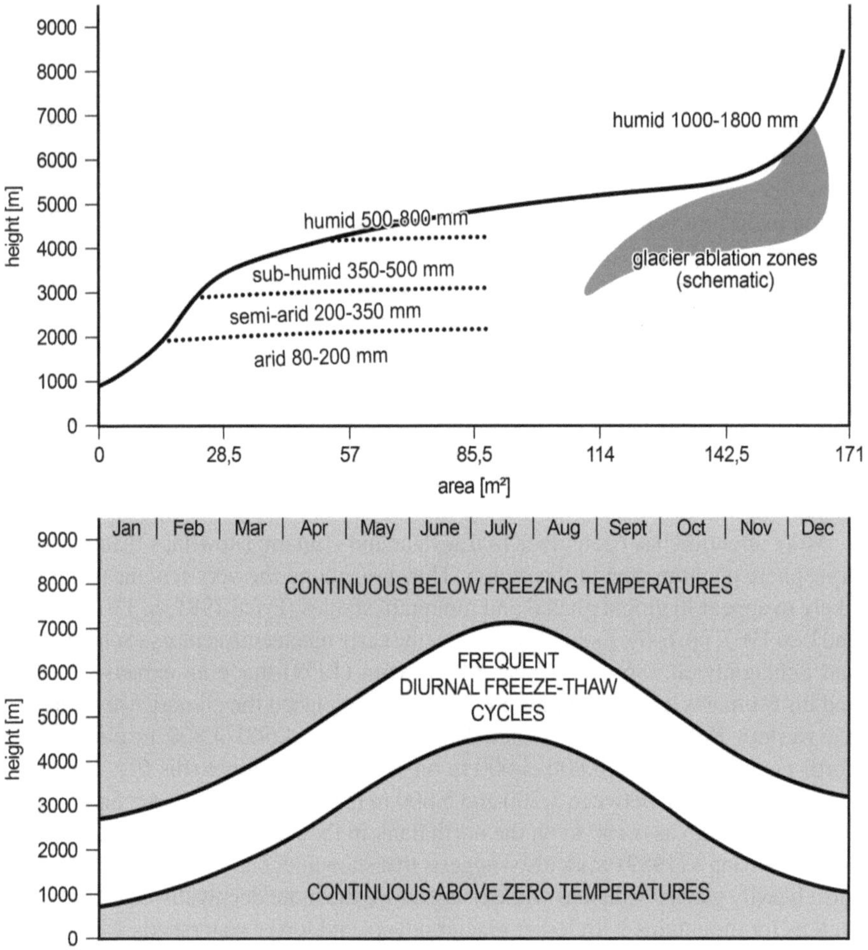


**Fig. 1.4** Schematic relations of precipitation to upper Indus basin hypsometry and glacier ablation zones (above); and relations of temperature and freeze-thaw cycles to elevation (below). Both need to be considered as controls over the seasonal and perennial snow and ice cover (After Hewitt 1993)

exceptionally rapid warming in which the glacier cover has certainly diminished. The literature on climate change and models of it presume snowlines and their equivalent on glaciers, known as the equilibrium-line altitude (ELA, Chap. 6), have been rising, some say dramatically in the HKH (FAO 2007; ICIMOD 2009). One set of estimates of changes since the Little Ice Age is for glacier basins in the Nepal Himalaya, based on snowlines from October 1992 photographs. They were considered to show changes of between 66 and 84 m (Kayastha and Harrisson 2008, p. 97).

However, in high mountains, the transient and climatic snowlines usually have a very irregular trace according to steepness of slopes, solar orientation, wind direction and other factors. Visible snow limits are extremely variable at any given place and time and from one year to the next. In the Karakoram, I find it hard or impossible

**Plate 1.9** Where are the snowlines? This view of the complex patterns of snow on the ground is in the upper Karambar Valley, Gilgit Basin, looking from north to south from 4,500 m below Qalandar Uwin Pass (4,600 m). It is a midsummer view around the elevations where the snowlines are said to occur in the western Karakoram (Hewitt late June 1998)

to decide with any confidence where the snowlines are, to within some hundreds of metres (Plate 1.9). Part of the problem is that year-to-year changes seem larger than any discernable trend. While the advent of high resolution and relatively frequent satellite imagery improves observable detail, it reveals a more complicated and confusing situation. The lower limits are highly irregular. Rather than anything resembling 'lines', there are patches and flags of snow and narrow vertical traces in chutes and gullies among bare slopes, features highly variable in themselves and spanning elevation ranges of as much as 500 and even 1,000 m. The chances are slim of finding end-of-summer limits that are at all 'linear' or at well-defined elevations. Surely this is not so surprising. Von Wissmann's and others' climatic snowlines are at elevations with the most frequent and all-year avalanching. On and below interfluves close to supposed snowlines, there is massive wind redistribution of snow. It is my sense that past estimates may have been grossly oversimplified, based on distant and travellers' impressions rather than detailed measurement and not providing a well-defined climatic signal.

Cogley (2011) rightly challenges a common assumption that the climatic snowline is essentially determined by temperature and often presumed to coincide with the 0 °C annual isotherm. What isotherm, if any, could represent snow limits in the Karakoram is not known, but temperatures measured, on and off glaciers, vary hugely at the same elevations and times. They reflect various topoclimatic and weather conditions, not simply elevation. Rather, thermal conditions appear to form an irregular patchwork across slopes that will change according snowfall temperatures, exposure to sun and wind, slope winds, sublimation and evaporative cooling and nocturnal heat loss. Daytime albedo of snow on the ground is itself a key and highly variable factor, as are surface melting, moisture percolation and refreezing. As noted, much of the snow around the presumed snowlines is avalanche delivered and/or wind redistributed.

Weather conditions towards the end of summer, when the upper limit of snow melting is expected to occur, vary greatly in the HKH and in different parts of the region. The late summer extent and fate of snow will depend on conditions described above, in which the occurrence of westerly and monsoon storms is highly variable, as are longer or shorter episodes of clear weather and intense melting.

It is also sometimes inferred or expected that the snowlines will coincide with a well-defined interface between snowfall and rainfall in precipitation events. It is true that, sometimes after a storm, the base of snow on the ground is a well-defined level but this is quite short lived, irrelevant for the climatic snowline and much less common in the high mountains than those around the North Atlantic shores where the notion has found favour. In summer storms, snowfall at elevations well below the prior limit of snow on the ground is normal. A return to sunny weather can lead to a rapid rise in the visible snow limits of a 1,000 m or more. Apparently this contrasts with tropical glaciers or, at least, those closer to the Equator, where little difference is reported in the snowfall–rainfall limit through the year (Kaser and Osmaston 2002, p. 23).

In the Karakoram, the limits of perennial and seasonal snow covers involve a broad zone of transitional and fluctuating conditions. Meanwhile, the clearly continuous, perennial snow areas lie well above von Wissmann's (1959) snowline elevations. Only above about 6,500 m do the forms and textures of rock walls become masked and dominated by snow and ice forms rather than bedrock textures. The latter prevail at the elevations commonly identified with snowlines.

It is not in doubt that the elevations available in the HKH mean there are large areas with a permanent snow cover and sub-zero temperatures. Focusing on just where their lower boundaries are, however, suggests an unwarranted and, perhaps, unattainable precision while ignoring the transitional and highly dynamic conditions that apply. Perennial or seasonal snow covers will be referred to here and, if necessary, snowfall limits. The term 'snowline' will be avoided and some of the other more or less imaginary boundary lines.

## References

Alam UZ (2002) Questioning the water wars rationale: a case study of the Indus Waters Treaty. Geogr J 168(4):341–353

Alford D (1985) Mountain hydrologic systems. Mt Res Dev 5:349–363

Archer DR, Fowler HA (2004) Spatial and temporal variations in precipitation in the upper Indus River: global teleconnections and hydrological implications. Hydrol Earth Sci Sys 8(1):47–61

Bajracharya SR, Mool PK, Shrestha BR (2008) Global climate change and melting of Himalayan glaciers. In: Ranade PS (ed) Melting glaciers and rising sea levels: impacts and implications. Icfai University Press, Hyderabad, pp 28–46

Bandyopadhyay J, Gyawali D (1994) Himalayan water resources: ecological and political aspects of management. Mt Res Dev 14(1):1–24

Barros AP, Joshi M, Putkonen J, Burbank DW (2000) A study of the 1999 monsoon rainfall in a mountainous region in central Nepal using TRMM products and rain gauge observations. Geophys Res Lett 27:3683–3686

Barros AP, Chiao S, Lang TJ, Burbank DW, Putkonen J (2009) From weather to climate: seasonal and interannual variability of storms and implications for erosion processes in the Himalaya. In: Willett SD, Hovius N, Brandon MT, Fisher DM (eds) Tectonics, climate, and landscape evolution, vol 398, Special papers. Geological Society of America, Boulder, pp 17–38

Barry RG (1992) Mountain weather and climate, 2nd edn. Routledge, London

BIG (Batura Investigations Group) (1979) The Batura Glacier in the Karakoram mountains and its variations. Sci Sin 22(8):958–974

Bocchiola D, Diolaiuti G (2012) Recent (1980–2009) evidence of climate change in the upper Karakoram, Pakistan. Theor Appl Clim. doi:10.1007/s00704-012-0803-y

Böhner J (2006) General climatic controls and topoclimatic variations in Central and High Asia. Boreas 35:279–295

Brown J, Ferrians OJ Jr, Heginbottom JA, Melnikov ES (1997) Circum-arctic map of permafrost and ground-ice conditions. Scale 1:10,000,000. United States Geological Survey, Washington, DC

Burbank DW, Leland J, Fielding E, Anderson RS, Brozovic N, Reid MR, Duncan C (1996) Bedrock incision, rock uplift and threshold hill slopes in the northwestern Himalayas. Nat 379:505–510

Burrard SG, Hayden HH (1933) A sketch of the geography and geology of the Himalayan mountains and Tibet, 2nd edn. Survey of India, New Delhi

Cheng G, Wu T (2007) Responses of permafrost to climate change and their environmental significance. Qinghai-Tibet Plateau. J Geophys Res 112:F02S03. doi:10.1029/2006JF000631

Cogley (2010) In: Singh VP, Singh P, Haritashaya UK (eds) Encyclopaedia of snow, ice and glaciers. Springer, Dordrecht

Cogley JG (2011) Present and future states of Himalayan and Karakoram glaciers. Ann Glaciol 52(59):69–73

Cornwell K, Norsby D, Marston R (2003) Drainage, sediment transport, and denudation rates on the Nanga Parbat Himalaya, Pakistan. In: Butler DR, Walsh SJ, Malanson GP (eds) Mountain geomorphology – integrating earth systems. Proceedings, 32nd Binghamton symposium in geomorphology. Geomorphology, vol 55, pp 25–43

Cronin (1989) In: Malinconico L Jr, Lillie RJ (eds) Tectonics of the Western Himalayas, Special Paper 232. Geological Society of America, Boulder

Cunningham A (1854) Ladak, physical, statistical and historical, with notices of the surrounding countries. Allen, London

Dainelli G (1922) Studi sul Glaciale: Relazione Scientifiche della Spedizione Italiana de Filippi, nell'Himalaia, Caracorum e Turchestan Cinese (1913–1914), vol 3+volume of plates. Zanichelli, Bologna

Dainelli G (1924–1935) Relazioni Scientifiche della Spedizione Italiana De Filippi nell'Himalaia, Caracorum e Turchestan Chinese (1913–1914). Series II, 10 vols. Zanichelli, Bologna

Dainelli G, Marinelli O (1928) Le Condizioni Fische Attuali. Relazioni Scientifiche della Spedizione Italiana de Filippi nell' Himalaja, Caracorum e Turchestan Cinese (1913–1914). Series 1, vol 8. Bologna, Zanichelli

Derbyshire E, Owen L (1990) Quaternary alluvial fans in the Karakoram mountains. In: Rachochi AH, Church M (eds) Alluvial fans: a field approach. Wiley, New York, pp 27–5

Derbyshire E, Owen L (1997) Quaternary glacial history of the Karakoram mountains and Northwest Himalaya: a review. Quatern Int 38/39:85–102

Derbyshire E, Jijun L, Perrot FA, Xu S, Waters RS (1984) Quaternary glacial history of the Hunza Valley, Karakoram mountains, Pakistan. In: Miller KJ (ed) International Karakoram project, vol 2. Cambridge University Press, Cambridge, pp 456–495

Ding Y (1992) Some glacio-micrometeorological features on the north side of Mount Qogir (K2), Karakoram mountains. Ann Glaciol 16:67–72

Drew F (1875) The Jammoo and Kashmir territories. A geographical account. Edward Stanford, London, reprinted (1980), Indus Publications, Karachi
Dyurgerov MB (2005) Mountain Glaciers are at risk of extinction. In: Huber UM, Bugmann HKM, Reasoner MA (eds) Global change and Mountain Regions: an overview of current knowledge. Kluwer, Dordrecht, pp 177–184
Dyurgerov M, Meier MF (2005) Glaciers and the Changing Earth System: A 2004 Snapshot, Occasional Paper 58. Institute of Arctic and Alpine Research, Boulder, 117p
FAO (Food and Agricultural Organization) (2007) Climate change causing species disappearance in mountain areas, Newsroom, UN, FAO, December, Rome. www.fao.org/newsroom/en/news/2007/1000722/index.html
Finsterwalder R (1960) German expeditions to Batura Mustagh and Rakaposhi. J Glaciol 3:57–108
Flohn (1959) In: Wissmann, H von (ed) Die heutige Vergletscherung und Schneegrenze in Hochasien mit Hinweisen auf die Vergletscherung der letzten Eiszeit. Akademie der Wissenschaften und der Literatur in Mainz. Abhandlungen der mathematisch-naturwissenschaftlichen Klasse 14
Forsythe N, Kilsby CG, Fowler HJ, Archer DR (2012) Assessment of runoff sensitivity in the Upper Indus Basin to interannual climate variability and potential change using MODIS Satellite Data Products. Mount Res Dev 32(1):16–29, http://www.bioone.org/doi/full/10.1659/MRD-JOURNAL-D-11-00027.1-aff1
French (2011) In: Singh VP, Singh P, Haritashaya UK (eds) Encyclopaedia of snow, ice and glaciers. Springer-Verlag, Dordrecht
Gansser A (1964) The geology of the Himalayas. Wiley Interscience Publishers, New York
Gansser A (1983) The morphogenetic phase of mountain building. In: Hsu KJ (ed) Mountain building processes. Academic, London, pp 221–228
Godwin-Austen HH (1864) On the Glaciers of the Mustagh Range (Trans-Indus). Proc Roy Geogr Soc 34:19–56
Godwin-Austen HH (1867) Notes on the Pangong Lake district of Ladakh, from a journal made during the Survey in 1863. J Roy Geogr Soc 37:343–363
Goudie AS et al (1984) The geomorphology of the Hunza valley, Karakoram Mountains, Pakistan. In: Miller KJ (ed) vol 2, pp 359–410
GP (Government of Pakistan) (1985) Atlas of Pakistan. Survey of Pakistan, Rawalpindi
Grove J (1988) The little ice age. Methuen, London
Haeberli W, Gruber S (2009) Global warming and mountain permafrost. In: Margesin R (ed) Permafrost soils, vol 16, Soil biology. Springer, Berlin, pp 205–218, Chap 14
Harris C, Murton JB (2005) Cryospheric systems: glaciers and permafrost, vol 242, Special publication. Geological Society, London
Haserodt K (1984) Abflussverhalten der Flüsse mit Bezügen zur Sonnenscheindauer und zum Niederschlag zwischen Hindukusch (Chitral) und Hunza-Karakorum (Gilgit, Nordpakistan). Mitteilungen der Geographischen Gesellschaft, München 96:29–36
Haserodt K (1989) Chitral (Pakistanischer Hindukusch) Strukturen, Wandel und Probleme eines Lebensraumes im Hochgebirge zwischen Gletschern und Wüste. In: Haserodt K (ed) Beiträge und Materialen zur Regionalen Geographie, vol 2. Institut für Geographie der Technischen Universität, Berlin, pp 182–233
Hewitt K (1968a) The freeze-thaw environment of the Karakoram Himalaya. Can Geogr 12:85–98
Hewitt K (1968b) Geomorphology of the mountain regions of the upper Indus Basin. 2 vols. Ph.D. dissertation, University of London, London
Hewitt K (1988) Catastrophic landslide deposits in the Karakoram Himalaya. Science 242:64–67
Hewitt K (1989) European science in High Asia: geomorphology in the Karakoram Himalaya to 1939. In: Tinkler KJ (ed) History of geomorphology; Hutton to Hack. Unwin Hyman, Boston, pp 165–203
Hewitt (1993) In: Shroder JF Jr (ed) Himalaya to the sea: geology, geomorphology and the quaternary. Routledge, London
Hewitt K (1998a) Catastrophic landslides and their effects on the Upper Indus streams, Karakoram Himalaya, northern Pakistan. Geomorphology 26:47–80

Hewitt K (1998b) Himalayan Indus streams in the Holocene: Glacier-, and Landslide-'Interrupted' fluvial systems Chap 1. In: Stellrecht I (ed) Karakorum-Hindu Kush-Himalaya: Dynamics of change Part I. Rudiger Koppe Verlag, Koln, pp 1–28

Hewitt K (1999) Quaternary moraines vs. catastrophic rock avalanches in the Karakoram Himalaya, Northern Pakistan. Quat Res 51(3):220–237

Hewitt K (2005) The Karakoram anomaly? Glacier expansion and the "elevation effect", Karakoram Himalaya. MT Res Dev 25:332–340

Hewitt K (2007) Tributary glacier surges: an exceptional concentration at Panmah Glacier, Karakoram Himalaya. J Glaciol 53(181):181–188

Hewitt K (2009a) Rock avalanches that travel onto glaciers and related developments, Karakoram Himalaya, Inner Asia. Geomorphology 103:66–79

Hewitt K (2009b) Catastrophic rock slope failures and late Quaternary developments in the Nanga Parbat-Haramosh Massif, Upper Indus basin, Northern Pakistan. Quat Sci Rev 28:1055–1069

Hewitt K (2011) Glacier change, concentration and elevation effects in the Karakoram Himalaya, Upper Indus Basin. Mount Res Dev 31(3):1–13, http://dx.doi.org/10.1659/MRD-JOURNAL-D-11-00020.1

Hewitt K (2010) Gifts and perils of landslides: catastrophic rockslides and related landscape developments are an integral part of human settlement along upper Indus streams. American Journal of Science 48:410–419

Hewitt K, Liu J (2010) Ice-dammed lakes and outburst floods, Karakoram Himalaya: historical perspectives and emerging threats. Phys Geogr 31(6):528–551

Hewitt K, Clague JJ, Deline P (2010) Catastrophic rock slope failures and mountain glaciers. In: Singh et al (eds), pp 112–126

Hewitt K, Clague JJ, Gosse J (2011) Rock avalanches and the pace of late Quaternary development of river valleys in the Karakoram Himalaya. Geol Soc Am Bull 14. doi:10.1130/B30341.1

Hilton I (2012) Exploring the third pole, thirdpole.net reader, Chinadialogue. http://184.172.177.95//~chinad/wp-content/uploads/2012/06/Combined-Final

ICIMOD (2009) Climate change in the Himalayas. Information Sheet #3/09, International Centre for Integrated Mountain Development, Kathmandu

Kalvoda J (1992) Geomorphological record of the Quaternary Orogeny in the Himalaya and the Karakoram, vol 3, Developments in earth surface processes. Elsevier, Amsterdam

Kamp U, Haserodt K (2004) Quaternary glaciations in the high mountains of northern Pakistan. In: Ehlers J, Gibbard PL (eds) Extent and chronology of glaciation, vol. 3, South America, Asia, Africa, Australia, Antarctica. Elsevier, Amsterdam, pp 293–311

Kaser G, Osmaston H (2002) Tropical glaciers. Cambridge University Press, Cambridge

Kaser G, Großhauser M, Marzeion B (2010) Contribution potential of glaciers to water availability in different climate regimes. ProcUS Nat Acad Sci 107(47):20223–20227

Kattelmann R (2003) Glacial lake outburst floods in the Nepal Himalaya: a manageable hazard? Nat Haz 28:145–154

Kayastha RB, Harrisson SP (2008) Changes of the equilibrium-line altitude since the Little Ice Age in the Nepalese Himalaya. Ann Glaciol 48:93–99

Khan MA, Treloar PT, Searle MP, Jan MQ (2000) Tectonics of the Nanga Parbat syntaxis and western Himalaya, vol 170, Geological society special Publication. Geological Society, UK

Kick W (1989) The decline of the last Little Ice Age in High Asia compared with that in the Alps. In: Oerlemans J (ed) Glacier fluctuations and climate change. Kluwer, Dordrecht, pp 129–142

Korup O, Clague JJ, Hermanns RL, Hewitt K, Strom AL, Weidinger JT (2006) Giant landslides, topography, and erosion. Earth Plan Sci Lett 261:578–589

Kotlyakov VM (ed) (1997) Atlas snezhno-ledovykh resursa mira [World atlas of snow and ice resources]. Russian Academy of Sciences, Institute of Geography, Moscow [in Russian]

Kreutzmann H (2011) Scarcity within opulence: water management in the Karakoram mountains revisited. J Mount Sci 8:525–534

Kuhle M (1998) Reconstruction of the 2.4 million km$^2$ late pleistocene ice sheet on the Tibetan Plateau and its impact on the global climate. Quat Int 45(46):71–108

Kuhle M (2006) The past Hunza Glacier in connection with a Pleistocene Karakoram ice stream network during the last ice age (Wurm). In: Kreutzmann H (ed) Karakoram in transition: culture, development and ecology in the Hunza Valley. Oxford University Press, Karachi, pp 24–48

Leland J, Reid MR, Burbank D, Finkel R, Caffee M (1998) Incision and differential bedrock uplift along the Indus River near Nanga Parbat, Pakistan Himalaya, from 10Be and 26Al exposure age dating of bedrock straths. Earth Plan Sci Lett 154:93–107
MacFarlane A, Sorkhabi RB, Quade J (1999) Himalaya and Tibet: mountain roots to mountain tops, vol 328, Special paper (Geological Society of America). Geological Society of America, Boulder
Malinconico L Jr, Lillie RJ (1989) Tectonics of the Western Himalayas, Special paper (Geological Society of America). Geological Society of America, Boulder, 232
Mayer C, Lambrecht A, Belo M, Smiraglia C, Diolaiuti G (2006) Glaciological characteristics of the ablation zone of Baltoro Glacier, Karakorum, Pakistan. Ann Glaciol 43:123–131
Mayewski PA, Jeschke PA (1979) Himalayan and transHimalayan glacier fluctuations since AD 1812. Arc Alp Res 11:267–287
Messerli B, Ives JD (eds) (1997) Mountains of the world: a global priority. Parthenon Publishing, New York
Middleton NJ (1986) The geography of dust storms in south-west Asia. J Clim 6:183–196
Moazzam SM (1966) Western disturbances, regional meteorological centre report. Pakistan Meteorological Service, Lahore
Mool PK, Bajracharya SR, Joshi SP (2001) Inventory of glaciers, glacial lakes, and glacial lake outburst flood monitoring and early warning systems in the Hindu Kush-Himalayan region: Nepal. International Centre for Integrated Mountain Development (ICIMOD), Kathmandu, Kathmandu and United Nations Environment Programme, Regional Resource Centre for Asia and the Pacific, 2 vols
Mooley DA (1957) The role of Western Disturbances in the production of weather over India during different seasons. Ind J Meterol Geophys 8:254–260
Norin E (1925) Preliminary notes on the late Quaternary glaciation of the North West Himalaya. Geogr Ann 7:165–194
Owen L (1996) Quaternary lacustrine deposits in a high-energy, semi-arid mountain environment, Karakoram mountains, Northern Pakistan. J Quatern Sci 11:461–483
Owen L (2006) Quaternary glaciation. In: Kreutzmann H (ed) Karakoram in transition: culture, development and ecology in the Hunza Valley. Oxford University Press, Karachi, pp 12–23
Owen L, England J (1998) Observations on rock glaciers in the Himalayas and Karakoram mountains of northern Pakistan and India. Geomorphology 26:199–213
Paffen KJ, Pillewizer W, Schneider HJ (1956) Forschungen im Hunza-Karakorum. Erdkd 10:1–33
Pant GB, Rupa K (1997) Climates of South Asia. Wiley, New York
Price LW (1981) Mountains and man: a study of process and environment. University of California Press, Berkeley
Qui J (2008) China: the third pole. Nature 454:393–396
Richardson SD, Reynolds JM (2000) An overview of glacial hazards in the Himalayas. Quat Int 65(66):31–47
Scherler D, Bookhagen B, Strecker MR (2011) Spatially variable response of Himalayan glaciers to climate change affected by debris cover. Nat Geosci 4:156–159. doi:10.1038/ngeo1068
Schlagintweit HA von, von Schlagintweit R (1860–1866) Results of a scientific mission to India and High Asia. 4 vols+Atlas, Truebner, London
Searle M (1991) Geology and tectonics of the Karakoram mountains. Wiley, New York
Shroder JF Jr (1993) Himalaya to the sea: geology, geomorphology and the quaternary. Routledge, London
Shroder JF Jr (1998) Slope failure and denudation in the western Himalaya. Geomorphology 26:81–106
Shroder, Bishop (2000) In: Khan MA, Treloar PT, Searle MP, Jan MQ (eds) Tectonics of the Nanga Parbat syntaxis and Western Himalaya, Special publication 170. Geological Society, UK
Shroder, Bishop (2010) In: Williams RS, Ferrigno JG (eds) Satellite image atlas of glaciers of the world: Asia, Professional paper 1386-F. U.S. Geological Survey and U.S. Government Printing Office, Washington, DC

Shroder et al (1989) In: Malinconico L Jr, Lillie RJ (eds) Tectonics of the Western Himalayas. Special paper 232. Geological Society of America
Singh VP, Singh P, Haritashaya UK (2010) Encyclopaedia of snow, ice and glaciers. Springer, Dordrecht
Strachey H (1853) Physical geography of Western Tibet. J Roy Geogr Soc 23:1–69
Tahirkheli RAK (1982) Geology of the Himalaya, Karakoram and Hindu Kush, Pakistan. University of Peshawar Geological Bulletin, Special Issue II, p 187
Tandong Y (ed) (2007) Map of the glaciers and lakes on the Tibetan Plateau and adjoining regions. Xi'an Cartographic Publishing House, Xi'an
Treloar PJ, Searle MP (eds) (1993) Himalayan tectonics, vol 74, Geological Society, Special publication. Geological Society, London
Visser PhC, Visser-Hooft J (1935–1938) Wissenschaftliche Ergebnisse der niederländischen Expeditionen in den Karakorum und die angrenzenden Gebiete in den Jahren 1922, 1925 und 1929–30. E.J. Brill, Leiden
Wadia DN (1931) The Syntaxis of the North-West Himalaya: its rocks, tectonics and orogeny. Rec Geolo Surv India 65:189–220
Wake CP (1989) Glaciochemical investigations as a tool for determining the spatial and seasonal variation of snow accumulation in the central Karakorum, northern Pakistan. Ann Glaciol 13:279–284
Williams RS, Ferrigno JG (eds) (2010) Satellite image atlas of glaciers of the world: Asia, vol 1386-F, U.S. Geological Survey Professional paper. U.S. Government Printing Office, Washington, DC
Wissmann, H von (1959) Die heutige Vergletscherung und Schneegrenze in Hochasien mit Hinweisen auf die Vergletscherung der letzten Eiszeit. Akademie der Wissenschaften und der Literatur in Mainz. Abhandlungen der mathematisch-naturwissenschaftlichen Klasse 14:1103–1431
Workman FB, Workman WH (1905) From Srinagar to the sources of the Chogo Lungma Glaciers. Geogr J 25:245–268
WWF for Nature-Pakistan (2008) Climate change in Northern Areas Pakistan: impacts on glaciers, ecology and livelihoods. World Wide Fund, Gilgit Conservation and Information Center (GCIC), NLI Colony, Jutial Gilgit, Internal Report, April
Yafeng S, Liu S, Shangguan D, Li D, Ye B (2006) Peculiar phenomena regarding climatic and glacial variations on the Tibetan Plateau. Ann Glaciol 43:106–111
Yang HA, An RZ (1991) Yarkant River basin (interior drainage area of Tarim Basin), vol 3, Glacier inventory of China. Science Press, Beijing
Zeitler PK (1985) Cooling history of the NW Himalaya, Pakistan. Tectonics 4:127–151
Zhang T, Barry RG, Knowles K, Ling F, Armstrong RL (2003) Distribution of seasonally and perennially frozen ground in the Northern Hemisphere. In: Phillips M, Springman SM, Arenson LU (eds) Proceedings of the eighth international conference on permafrost, A.A. Balkema, pp 2017–2076
Zhao L, Wu Q, Marchenko SS, Sharkhuu N (2010) Thermal state of permafrost and active layer in central Asia during the international polar year. Permafrost Periglac Processes 21:198–207

## Annex 1: Inventories of High Asian Glaciers and Related Resources

Dainelli G (1924–35) Relazioni Scientifiche della Spedizione Italiana De Filippi nell'Himalaia, Caracorum e Turchestan Chinese (1913–1914). Series II, 10 vols Zanichelli, Bologna
Field WO (ed) (1975) Mountain glaciers of the northern hemisphere, Part III Glaciers of South Asia.) V.1+Atlas, Cold Regions Research and Engineering Laboratory (CREEL), Hanover, New Hampshire

Chap. 2 Hindu Kush, (Horvath E, 1970), pp 361–370
Chap. 3 Karakoram (Mercer J, 1968), pp 371–410
Chap. 4 Himalaya (Mercer JH, 1968), pp 411–448
Chap. 5 China (Lehr P, Horvath E. 1970), pp 449–477

Hagg W, Severskiy IV, Young G (eds) (2009) Assessment of snow, glacier and water resources in Asia. IHP/HWRP, Koblenz

ICIMOD (2007) Inventory of glaciers, glacial lakes and identification of potential glacial lake outburst flood GLOFs affected by global warming in the mountains of Himalayan Region. DVD ROM International Centre for Integrated Mountain Development, Kathmandu

Kaul MK (1999) Inventory of the Himalaya glaciers: a Contribution to the International Hydrological Programme GSI Special Publication. No.34, Indian Geological Survey, Calcutta

Kotlyakov VM (ed) (1997) Atlas snezhno-ledovykh resursa mira [World atlas of snow and ice resources]. Russian Academy of Sciences. Institute of Geography, Moscow [in Russian]

Kuhle M (2004) The Pleistocene Glaciation in the Karakoram mountains: reconstruction of past glacier extensions and ice thicknesses. J Mt Sci 1(3):17–298

LIGC (1984) A bibliography of the glaciology and cryopedology in China and its adjacent districts (1820–192), Lanzhou Institute of Glaciology and Cryopedology, Academia Sinica, Gansu People's Publishing House (in Chinese and English)

McSaveney MJ (2002) Recent rockfalls and rock avalanches in Mount Cook National Park, New Zealand. In: Evans SG, Degraff JV (eds) Catastrophic landslides: effects, occurrence, and mechanisms. Geological Society of America, Reviews in Engineering Geology, vol XV, pp 35–70

Mool PK, Bajracharya SR, Joshi SP (2001) Inventory of glaciers, glacial lakes, and glacial lake outburst flood monitoring and early warning systems in the Hindu Kush-Himalayan region. International Centre for Integrated Mountain Develo-pment (ICIMOD), Kathmandu and United Nations Environment Programme, Regional Resource Centre for Asia and the Pacific, 2 vols

Raina VK (2009) Himalayan glaciers. A state-of-art review of glacial studies, glacial retreat and climate change. Ministry of Environment and Forests, India. G.B. Pant Institute of Himalayan Environment & Development, Kosi-Katarmal, Almora

Raina VK, Srivastava D (2008) Glacier Atlas of India, Geological Society of India, Bangalore

Roohi RP, Ashraf A, Bajracharya S, Amjad S, Naz R (2005) Inventory of glaciers, glaciers, glacial lakes the identification of potential glacial lake outburst floods affected by global warming in the Mountains of Himalayan Range, Pakistan, International Centre for Integrated Mountain Development (ICIMOD), Kathmandu Nepal & PARC, Pakistan (CD)

Sangewar CV, Shukla SP (eds) (2009) Inventory of the Himalayan glaciers: a contribution to the International Hydrological Programme. Updated edn. Geological Survey of India, Kolkata. (Special Publication 34)

SIHP (1990) Classified bibliography of Upper Indus Basin and South Central Asian Mountain systems, + select bibliography on glaciers, snow, and high mountain hydrology. Snow and Ice Hydrology Project, Final Report, v. IV, WAPDA/WLU/ICIMOD, Wifrid Laurier University, 125p

Tandong Y (ed) (2007) Map of the glaciers and lakes on the Tibetan Plateau and adjoining regions. Xi'an Cartographic Publishing House, Xi'an, China[6]

Tsevetkov DG, Osipova GB, Xie Z, Wang Z, Ageta Y, Baast P (1998) Glaciers of Asia: In: UNSECO (1998) Into the second century of worldwide glacier monitoring: prospects and strategies. UN Educational, Scientific and Cultural Organization, Paris, pp 177–196

Williams RS, Ferrigno JG (eds) (2010)[7] Satellite image atlas of glaciers of the world: Asia. Professional paper 1386-F, U.S. Geological Survey, U.S. Government Printing Office, Washington DC

---

[6]This includes a 28-page pamphlet with statistics for glaciers and lakes and distribution by mountain ranges, subregions and river basins.

[7]Most chapters were written in the late 1970s or early 1980s. There has been some updating of these and some more recent supplementary sections added.

F-1 Glaciers of the Former Soviet Union (Kotlyakov et al.), pp 1–93
F-2 Glaciers of China (Yafeng et al.), pp 127–166
F-3 Glaciers of Afghanistan (Shroder J, Bishop M), pp 167–199
F-4 Glaciers of Pakistan (Shroder J, Bishop M), pp 201–257
F-5 Glaciers of India (Vohra CH et al.), pp 259–291
F-6 Glaciers of Nepal (Higuchi K et al.), pp 293–320
F-7 Glaciers of Bhutan (Shuji Iwata), pp 321–334

Wissmann H von (1959) Die heutige Vergletscherung und Schneegrenze in Hochasien mit Hinweisen die Vergletscherung der letzten Eiszeit Akademie der Wissenchaften und der Literatur in Mainz Nr. 14:1103–1431

WWF-Nepal (2005) An overview of glaciers, glacier retreat, and subsequent impacts in Nepal, India and China. World Wildlife Fund, Nepal Program, Kathmandu

WWF–Pakistan (2008) Climate Change in Northern Areas of Pakistan: Impacts on Glaciers, Ecology, and Livelihoods. Gilgit Pakistan: World Wildlife Fund, Gilgit Conservation and Information Center (GCIC), Internal report

Yang HA, An RZ (1991) Yarkant River Basin (interior drainage area of Tarim Basin). Glacier inventory of China, vol 3. Ice-dammed lakes and outburst floods. Science Press, Beijing

# Chapter 2
# Snow, Ice and Verticality in the Karakoram

**Abstract** The distribution of perennial snow and ice in the Karakoram Himalaya is examined and its area–altitude relations. The presence and extent of snow and ice are shown to depend upon, and be positively correlated with, interfluve heights. The elevations and extent of the highest altitude terrain are of decisive significance. The size, length and lowest reach of glaciers increase as elevation increases up to the highest watersheds. In the Central Karakoram, the 'glaciation level', or minimum elevation needed to generate a glacier, is found at about 5,250 m on north-facing slopes and 5,500 m on south-facing slopes. At the western margins, the averages are 4,600 m and 5,200 m, respectively. They rise eastwards by about 1,200 m and 900 m to the highest glaciation levels found in the eastern margins of the Karakoram. A main set of the 42 largest valley glaciers is introduced, with basin areas exceeding 130 km$^2$ and ice streams over 16 km in length. These have exceptional elevation ranges, five spanning more than 5,000 m and 34 more than 3,000 m. Their long profiles exhibit two main features. Most of the vertical descent is accomplished in less than 10 % of ice stream lengths, mainly in icefalls in the upper parts of the basins. However, their longest sections, in the middle and lower reaches, are of relatively gentle gradient. Some 85 % of main ice stream areas lie between 4,000 and 6,000 m, the critical elevation zone in terms of ice cover. This must be balanced against the extreme high elevations of their watersheds and also the exceptionally low termini of many Karakoram glaciers compared to most in the Greater Himalayan Region.

**Keywords** Verticality • Elevation range • Area–altitude distributions • Glaciation limit • Rock glacier thresholds • Glacier long profiles

## 2.1 Glaciers and the Perennial Snow and Ice Cover

Most of the Karakoram ice cover and its largest glaciers are in the highest parts of the range, especially the main axis, aptly named the Mustagh (= 'ice mountain') Karakoram. Here the perennial snow and ice cover exceeds 70 % (Fig. 2.1). The Rakaposhi–Haramosh

K. Hewitt, *Glaciers of the Karakoram Himalaya: Glacial Environments, Processes, Hazards and Resources*, Advances in Asian Human-Environmental Research,
DOI 10.1007/978-94-007-6311-1_2, © Springer Science+Business Media Dordrecht 2014

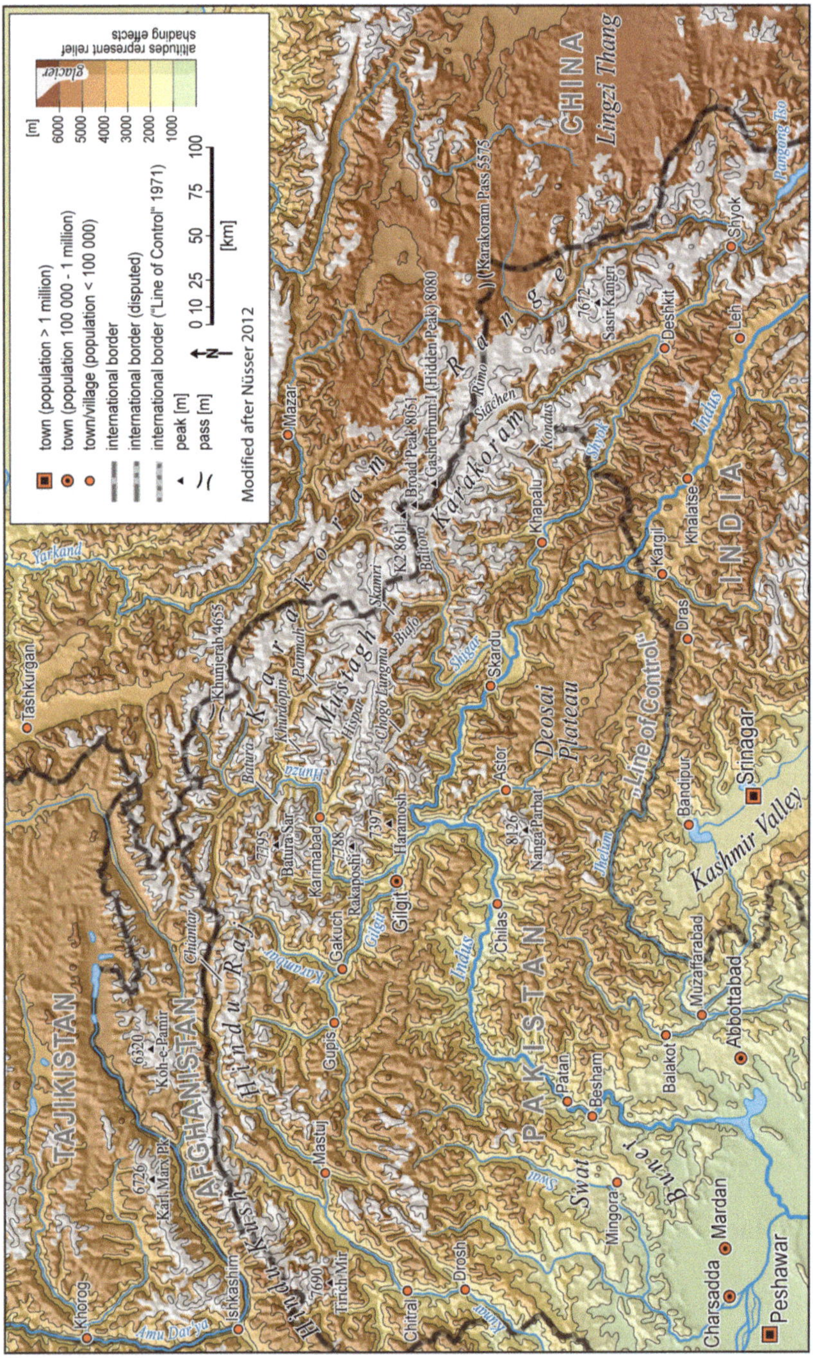


**Fig. 2.1** Map of glacier cover and principal basins and sub-ranges of the Karakoram (modified after Nüsser 2012)

Massif is south of the main axis but shares its characteristics of great height, ruggedness and ice cover. The main glaciers commence on the highest watersheds, but the larger ice masses spread outward and down to relatively low elevations. They form a nearly continuous glacierised zone over a distance of some 400 km and from 20 to 70 km in width. This Mustagh zone comprises about 12,500 $km^2$ out of the total Karakoram perennial snow and ice cover of 20,500 $km^2$.[1]

All of the larger glaciers are in the highest headwaters of the Indus and Yarkand Rivers. The greatest concentration surrounds the highest K2 (8,610 m) Massif where the longest ice streams originate: Siachen (76 km), Biafo (66 km) and Baltoro (62 km).[2] These and several others comprise most of the biggest valley glaciers outside high latitudes.[3] On the Karakoram north flank, glaciers draining to the Shaksgam and Eastern Yarkand tributaries include the largest in China, the Skamri or *Yengisogat Binchuan*, 50 km in length and over 500 $km^2$ in area (Yafeng et al. 2010).

Several clusters of intermediate and small Karakoram glaciers occur south of the main axis, between it and the courses of the Shyok, Middle Indus and Gilgit Rivers. Others are to the northwest, mainly in the Shimshal and Chapursan Basins. The perennial snow and ice cover of these mountain areas varies from 20 % to 50 % (Dainelli and Marinelli 1928, Plate XII), and their glaciers add another 6,500 $km^2$ or so to the Karakoram total.

The larger Karakoram glaciers are exceptional for the subtropics and compared with most of High Asia (Table 2.1). They are of special interest in themselves, because they comprise such huge stores of fresh water and make significant contributions to the Indus or Yarkand Rivers. Water yields are as much as ten times greater than their share of the whole basin area.

Many of the larger glaciers have received little attention and the smaller ones none at all. Taken alone, however, even the latter add up to greater concentrations of ice than some of the most intensively studied systems elsewhere. And the small glaciers are usually more crucial to the livelihoods of people living in the mountain. In part, it is because the valleys of the lesser ranges provide more land suitable for settlement and partly because local communities are better able to exploit waters

---

[1] I estimate that not more than 4,500 $km^2$ of Karakoram perennial snow and ice is in the Yarkand drainage. Yafeng et al. (2010, F143) find the glacier cover on the north, China flank to be 6,262 $km^2$. It seems high, but opinions vary as to how far east the Karakoram extends, and their estimate seems to include the Aghil Range and others adjacent to the Karakoram in the Yarkand Basin.

[2] Chiantar Glacier, at the head of the Yarkhun–Chitral River, is often placed in the Hindu Raj (von Wissmann 1959) but is treated as a 'Karakoram' Glacier here. Its headwaters are in the western extremity of the range. Also in the upper Indus Basin, on the Chitral flank of the Tirich Mir range, are three glaciers of intermediate size: the Tirich, 21 km long; Atrak, 21 km; and Kotgaz, 16 km (ibid, p. 135; Shroder and Bishop 2010, F 211–212). Two Nanga Parbat glaciers, the Rupal (Toshain), at 15.5 km, and Rakhiot, 15 km, are just within the intermediate category. A few glaciers of intermediate size drain to the trans-Himalayan Indus from the Greater Himalayan range to the southwest of the Karakoram.

[3] Only the Fedchenko Glacier in the Pamir Range, which is 77 km long and 652 $km^2$ in area, exceeds all Karakoram Valley glaciers in length, but seven of the latter have larger basins (Kotlyakov et al. 2010).

**Table 2.1** Some dimensions of the fifteen largest glaciers in the Karakoram, including three in Yarkand Basin (= Shaksgam River), sorted by basin area

| Glacier | Basin area ($km^2$) | Length (km) | River basin |
|---|---|---|---|
| Siachen | 1,400 | 75 | Shyok–Indus |
| Baltoro | 1,270 | 62 | Braldu–Indus |
| Biafo | 855 | 68 | Braldu–Indus |
| Hispar | 785 | 53 | Hunza–Indus |
| Batura | 710 | 59 | Hunza–Indus |
| Chogo L. | 690 | 47 | Shigar–Indus |
| Panmah | 680 | 44 | Braldu–Indus |
| Rimo | 612 | 45 | Shyok–Indus |
| Skamri | 510 | 40 | Shaksgam–Y. |
| Kondus–K. | 490 | 36 | Shyok–Indus |
| Chiantar | 436 | 51 | Chitral–Indus |
| Braldu | 430 | 36 | Shaksgam–Y. |
| Khurdopin | 415 | 41 | Hunza–Indus |
| Sarpo Largo | 390 | 32 | Shaksgam–Y. |
| Virjerab | 389 | 36 | Hunza–Indus |

from small glaciers (Chap. 13). The latter appear more sensitive to climate fluctuations adding to their significance for mountain societies. The lack of investigations limits what can be said, but an attempt to increase awareness is made below.

## 2.2 Verticality: The Primary 'Himalayan' Dimension

In a high mountain context, the elevations at which glaciers exist and elevation ranges they span are critical in their maintenance, morphology and behaviour (Plate 2.1). The Karakoram is singled out as having the greatest array of high peaks, elevated mass and deep dissection in the whole Himalayan region and said to include the 'steepest places on Earth' (Miller 1984).[4] There are four peaks over 8,000 m and 71 recognised summits between 7,000 and 8,000 m. As noted, the bulk of the ice cover originates in the highest and steepest parts of the mountain range.

In high mountains, the consequences of elevation relate to, but differ from, free air phenomena or elevated plateaux.[5] Environmental relations between elevation and topography are as important as their absolute dimensions and need to be addressed together. The term 'verticality' conveys this. In particular, it identifies issues 'Relating to or composed of elements *at different levels*' and 'of, constituting, or resulting in, *vertical combination*' (Dictionary.com 2010, *emphases added*). Verticality includes conditions defined by the 'vertical zonation' of Klimek and Starkel (1984), the

[4]While these comments are often made, it seems the greatest vertical span of a rock face is the nearby south-facing Rupal Wall of Nanga Parbat and the highest measured elevation range in the Annapurna Region of Nepal (Dr. M. Nüsser, personal communication).

[5]The words elevation, altitude and height are used interchangeably, except where 'elevation' refers to the act of lifting.

**Plate 2.1** Bualtar Glacier descends northwards from Minapin Peak (7,266 m) in the Rakaposhi Range, a fall of some 4,965 m in 22 km to its snout. The extreme steepness and elevation ranges of larger Karakoram glacier basins are indicated, and a sense of the great vertical changes in conditions from the debris covered ice of lower tongue to precipitous, avalanched walls at the head (Hewitt June 1997)

'altitudinal belts' of Ives et al. (1997, p. 7) and the 'elevation effect' and 'altitudinal organization' of Hewitt (1993, 2005). While these terms tend to privilege horizontal zoning, verticality directs attention to spatial–physical relationships in which gravity is of foremost significance. It reinforces the sense of connections up- and downslope, quite as important as the distinctive elevation belts evident in high mountains.

Phenomena associated with verticality also increase in importance and intensity as elevation and, especially, elevation range and slope angles increase. In the Karakoram, these drive key aspects of glacial and other cold-related or cryosphere hydrological and geomorphic processes. Geographically, the scope and diversity of conditions relating to verticality tend to increase towards lower latitudes. Studies of cryogenic features and vegetation zones suggest the subtropics are where the widest diversity and strongest gradients occur (Troll 1954; Ives et al. 1997). It is partly because of greater variety in the lower elevations, as well as the requirement of very high elevations if the mountains reach into cold climates. In the Himalaya–Karakoram–Hindu Kush (HKH), conditions and ecosystems at lower as well as higher elevations vary markedly according to prevailing weather systems, degrees of rain shadowing, topography, steepness and orientation.

In the Karakoram, great height, elevation range and the sheer extent of high altitude terrain are decisive for the distribution of snow and ice. Such conditions are, therefore, pivotal throughout this book. Verticality will be looked at first as it relates to relevant dimensions of glacierised areas and, especially, glacier basins.

## 2.3 A Main Set of Glaciers

With respect to size, Karakoram glaciers will be treated in four groups. The first three groups are valley glaciers. Those called 'large' have main ice streams longer than 40 km, which also involve basin areas exceeding 400 km$^2$. There are twelve of these in total. 'Intermediate' glaciers are between 15 and 40 km in length, and 35 have been identified. Together, these large and intermediate ice masses comprise about two-thirds of the Karakoram glacier cover. The remaining valley glaciers are 'small', between 2 and 15 km long. There are probably as many as 2,000 of these, although they comprise the lesser share of total ice cover. By far the most numerous is a fourth group of 'minor' ice masses, less than 2 km in length or diameter, probably more than 7,000 in all (Williams and Ferrigno 2010). Most of these are not strictly valley glaciers, but occur in a wide variety of forms addressed in Chap. 3. They are scattered below lesser peaks and watersheds with limited areas in the perennial snow zone, or high up within the perennial frozen zone of larger glacier basins, but disconnected from main valley glaciers.

To help define some basic dimensions of glacierisation, a reference set of 42 glaciers has been chosen (Table 2.2). They are all of intermediate or large size and have basins that are 130 km$^2$ in area or larger, up to the maximum of Siachen Glacier at 1,400 km$^2$. These basins comprise about 75 % of the perennial snow and ice cover of the Karakoram. Being much thicker than the majority of smaller glaciers, they store a larger fraction of regional ice mass. They will serve, firstly, to establish area, verticality and environmental relations of the glaciers and their basins. The 'glacial zone' is treated as all areas above glacier termini. It includes, in addition to parts covered by glacier ice, seasonally thawed off-ice areas surrounding the lower glacier tongues and perennially frigid areas above, mostly too steep to support build-up of snow and glacier ice.

### 2.3.1 *Available Relief and Elevation Range of Glacier Basins*

As a whole, the Karakoram glacial zone spans more than 6,300 m vertically, that is, from the summit of K2 (8,610 m) to the lowest glacier termini which, in the Hunza Valley, reach down to, and occasionally below, 2,300 m. No individual basin spans the whole elevation range, but some are exceptional with over 5,000 m (Fig. 2.2; Table 2.3). At least 34 fall more than 3,000 m.[6]

There are dozens of small Karakoram glaciers with basins spanning 2,500–3,500 m and some that rival the larger glaciers. In the Hunza Valley, Pisan Glacier falls 5,388 m in just 11 km. Compare this with the largest valley glacier in western Eurasia, the Aletsch in the Swiss Alps, with a maximum descent of 2,540 m in 23 km. The major glaciers of the Alaska–Yukon ranges are much larger than those of the

[6]This refers to the elevation difference between termini and highest points on the basin watersheds. Few of the main connected glaciers are continuous over the full span (Chap. 3).

**Table 2.2** Locations and dimensions of the main set of 42 Karakoram glaciers, of which seven drain to the Yarkand from Shaksgam River[a]

| Glacier | Lat.: °N; Long: °E | Basin area ($km^2$) | Length (km) | River basin |
|---|---|---|---|---|
| Siachen | 35 28; 77 01 | 1,400 | 75 | Shyok–Indus |
| Baltoro | 35 44; 76 31 | 1,270 | 62 | Braldu–Indus |
| Biafo | 35 56; 75 36 | 855 | 68 | Braldu–Indus |
| Hispar | 35 04; 75 18 | 785 | 53 | Hunza–Indus |
| Batura | 36 33; 74 37 | 710 | 59 | Hunza–Indus |
| Chogo Lungma | 35 56; 75 06 | 690 | 47 | Shigar–Indus |
| Panmah | 35 56; 75 56 | 680 | 44 | Braldu–Indus |
| Rimo | 35 26; 77 24 | 612 | 45 | Shyok–Indus |
| Skamri | 36 04; 76 08 | 510 | 40 | Shaks–Yark'd |
| Kondus–K. | 35 28; 76 39 | 490 | 36 | Shyok–Indus |
| Chiantar | 36 45; 73 43 | 436 | 51 | Chitral–Indus |
| Braldu | 36 08; 75 51 | 430 | 36 | Shaksgam–Y. |
| Khurdopin | 36 13; 75 27 | 415 | 41 | Hunza–Indus |
| Sarpo Laggo | 35 54; 76 17 | 390 | 32 | Shaks–Yark'd |
| Virjerab | 36 13; 75 38 | 389 | 36 | Hunza–Indus |
| Terong NS | 35 31; 77 24 | 355 | 26 | Shyok–Indus |
| Yashuk Y. | 36 42; 74 18 | 330 | 20.5 | Chapursan–Indus |
| Hasanabad | 36 25; 74 33 | 320 | 23 | Hunza–Indus |
| Bilafond | 35 16; 76 52 | 290 | 19 | Saltoro–Indus |
| N. Shukpa | 34 49; 77 52 | 280 | 25 | Shyok–Indus |
| Charakusa | 35 30; 76 28 | 265 | 18 | Hushe–Indus |
| Kukuar | 36 31; 74 12 | 225 | 22 | Hunza–Indus |
| Toltar–Bal. | 36 29; 74 24 | 205 | 17 | Hunza–Indus |
| Yazghil | 36 23; 75 23 | 220 | 29 | Shimshal–Indus |
| Sosbun | 35 50; 75 35 | 215 | 16 | Braldu–Indus |
| Karambar | 36 36; 74 09 | 205 | 23 | Gilgit–Indus |
| Barpu | 36 08; 74 51 | 198 | 29 | Hunza–Indus |
| N. Gasherbr | 35 50; 76 41 | 190 | 26 | Shaksgam–Y. |
| Sherpigang | 35 21; 76 55 | 190 | 20 | Saltoro–Indus |
| Ghondo'ro | 35 35; 76 22 | 180 | 21 | Shyok–Indus |
| Momhil | 36 29; 75 03 | 175 | 25 | Shimshal–Indus |
| Singhi | 35 40; 77 00 | 170 | 24 | Shaksgam–Y. |
| Malangutti | 36 29; 75 13 | 170 | 22 | Shimshal–Indus |
| Urdok | 35 45; 76 45 | 162 | 26 | Shaksgam–Y. |
| Gyong | 35 08; 76 59 | 160 | 16 | Saltoro–Indus |
| Chong Khum | 35 12; 77 33 | 155 | 21 | Shyok–Indus |
| Sokha Lungma | 35 56; 75 26 | 150 | 15.5 | Basha–Indus |
| Bualtar | 36 08; 74 44 | 140 | 22 | Hunza–Indus |
| S. Shukpa | 34 43; 77 47 | 135 | 21 | Shyok–Indus |
| Gharesa | 36 15; 74 58 | 135 | 22 | Hunza–Indus |
| Kutiah | 35 48; 74 58 | 130 | 18.5 | Stak–Indus |
| Kyagar | 35 39; 77 10 | 130 | 19.5 | Shaksgam–Y. |
| **Total** | | **15,542** | **1,312** | |
| **Mean** | | **370.0** | **31.2** | |

[a]The northern part of the Rimo drains across the watershed to the eastern Yarkand River

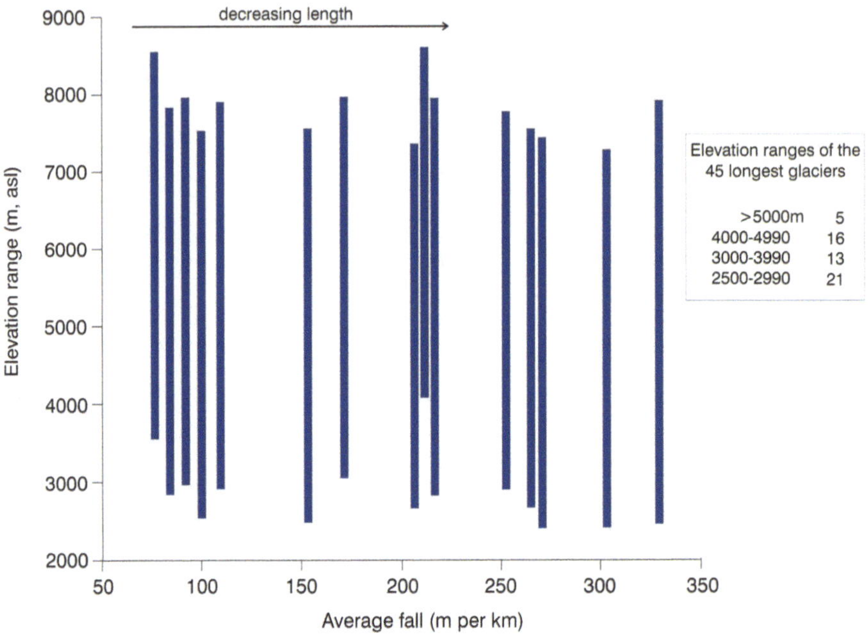


**Fig. 2.2** Elevation ranges of the largest Karakoram Glaciers

**Table 2.3** Elevations, elevation range and gradients of the main set of Karakoram glaciers

| Glacier | Lowest elevation (m asl) | Highest elevation (m asl) | Gradient (m per km) |
|---|---|---|---|
| Siachen | 3,600 | 7,742 | 55 |
| Baltoro | 3,600 | 8,610 | 81 |
| Biafo | 3,070 | 7,285 | 62 |
| Hispar | 2,950 | 7,885 | 93 |
| Batura | 2,570 | 7,795 | 86 |
| Chogo Lungma | 2,760 | 7,458 | 100 |
| Panmah | 3,500 | 7,285 | 86 |
| Rimo | 4,850 | 7,455 | 74 |
| Skamri | 3,910 | 7,265 | 84 |
| Kondus–K. | 3,320 | 7,665 | 136 |
| Chiantar | 3,600 | 6,245 | 52 |
| Braldu | 3,540 | 6,477 | 73 |
| Khurdopin | 3,200 | 7,760 | 111 |
| Sarpo Laggo | 3,900 | 7,360 | 124 |
| Virjerab | 3,540 | 6,477 | 74 |
| Terong NS | 3,975 | 7,385 | 131 |
| Yashuk Y. | 3,380 | 6,928 | 173 |
| Hasanabad | 2,530 | 7,762 | 267 |
| Bilafond | 3,780 | 7,428 | 192 |
| N. Shukpa | 4,670 | 7,672 | 120 |

(continued)

**Table 2.3** (continued)

| Glacier | Lowest elevation (m asl) | Highest elevation (m asl) | Gradient (m per km) |
|---|---|---|---|
| Charakusa | 3,480 | 7,282 | 211 |
| Kukuar | 2,880 | 7,168 | 185 |
| Toltar–Bal. | 2,950 | 7,785 | 252 |
| Yazghil | 3,230 | 7,852 | 159 |
| Sosbun | 3,545 | 6,462 | 182 |
| Karambar | 2,870 | 7,168 | 172 |
| Barpu | 2,835 | 7,453 | 159 |
| N. Gasherbr | 4,230 | 8,047 | 167 |
| Sherpigang | 3,510 | 7,742 | 212 |
| Ghondo'ro | 3,435 | 7,163 | 178 |
| Momhil | 2,895 | 7,885 | 200 |
| Singhi | 4,520 | 7,245 | 114 |
| Malangutti | 2,910 | 7,885 | 226 |
| Urdok | 4,260 | 8,068 | 147 |
| Gyong | 4,140 | 6,620 | 155 |
| Chong Khum | 4,700 | 7,530 | 135 |
| Sokha Lungma | 3,255 | 6,413 | 204 |
| Bualtar | 2,300 | 7,266 | 169 |
| S. Shukpa | 4,500 | 7,513 | 143 |
| Gharesa | 3,500 | 7,728 | 197 |
| Kutiah | 2,800 | 7,397 | 271 |
| Kyagar | 4,740 | 7,245 | 134 |
| **Total** | **147,730** | **310,856** | **6,146** |
| **Mean** | **3,517.4** | **7,401.3** | **146.3** |

Karakoram, and some that descend to or near sea level have comparable elevation ranges, but their interfluves are not so high.[7] The highest Karakoram elevations are matched in some other parts of the Himalaya, but few glaciers descend as low or have comparable elevation ranges. Two of the larger Everest region glaciers, the 12 km long Khumbu in Nepal and the 17 km long Kangshung on the Tibetan side, commence on Mount Everest (8,848 m), higher than any Karakoram glacier. But they terminate at 4,800 and 4,560 m, respectively, a vertical span of just over 4,000 m. The Pisan is shorter than either, of course, and its neighbour, the 16 km Minapin Glacier, spans 4,920 m vertically. Since lower elevations are immediately available for Everest glaciers, the main constraint must be the balance between nourishment and melting of the ice, not available relief.

In the Karakoram, on the other hand, elevation range of the larger glaciers is constrained more by available relief, or basin morphology, than nourishment. Most have long, low-angle tongues (Plate 2.2). The valleys in front of them are mostly of relatively low gradient too. Advances or retreats of some kilometres may have small effect on elevation range, meaning terminus elevations are fairly insensitive to changes in glacier mass (Plate 2.3). Similar reasons explain why termini towards

[7] The highest point is Mount McKinley (6,193 m).

**Plate 2.2** The broad, relatively low-angle main ice stream of Panmah Glacier, Nobonde Sobande branch, is shown between 4,300 and 5,300 m. It is about 2 km wide and 20–30 km above the terminus and typical of the main ice mass areas of the larger Karakoram Valley glaciers. The highest peak shown is Bobisghir (6,415 m) in the centre background. The highest point in Panmah Basin is Baintha Brakk (7,285 m) and total elevation range 3,785 m

**Plate 2.3** Terminus area of Panmah Glacier showing relatively low-angle tongue and river flats or outwash plain. A retreat of some 2 km between the 1920s and 2005 changed the glacier's elevation span by barely 100 m (Chap. 12)

the eastern limits of the Karakoram, and on its northern Yarkand flank, are *not* among the lowest, regardless of glacier size. The North Gasherbrum and Urdok Glaciers, for example, descending from K2 have elevation ranges of roughly 3,800 m. They terminate at just over 4,200 m. However, even if they grew by many kilometres, they would not reach as low as, say, 4,000 m or not without considerable deepening of the Shaksgam River valley. Compare them with Karambar Glacier in the western Karakoram and of similar length. Its highest point is just 7,168 m, but a terminus at 2,870 m on the upper Gilgit (Ishkoman) River gives a fall of almost 4,300 m.

The longest glacier, the Siachen, has a terminus at 3,880 m, almost 1,500 m higher than the lowest in Hunza, essentially because it is 400 km upstream of the lowest sections of the Indus River in the Karakoram. Higher in the Shyok Basin, a huge glacier like the Rimo terminates at 5,000 m.

In this regard, it is useful to compare Karakoram glaciers with those on nearby Nanga Parbat (8,125 m). Maximum elevations and available relief are even greater than in the nearby Hunza Basin, and Nanga Parbat's west flank is as steep or steeper (Hewitt 2006). The Indus is incised to a greater depth. Yet, no glacier reaches as low down as in Hunza. The Rakhiot Glacier descends almost 4,500 m, beginning at the summit, to terminate at 3,070 m. However, below where the glacier ends, the valley continues to fall steeply another 2,000 m to the Indus for a total drop of over 7,000 m in 21 km. The glacier's vertical span is constrained by ice supply and not by available relief.

This can also be put in perspective by considering the largest glaciation(s) of the Quaternary. During maximum glaciations, Nanga Parbat ice moved well down into the Indus valley, the depositional evidence suggesting a total vertical span of at least 7,300 m (Shroder et al. 1989; Searle 1991, his Chap. 11; Kuhle 2006; Hewitt 2009). It was and remains a greater range than any other valley glacier in the world – with one possible exception! A continuous 'Indus Glacier' from K2 through the Nanga Parbat–Haramosh Massif would have had a greater vertical span, or around 8,000 m. However, there is uncertainty about whether or when this existed (Shroder 1993).

## 2.4 Interfluves and Glaciation Limits

Topography is, ultimately, a legacy of the interaction of tectonics and erosion processes working against rock resistance. Substantial changes in relief usually take many thousands of years, if not millions. At any given time, available relief or the height of interfluves and valley floors are relatively inflexible constraints and may be regarded as essentially fixed. They intervene in the atmospheric processes that supply and remove snow and ice. In high mountains, along with orientation of watersheds, interfluve heights exercise a critical influence over the incidence and scale of glacial phenomena.

Interfluve elevations can also be used to identify thresholds critical for cryosphere developments that are, otherwise, hard to observe in rugged terrain. Five threshold features for Karakoram glaciers are identified here:

(i) *The glaciation level or threshold*: the lowest interfluve heights which support glaciers at, descending from, or at some distance *below them*. The latter needs emphasis because of the role of avalanche nourishment. The 'glaciation level' is a widely employed limit (Ostrem 1966; Benn and Evans 1998, pp. 41–43).
(ii) *The glacier-free zone*: the highest interfluve(s) *without* glaciers. In the Karakoram context, this means having no glacier ice at and descending from the interfluve *or* displaced downslope by avalanche nourishment.
(iii) The *highest suitable peak* supporting, or *not* supporting, a glacier, a term introduced by Ostrem (1966).
(iv) *The lowest reach of glacier termini.*
(v) *The greatest elevation span of glacier ice*: the glacier(s) with the largest vertical or relief range in given valleys or sub-ranges.

To arrive at some values, interfluve elevations were measured along four north–south transects. They are based on crest line or arête heights sampled in multiple short scans across interfluves using satellite imagery and, where available, good contour maps. The complexity of Karakoram interfluve topography, including ridge orientation and microclimates, means it is best to sample interfluves fairly frequently in given areas. The figures determined here involve 10–15 local ridge heights and cross-ridge conditions, along one or more interfluves within a 5 km radius at intervals of 50 km along each N–S transect. In fact, just the highest local peak proved sufficient to define the glacier level to within 200 m, a tiny percentage of available relief. Variability arises less from how one measures interfluve heights than from how terrain influences the complex relations between wind direction and deposition, scouring and redistribution by wind action and exposure to sunshine (Barry 1992, p. 82). As with the problems of 'snowlines', there is considerable local variability, but the interfluve heights themselves are fixed.

The observations identify the absence or variable local presence of perennial ice masses. The limits given are mainly defined from lesser offshoots or ranges surrounding the high Karakoram. In much of the high Mustagh Karakoram, a glaciation limit cannot be determined as such because local interfluves are well above that limit and support glaciers.

### *2.4.1 Illustrative Transect, East Central Karakoram*

A transect in the East Central Karakoram serves to show the threshold levels identified and some of the issues they raise (Table 2.4). Ridge heights were sampled and relevant glacial phenomena recorded in a belt roughly 100 km long and up to 40 km wide, from Shaksgam River through the K2 Massif southwards to the Indus River.

Along this transect, the lowest interfluves supporting a glacier, or glaciation levels, cluster around 5,250 m on north-facing slopes and 5,500 m on south-facing

**Table 2.4** Glacier thresholds based on a transect north from 2,465 m on the Indus River through the Shimshak Mountains and K2 Massif to Shaksgam River at 4,200 m (Lat: 35°16′-55′; Long: 76°15′). Where they involve particular glacier basins, they are specified

| Cryosphere threshold | Elevations (m) | |
|---|---|---|
| *Glaciers* | | |
| (i) Glaciation level | Average | Extremes |
| North faces | 5,250 | 5,050 |
| South faces | 5,500 | 5,350 |
| (ii) Glacier-free zone (at and below) | | |
| North faces | 5,050 | |
| South faces | 5,350 | |
| (iii) Highest 'suitable' peak | 8,610 (K2) | |
| (iv) Lowest glacier termini | 3,050 (Biafo Glacier) | |
| | 3,400 (Baltoro Glacier) | |
| | 4,200 (N. Gasherbrum Glacier, Shaksgam, Karakoram north slope) | |
| (v) Glacier elevation spans | | |
| North faces | 4,460 (K2 Glacier–Shaksgam) | |
| West faces | 5,210 (Baltoro Glacier) | |
| South faces | 4,345 (Kondus–Kaberi Glacier) | |
| East faces | 3,820 (N. Gasherbrum Glacier–Shaksgam) | |
| Greatest span | (8,610–3,400) = 5,210 | |

slopes. Extreme values show that a few glaciers occur with interfluves as low as 5,050 m but, in some south-facing slopes, are not found if interfluves are below 5,500 m. A glacier-free zone is defined at and below 5,350 m on southerly slopes and 5,050 m on northerly ones.

Where interfluves are at the glaciation level and a few hundred metres above it, only minor glaciers develop, rarely more than 1 km in diameter. The relation between ice mass elevations themselves and glaciation level is not straightforward. Minor glaciers can start at and hug the interfluve. Small aprons or flag-like tongues of snow and ice may delineate the watershed, usually on the lee flank. Equally common, however, is to find *no* glaciers at the actual interfluve. The ice masses are displaced well downslope, usually being avalanche fed, sometimes wind fed. In this transect, again according to orientation, glacier ice may commence 500–1,000 m *lower* than the glaciation limits, and some minor glaciers terminate below 4,400 m.

For many interfluves close to the glaciation level, glacier ice is interspersed with rock glaciers and in complex relations to them. Separating the two is complicated by local topography, wind and avalanche nourishment, debris covers on the ice and abundant talus. Given the complexities and neglect of the considerable numbers of rock glaciers, these questions will be addressed in a separate chapter (Chap. 11).

Ostrem's (1966) notion of a 'highest suitable mountain without a glacier' (Sugden and John 1976, p. 92; Benn and Evans 1998, p. 42) needs qualification. In the Karakoram, there are vast areas of rock wall and countless summits above the glaciation limit that do not have glaciers on them. There are plenty of summits 'without space for a glacier'. However, there is no evidence for an upper limit to glacier

**Plate 2.4** A view of K2 (8,610 m) from Concordia (4,600 m), showing the vast extent of rock walls and small cover of glacier ice on this flank, which faces the prevailing winds (Hewitt 2005, see text)

formation. For example, in classic views of K2 from the south, glacier ice is virtually absent from the last 2,500 to 3,000 m below the summit (Plate 2.4). The north and northeast flanks tell a different story. Ice aprons and small hanging glaciers are within 300–600 m of the summit, therefore above 8,000 m. They reflect prevailing wind direction and wind-deposited snow on the lee slopes. Meanwhile, the base of the peak is completely enveloped in glaciers that derive most nourishment from avalanches down its rock faces. Much the same applies to nearly all peaks above 7,000 m. Windward flanks are commonly bare of ice but leeward flanks draped by perennial snow and minor ice masses. Large snow cornices and aprons at these elevations show that wind action is more critical for snow concentration than precipitation.

Nevertheless, the K2 situation shows that enough snow for glacier nourishment is available up to the highest elevations. This contrasts with Ostrem's (1966) idea, deduced mainly from conditions in Europe and North America. It may apply elsewhere in the Himalaya, reports suggesting some of the highest peaks lie in frigid deserts (Harper and Humphrey 2003). No glaciers are found on or below them because snowfall is absent or insufficient to sustain ice masses. In the Karakoram, absence of glaciers at any place above the glaciation limit is due to steep walls and high winds, not lack of snow.

As a rule, basins with higher elevations not only have greater ice covers but their glaciers penetrate to lower elevations. Glaciers surrounding massifs culminating above 7,500 m are much more extensive than around those culminating between,

**Table 2.5** Glaciation limits and other relations of elevation and terrain in metres for different south–north transects across the Karakoram

| Cryosphere threshold | Darkot–Karambar Pass | | Hunza–Bagrot–Sost. | | K2–Shimshak Mts | | Rimo–Chang Chenmo | |
|---|---|---|---|---|---|---|---|---|
| | Lat: 36°30′–54′ | | Lat: 36°00′–44′ | | Lat: 35°16′–55′ | | Lat: 34°30′–35°30′ | |
| | Long: 73° 31′ | | Long: 74° 32′ | | Long: 76°15′ | | Long: 77° 30′ | |
| (i) Glaciation level | Average | Extreme | Average | Extreme | Average | Extreme | Average | Extreme |
| North faces | 4,600 | 4,500 | 4,900 | 4,550 | 5,250 | 5,050 | 5,800 | 4,750 |
| South faces | 5,200 | 5,150 | 5,350 | 5,300 | 5,500 | 5,350 | 6,100 | 5,930 |
| (ii) Glacier-free zone | | | | | | | | |
| North faces | 4,720 | | 4,800 | | 5,000 | | 5,960 | |
| South faces | 5,300 | | 5,200 | | 5,350 | | 6,020 | |
| (iii) Highest 'suitable' point | 6,860 | | 7,794 | | 8,610 | | 7,742 | |
| (iv) Lowest glacier terminus | 2,750 | | 2,310 | | 3,050 | | 4,600 | |
| Greatest span | 6,860–2,900 | | 7,794–2,700 | | 8,610–3,400 | | 7,455–4,850 | |

say, 6,500 and 7,000 m. The K2 Massif not only supports the most extensive glaciers in the region but Baltoro descends to 3,600 m and Kondus–Kaberi to 3,300 m, more than 2,500 m below the glaciation level.

### 2.4.2 *Glaciation Thresholds Across the Karakoram*

Comparing the results of three other transects, it is shown that glacier limits and thresholds vary across the range from north to south and east to west (Table 2.5). At the western margins, the glaciation level averages 4,600 m for north-facing slopes and 5,200 m for south-facing slopes. Between these and the eastern margins of the Karakoram, it rises by some 1,200 m and 900 m, respectively, or more than 1,000 m on average.

In each of the transects, perennial snow and glaciers tend to form at slightly lower elevations on east-facing slopes than west-facing slopes, and except at the highest levels, as noted, northeast slopes have the lowest descent or reach of glaciers as well as lower glaciation limits. It may reflect a combination of shading from sunlight and blow-over of snow from the dominant westerly airflow. However, in some parts of the Karakoram, these limits are lower on westerly slopes, while, elsewhere in the High Asian mountains, glaciers may commence at lower elevations on southerly slopes (Williams and Ferrigno 2010). It is a reminder that sources and quantities of snowfall and, especially, its seasonal occurrence can override the role of sun angle.

In the Karakoram, where the highest elevations occur, north versus south relations of glacierisation are reversed compared to lower watersheds and the lesser ranges. Up to about 6,700 m, the glacier cover and reach of the glacier termini are generally greater on northerly slopes. For watersheds above 6,700 m, however, glaciers on *southerly slopes* tend to be longer and to reach the lowest elevations. These relations of orientation, elevation span and descent of glaciers seem to reflect conditions critical for glacier maintenance. As noted earlier, the lowest penetration of glacier termini also depends on the depth of valley incision in different tributaries of the Indus and Yarkand.

Except for the far west of the Karakoram, all the 42 largest glaciers come from watersheds that rise above 7,000 m (Table 2.3). For these, there are no substantial differences in length or ice area in comparing pairs of glaciers that flow in almost opposite directions, for example, Biafo and Hispar, or Baltoro and Siachen, or Batura and Chiantar. Yet, these pairs of glaciers will be shown to be significantly different in morphology, nourishment and behaviour. For instance, one of each pair is predominantly avalanche nourished, the other by direct snow accumulation, also leading to differences in debris-mantled ice and flow stability (Chap. 3).

## 2.5 Glacier Long Profiles

The great elevation range of Karakoram glaciers goes along with steep gradients (Table 2.3). These tend to decline with glacier length, but in a very noisy way, a further reflection of constraints of available relief, ruggedness and orientation or how the snow and ice cover itself is distributed by altitude. The long profiles of three of the major ice streams with the large elevation ranges will serve to introduce the point (Fig. 2.3).

Two characteristics of these long profiles are of interest. Firstly, most of the vertical descent, some 2/3 to 3/4, is accomplished in less than 1/10 of ice stream lengths, in some of the steepest tributaries, less than 1/50th. This occurs almost entirely in the uppermost basin areas and directs attention to the prevalence of steep rock walls and ice falls there. Secondly, however, by far the longest sections lie in mid-reaches with relatively gentle gradients. Some 85 % of these main ice stream surfaces plot between 4,000 and 6,000 m and about half between 4,500 and 5,500 m. It means the greatest extent occurs in a relatively narrow part of total elevation range, although the highest *and* lowest elevations lie well above or below these midsections. In intermediate and small ice masses, steep sections usually comprise a greater part of long profiles although, typically, many of their lower tongues also have relatively gentle slope, while more than half of their descent involves steep gradients and series of extensive ice falls (Plate 2.5).

It will be necessary to address questions that arise from the large elevation range and steepness of a seemingly small fraction of the profiles, conditions

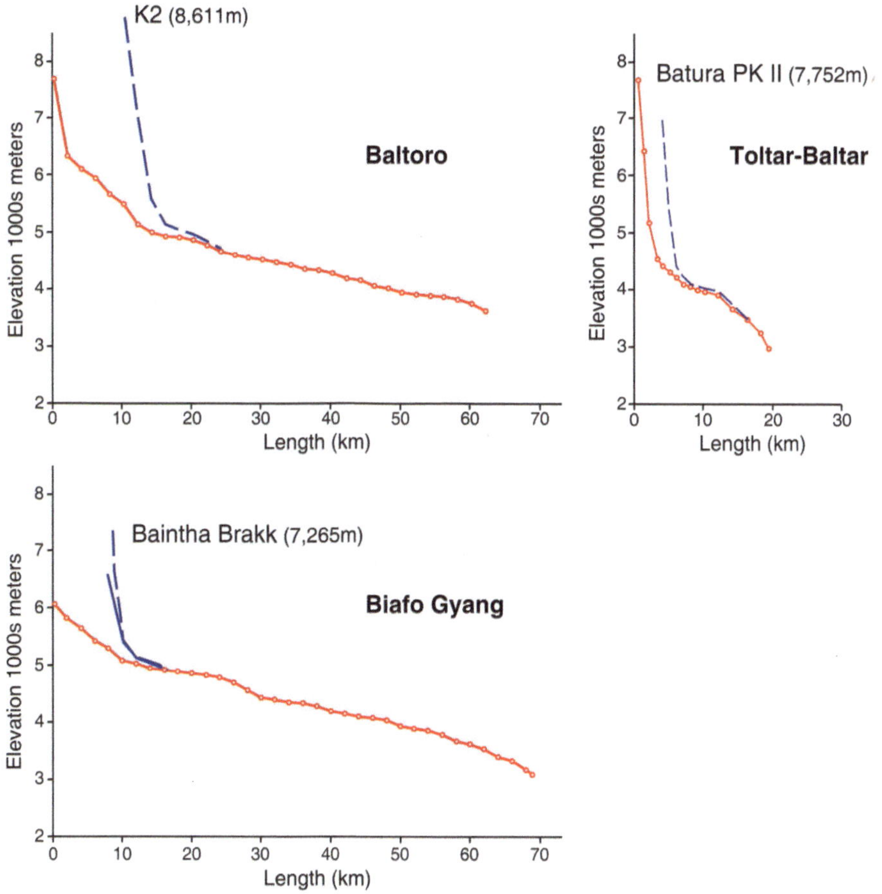


**Fig. 2.3** Long profiles of three large glaciers and their major tributaries

above 5,000 m about which so little is known. The significance of ice tongues below, say, 4,000 m, which most investigations have actually focused on, seems diminished although distinctive because of the unusually low penetration of ice for the HKH region.

## 2.6 Steepland Properties

> Many of the mountains far above the snow-line have such sharp slants and perpendicular faces that snow will not lodge on them, and these remain bare of snow at all times.... (Workman and Workman 1905, p. 261)

**Plate 2.5** Kaberi Glacier, a tributary of Kondus Glacier, EC Karakoram, showing its extensive, debris-mantled lower ice stream of relatively gentle slope and precipitous upper source areas around Chogolisa (7,665 m). It is viewed from 3,800 m and 10 km above the terminus (Hewitt July 1998)

Glacier long profiles only hint at the extent and exceptional steepness of the higher terrain in these basins. Icefalls are common and will be shown to account for the greater part of their vertical descent. Rock walls make up most off-ice areas above and below the perennial snow zone. An example will highlight something of the steepness involved.

### 2.6.1 *Off-Glacier Slopes*

The relatively good-quality 1937 Shipton (1938, 1940) survey and topographical map of Biafo Glacier Basin makes it possible to determine and sample slope angles at higher elevations.[8] Random samples of slope angles surrounding ice masses

[8]The original survey sheet held in the Royal Geographical Society's map library was used, being much more detailed than the published version (details, Hewitt 1968, I, Table 8.1, and II, Appendix 3).

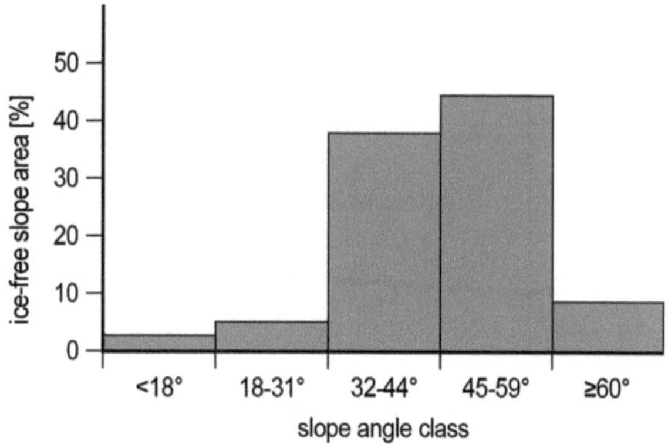


**Fig. 2.4** Distribution of slope angles in off-glacier terrain of Biafo Glacier

indicate a prevailing steepness but also considerable variety. Important features that emerged were as follows:

(i) Angles in excess of 32° predominate almost everywhere away from the glacier surfaces.
(ii) Mean slope angle increases with elevation.
(iii) Above 6,000 m angles greater than 45° apply almost everywhere and those over 55°predominate.
(iv) The largest class of off-glacier slopes is between 55° and 60°.
(v) While the steepest values are more frequent at higher elevations, some occur throughout, meaning an increased spread of angles towards lower elevations, not an absence of steep cliffs there.
(vi) Northerly slopes have somewhat steeper average angles compared to southerly. Angles less than 45° are more frequent on the latter, those greater than 60°more frequent on northerly slopes.
(vii) Of the two predominant rock types, average slope on the granitics is some 13° steeper than for metamorphics. This cannot be attributed only to lithological differences. Granitics mainly outcrop above 5,000 m elevation where angles are much steeper in both rock types. Metamorphics are more prevalent below about 4,800 m where the only slopes in unconsolidated material occur.
(viii) Slopes with lowest angles are most frequent between about 3,500 and 4,800 m, representing talus build-ups which only occur below the snowline and relatively limited areas of lateral moraine and kame terrace deposits along glacier margins.

Almost 50 % of the off-glacier terrain lies between 45° and 60° and 80 % between 32° and 60° (Fig. 2.4). Most of it is so steep it cannot support glacier ice because snow cannot build up there. Such steepness also means that *actual* off-ice surface areas are two or more times greater than their map area, which is likely to be critical for interactions with atmospheric conditions, and in the downslope cascade of snow, ice, meltwater and debris.

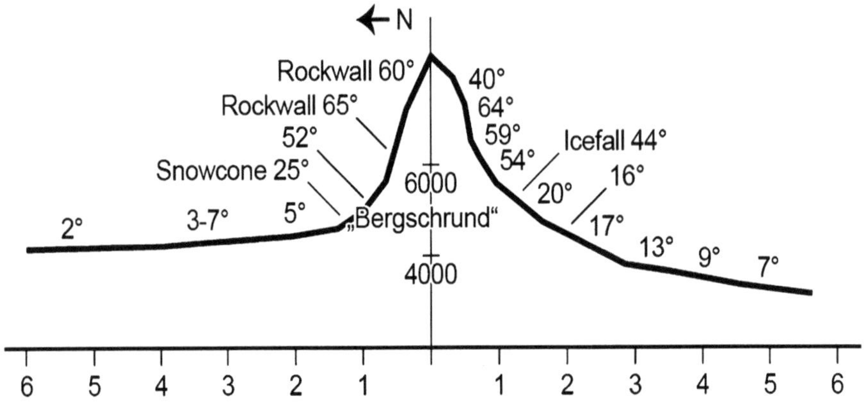


**Fig. 2.5** Slope profiles around Baintha Brakk, Biafo Basin

Two profiles from the highest part of the Biafo watershed around Baintha Brakk (7,285 m) indicate what to expect in the steepest areas of the perennial snow zone (Fig. 2.5). Rock walls prevail and although veneers of snow and ice are present everywhere above 6,000 m, there is almost no glacier cover except near the base of the cliffs. Slopes are generally steeper than 55°, and areas steeper than 65° are not unusual. Avalanche flutings can develop at angles around 40–45°. Areas less than about 55° but steeper than 25° involve complicated transitions between rock wall, ice flutings, wind-shaped cornices, avalanche cones and aprons merging into glacier ice. These various elements can be difficult to differentiate or identify as more than surface forms. Nevertheless, they recur over vast areas of the upper glacier basin areas in the Karakoram (Chap. 3).

While glacier ice can stand vertically in terminal cliffs and crevasse walls, the data here suggest ice streams rarely occur with slopes steeper than about 20°. It is possible that some glacier ice occurs beneath cones and aprons of avalanched and windblown snow with surface slopes between 20° and 35°. The only obvious glacier sections occur as icefalls characterised by intense crevassing and with surface slopes between about 7° and 20°. It is of note that, like so many of these mountains, Baintha Brakk appears massive from certain viewpoints, but others show it actually consists of remarkably narrow, flake-like interfluves between steep bounding walls (Plates 2.6, 2.7, and 2.8).

## 2.7 Glaciers and Regional Hypsometry

The area–altitude distribution of perennial snow and ice needs to be situated within the overall hypsometry of the Karakoram. More than 80 % of the perennial snow and glacier ice cover occurs between 4,000 and 6,000 m (Fig. 2.6). Only about 10 %

**Plate 2.6** The south aspect of Baintha Brakk (7,285 m) viewed from 4,100 m on Biafo Glacier giving an impression of a massive peak (Hewitt 2010)

**Plate 2.7** Baintha Brakk, the right-hand peak, viewed from the east on the upper Choktoi tributary of Panmah Glacier, showing how narrow the mass is (Hewitt 2005)

**Plate 2.8** Baintha Brakk viewed from the west and Hispar Pass, showing the precipitous north flank draped with snow and ice (cf. Fig. 2.5) and narrow, highest pinnacles of rock (Hewitt 2006)

of the whole trans-Himalayan upper Indus Basin is involved, about 30 % of the Indus flank of the Karakoram, and more than 50 % of its Yarkand flank. Conversely, almost 90 % of the two basins between 4,000 and 6,000 m have no perennial snow or glacier ice. This comprises lesser mountains and the high, relatively subdued, arid basin areas in the eastern part adjacent to the Tibetan Plateau. The bulk of glacier masses are found only where the highest mountains occur and descending into their immediate surrounds. Most of the ice is situated in narrow zones immediately beneath mountains that rise above 6,500 m and, overwhelmingly, above 7,500 m. This makes the true high mountain areas profoundly important in flows of the two rivers, despite a small share of the entire basin. In the hypsometric profile, the great array of high interfluves reduces to a single narrow spike, their plan areas being relatively tiny. Their significance in relation to large glaciers and the most concentrated ice cover is out of all proportion to their share of the whole Indus and Yarkand Basins.

In sum, verticality involves some seemingly contradictory features in the perennial snow and ice cover. Most of it, and the largest glacier basins, are associated with the highest watersheds. Watersheds above 6,000 m are decisive for a substantial ice cover and above 7,000 m for the larger ice masses. They are associated with glaciers having the greatest elevation ranges and lowest reach in their part of the mountains. Yet, even in these areas, the largest part of the snow and ice cover is confined, vertically, within a relatively narrow band, or less than a third of basin elevation ranges. In terms of actual glacier area and ice storage, the zone between 4,000 and 6,000 m is crucial. The next chapter explores this by looking more closely at the extent and character of glacier ice.

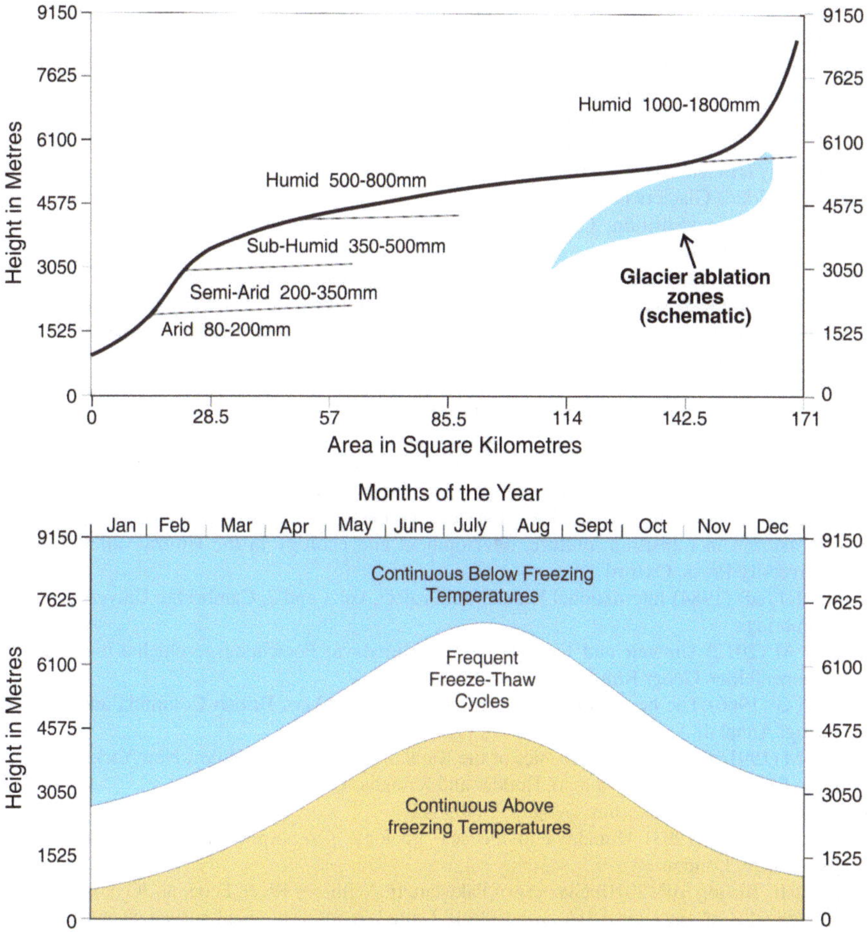


**Fig. 2.6** Schematic relations of precipitation to upper Indus basin hypsometry and glacier ablation zones (above); and relations of temperature and freeze-thaw cycles to elevation (below). Both need to be considered as controls over the seasonal and perennial snow and ice cover (After Hewitt 1993)

# References


Barry RG (1992) Mountain weather and climate, 2nd edn. Routledge, London

Benn DI, Evans DJA (1998) Glaciers and glaciation. Hodder Arnold Publication, London

Dainelli G, Marinelli O (1928) Le condizioni fische attuali. Relazioni Scientifiche della Spedizione Italiana de Filippi nell' Himalaja, Caracorum e Turchestan Cinese (1913–1914) Series 1, vol 8. Bologna, Zanichelli

Dictionary.com (2010) Definition of 'Verticality' http://dictionary.reference.com/browse/verticality?s=t. Accessed Oct 2010

Harper JT, Humphrey NF (2003) High altitude Himalayan climate inferred from glacial ice flux. Geophys Res Lett 30(14):1764. doi:10.1029/2003GL017329

Hewitt K (1968) Geomorphology of the mountain regions of the upper Indus Basin, 2 vols. Dissertation, University of London
Hewitt K (1993) Altitudinal organization of Karakoram geomorphic processes and depositional environments. In: Shroder JF Jr (ed) Himalaya to the sea: geology, geomorphology and the Quaternary. Routledge, New York, pp 159–183
Hewitt K (2005) The Karakoram anomaly? Glacier expansion and the 'elevation effect', Karakoram Himalaya, Inner Asia. Mt Res Dev 25(4):32–40
Hewitt K (2006) Glaciers of the Hunza Basin and related features. Ch 5. In: Kreutzmann H (ed) Karakoram in transition: Culture, development and ecology in the Hunza Valley. Oxford University Press, Oxford
Hewitt K (2009) Catastrophic rock slope failures and late Quaternary developments in the Nanga Parbat-Haramosh Massif, Upper Indus basin, northern Pakistan. Quat Sci Rev 28:1055–1069
Ives JD, Messerli B, Spiess E (1997) Mountains of the world: a global priority. In: Messerli B, Ives JD (eds) Mountains of the world: a global priority. Parthenon, New York
Klimek K, Starkel L (1984) Vertical zonality in the southern Khangai Mountains (Mongolia), vol 136, Prace Geograficzne. PAN Institute Geography, Wroclaw
Kotlyakov et al (2010) In: Williams RS, Ferrigno JG (eds) (2010) Satellite image atlas of glaciers of the world: Asia. Professional paper 1386-F. U.S. Geological Survey and U.S. Government
Kuhle M (2006) The past Hunza Glacier in connection with a Pleistocene Karakoram Ice Stream network during the Last Ice Age (Wuerm). Chap 4. In: Kreutzmann H (ed) Karakoram in transition: culture, development and ecology in the Hunza Valley. Oxford University Press, Oxford
Miller KJ (ed) (1984) International Karakoram Project, vol 1 and 2. Cambridge University Press, Cambridge
Nüsser M (2012) Umwelt und Entwicklung im Himalaya: Forschungsgeschichte und aktuelle Themenfelder. Geogr Rundsch 64(4):4–9 and suppl
Ostrem G (1966) The height of the glaciation limit on northern British Columbia and Alberta. Geogr Annal 48A:126–138
Searle M (1991) Geology and tectonics of the Karakoram Mountains. Wiley, New York
Shipton EE (1938) Blank on a map. Hodder and Stoughton, London
Shipton EE (1940) Karakorum. Geogr J 95:409–427
Shroder JF (ed) (1993) Himalaya to the sea: geology, geomorphology and the Quaternary. Routledge, London
Shroder JF, Bishop MP (2010) Glaciers of Pakistan. In: Williams RS Jr, Ferrigno JG (eds) Satellite image atlas of glaciers: ASIA, vol 1386-F, United States geological survey, Denver, professional paper. U.S. G.P.O, Washington, DC, pp 201–257
Shroder JF, Khan MS, Lawrence RD, Madin IP, Higgins SM (1989) Quaternary glacial chronology and neotectonics in the Himalaya or northern Pakistan. In: Malinconico LL Jr, Lillie RJ (eds) Tectonics of the Western Himalayas, vol 232, Geological Society of America, special paper. Geological Society of America, Boulder, pp 275–294
Sugden DE, John BS (1976) Glaciers and landscape. Edward Arnold, London
Troll C (1954) Über das Wesen der Hochgebirgsnatur. Jahrbuch Deutscher Alpenverein 80:142–147
Wissmann, H von (1959) Die heutige Vergletscherung und Schneegrenze in Hochasien mit Hinweisen auf die Vergletscherung der letzten Eiszeit. Akademie der Wissenschaften und der Literatur in Mainz. Abhandlungen der mathematisch-naturwissenschaftlichen Klasse 14:1103–1431
Williams RS Jr, Ferrigno JG (2010) Satellite image atlas of glaciers: ASIA, vol 1386-F, United States geological survey, Denver, professional paper. U.S. G.P.O, Washington, DC
Workman FB, Workman WH (1905) From Srinagar to the sources of the Chogo Lungma Glaciers. Geogr J 25:245–268
Yafeng S, Mi D, Yao T, Zeng Q, Liu C (2010) Glaciers of China. In: Williams RS Jr, Ferrigno JG (eds) Satellite image atlas of glaciers: Asia, vol 1386-F, United States geological survey, Denver, professional paper. U.S. G.P.O, Washington, DC, pp 127–166

# Chapter 3
# Karakoram Glaciers: Types and Terrain

**Abstract** This chapter focuses on the glaciers themselves, ice morphology and distribution and relations to terrain in their basins. To identify the relative extent of zones with particular conditions, criteria are outlined to define terrain elements on and off the ice. Estimates are given of their share of basin areas for the 42 largest glaciers and provide a basis for the investigation of glacier mass balance and landforms in later chapters. The importance of verticality is further reinforced. Relative distributions of perennially frozen, snow-covered areas and seasonally snow-free and thawed surfaces are underscored. These prove to be distinct from conventional glacier accumulation and ablation zones in some important respects, mainly due to complications arising from the extent of off-glacier rock walls and of snow redistribution by wind and avalanches. Icefalls, supraglacial debris mantles and ice-margin deposits are of particular interest. Glacier ice itself is shown to comprise, on average, one-third of basin areas above glacier termini and about 45 % of perennial snow and ice areas. Rock walls make up more than 65 % of the latter and 60 % of whole basin areas. Considerable differences emerge between basins, as illustrated by comparing Baltoro, Biafo and Toltar–Baltar Glaciers. They support an argument for revisiting older classifications of Karakoram Glaciers into Mustagh, Turkestan and Alpine types, which emphasise differences in nourishment, especially the relative shares of avalanche-fed ice. A fourth 'wind-fed' class, widely present in the region, is added. With certain revisions, these classes help to identify key conditions in the Karakoram. The four types are also distinguished by the proportions of terrain elements in their basins, notably rock walls and the extent or absence of accumulation zones. Classifications by size, morphology, climatic and thermal regimes are also examined. They raise some unique questions and reinforce a sense of the distinctiveness and diversity of ice masses in the Karakoram.

**Keywords** Glacier size, distribution and classification • Alpine-, Mustagh- and Turkestan-type glaciers • Baltoro Glacier • Toltar–Baltar Glacier • Biafo Glacier • Terrain elements • Rock walls • Debris-covered glaciers

K. Hewitt, *Glaciers of the Karakoram Himalaya: Glacial Environments, Processes, Hazards and Resources*, Advances in Asian Human-Environmental Research,
DOI 10.1007/978-94-007-6311-1_3, © Springer Science+Business Media Dordrecht 2014

## 3.1 Introduction

The glaciers of the Karakoram and the types of ice mass found are of foremost interest here. Important features are revealed by looking first at the terrain and environments in glacier basins. In the absence of any glaciological observations from most of the glaciers, and only incomplete ones for individual cases, their visible snow, ice- and land-surface or terrain features offer a starting point. The largest, 1,400 km$^2$ basin of Siachen Glacier, for example, reveals some of the conditions needing to be addressed. Its high, northeast source areas seem to comprise continuous, perennial snow covers familiar in glacier accumulation zones elsewhere. They merge across the interfluves with the headwaters of the Rimo Glacier which this part of Siachen resembles. The upper northwestern source areas also consist of wide accumulation basins, but surrounded by steep, avalanched interfluves. The southwestern tributary basins are extremely rugged, with very few, minor snow basins and extensive rock walls at the head. Main ice streams quickly become covered in heavy supraglacial debris. Such variety is common within and between the 42 largest basins introduced earlier and points to the need to identify the range of conditions and zones present. It is equally important to characterise the extensive off-ice areas within glacier basins and how they relate to the ice cover.

## 3.2 Glacier Basin Terrain

> There is a fundamental difference between glacier surface forms affected by surface melting and those not so affected.... (Sugden and John 1976, p. 71)

Firstly, perennially frozen and snow-covered areas are distinguished from those where glacier ice is seasonally exposed and subject to surface melting. Terms used in mass balance studies for the *accumulation* and *ablation zones* can be applied initially. The former represents the area with a net annual gain of mass to the glacier, derived from direct snowfall and comprising the upper or source area in most valley glaciers. The ablation zone is defined by a net loss of ice, mainly due to summer surface melting of ice transferred from higher areas. Mass balance will be examined in the next three chapters, but surface conditions and related processes in these two zones are distinct and well defined. Their relative shares of basin areas turn out to be good indicators of fundamental aspects of glacier behaviour and environments. Some major differences do emerge from accumulation and ablation zones as usually reported (Table 3.1).[1]

Secondly, a distinction arises between glacierised areas in the strict sense, where active glacier ice occurs and where it does not occur. It is essential to establish the extent of glacier ice and by elevation zones. However, off-ice terrain in glacier basins is also

[1]There are promising developments using satellite imagery for this purpose and the potential of different spectral signatures to record otherwise invisible temperature and energy fluxes, for example, or to highlight or remove particular topographical dimensions (Bishop and Colby 2011). They remain to be applied to most Karakoram glaciers.

**Table 3.1** Classes of terrain, or land surface and topography, identified and of interest to characterise Karakoram Glacier basins

| Feature | Definition or units |
|---|---|
| Basin | Glacier name |
| Location | Latitude and longitude (low to mid-basin location) |
| | River basin |
| Main glacier | |
| Length | Terminus to furthest watershed point of longest ice stream (km) |
| Areas: | $km^2$ |
| Basin | Entire watershed above glacier terminus |
| PSI | Perennial snow and ice: all glacier cover and frigid snow zone |
| Glacier ice | All above the terminus |
| Main connected Gl. | Main glacier system |
| 'Source' zone | Conventional accumulation zone, plus slopes, esp. frozen rock walls, that descend to glacier surface |
| Accumulation zone | Conventional firn basins and streams; direct snow-fed parts of glaciers |
| Rock walls | Frozen, usually with frequent avalanches and steeper than 35° |
| Ablation Zone | Conventional glacier ice surfaces ablated during part of the year |
| Clean/dusty/dirty | Ablation zone surfaces of clean ice, or with thin dirt, fine-grained debris, dust, and scattered coarser debris |
| Debris covered | Heavy supraglacial debris usually with abundant coarse clasts |
| Elevations: | |
| Highest | Highest point on basin watershed |
| Lowest | Terminus height |
| Range | Span between highest and lowest |
| Disconnected tributaries | |
| Sources | |
| Rock walls | Frozen, usually with frequent avalanches |
| Glacier ice | Conventional firn basins and ablation zone ice streams and ablation zones |
| Glacier cover total | Main and disconnected glaciers in the basin |
| Ice- and snow-free areas | Areas off the glacier that are seasonally thawed and snow-free |
| Rock walls | Seasonally thawed and snow free |
| Ice-margin complex | Ablation zone lateral margins with conspicuous deposits |
| Steep rock walls: | |
| Frigid zone | Glacier source zone rock walls |
| Seasonal snow-free | Rock walls below source zone |
| Total | All rock walls in basin |

of great interest, including both seasonally thawed, snow-free areas and the perennially frozen ones. It will be shown that they make up the larger fraction of most basin areas in the Karakoram and prove to be highly significant in glacier maintenance.

Thirdly, glacierised basin areas involve a main, connected glacier system or network and disconnected tributary ice masses. The latter may be minor, or quite substantial. One of the detached tributary basins of Siachen, for example, contains the North and South Terong Glaciers with a maximum ice stream 25 km long, and glacierised area of 350 $km^2$, or 20 % of the basin above the Siachen terminus. It is larger than a majority of the other main set glaciers. For this reason, it is treated as an independent glacier. By contrast, although Baltoro Glacier basin has countless minor masses of disconnected tributary ice, they make up a tiny fraction of the basin area or the glacier cover (3.2.2). Other influential types of terrain are rock walls, debris-covered glacier surfaces, icefalls and sets of ice-margin deposits.

Area extent alone is not necessarily the most critical variable, but the larger the extent of a given type of terrain, the greater the need to give careful consideration to what roles it may be indicative of or play. In the literature, no doubt in fact, some relatively small terrain components have importance out of proportion to their share of basin areas, including some expressions of glacier dynamics that are barely visible at all. Ice-margin depositional features along ablation zones comprise rarely as much as 3 % of existing glacier basins but have received by far the greatest attention. This relates to their importance in reconstructing glacial processes and landforms, Quaternary and contemporary glacier change. These and some other features will be addressed in relation to relevant concerns in later chapters.

The various zones were identified and their dimensions estimated with the aid of detailed contour maps, where available, and visual classification of recent, usually late summer, satellite imagery. The classes were first developed for glaciers where I have spent extended periods of field work and carried out morphological mapping (Hewitt 1968; Hewitt et al. 1989; SIHP 1990).

Technical difficulties with precise delineation or separation of conditions are encountered even on the ground, and with the highest quality imagery. They arise particularly with the boundaries between accumulation and ablation zones, and ice edges generally. In much of the source zones, the latter are obscured by avalanched and windblown snow. It is virtually impossible from available imagery to determine the number and extent of minor ice masses scattered across steep interfluves, detached from the main glacier or obscured by avalanched- and windblown snow. They are a small part of total source areas but important indicators of processes there (Chap. 8). Problems arise along ablation zone margins with heavy debris covers that can make it difficult or impossible to identify transitions from slow-moving to stagnant ice. Transitions from thick and thin debris on active ice, as described below, are impossible to define accurately without ground inspection and difficult even then.

In spite of these limitations, a sense of the orders of magnitude and relative significance of various terrain elements is obtained, and some compelling results easily missed otherwise. A strong sense emerges of the regional scope and scale of the elements making up the glacier basins. They can be introduced by looking at the three distinctly different Karakoram Glaciers whose long profiles were discussed in the last chapter (Table 3.2).

### *3.2.1 Biafo Glacier*

Biafo is the second longest Karakoram Glacier and third largest in area (Fig. 3.1). The extent and share of glacier ice is unusually high for the Karakoram, comprising 63 % of the basin. There is a huge accumulation zone in the conventional sense, almost three times larger than the ablation zone (Plate 3.1), close to the proportions

**Table 3.2** Terrain profiles for Baltoro, Biafo and Toltar–Baltar Glaciers, Central Karakoram

| | Biafo | Baltoro | Toltar–Baltar |
|---|---|---|---|
| Glacier | $km^2$ (% basin) | $km^2$ (% basin) | $km^2$ (% basin) |
| Lat.; long. | 35°55′N; 75°40′E | 35°46′N; 76°15′E | 36°27′N; 74°24′E |
| Length (km) | 68 | 62 | 17 |
| Elevation range (m) | 4,215 | 5,210 | 4,835 |
| Gradient (metres per km) | 62 | 81 | 252 |
| Basin area | 855 | 1270 | 205 |
| Perennial snow and ice | 645 (75) | 910 (72) | 150 (73) |
| Total glacier | 540 (63) | 535 (42) | 49 (24) |
| Main glacier | 460 | 495 | 38 |
| Source zone | 480 (58) | 610 (50) | 85 (37) |
| Firn basins | 330 | 180 (12) | 3 (1.5) |
| Ablation zone | 130 (15) | 260 (21) | 35 (18) |
| Clean/dusty ice | 100 (12) | 175 (13) | 15 (8) |
| Heavy debris | 30 (4) | 85 (7) | 20 (10) |
| Off-ice, seasonally snow-free | 210 (25) | 360 (28) | 56 (27) |
| Rock walls | 385 (40) | 780 (60) | 151 (72) |
| Frigid zone | 155 (15) | 430 (33) | 101 (49) |
| Seasonally snow-free | 210 (25) | 350 (27) | 50 (23) |

reported for most valley glaciers in the literature (Benn and Evans 1998). The perennial snow and ice cover comprise 75 %. The area of steep rock walls, whose angles were described earlier, is about 40 % of the whole basin and 20 % of the perennial snow zone. Both are high compared to the most well-known valley glaciers, but low for the Karakoram. The expanses of rock wall at higher elevations are associated mainly with tributaries such as Baintha Lukpar whose characteristics differ from the main Biafo. The ablation zone area with a heavy debris mantle is 30 $km^2$, a sizeable area but, at 4 % of glacier area, much less than most Karakoram valley glaciers.

## 3.2.2 *Baltoro Glacier*

Baltoro is one of the highest relief basins spanning 5,210 m from K2 (8,611 m) to the snout (Dyhrenfurth et al. 1939). Although a near neighbour of Biafo, it has some very different characteristics. The basin is 400 $km^2$ larger, but the area of the glacier slightly smaller, and barely 42 % of the whole.[2] Perennial snow and ice have almost

[2]This was estimated from a 1:100,000 map (Yoshizawa and Miyamori 1988) and 2006–2009 Landsat imagery. In the literature, values for the glacier area vary considerably. Schneider (1962–1963, p. 56) gives 754 $km^2$, the Snow and Ice Hydrology Project (Hewitt in SHIP 1990, p. 16) gave 756.3 $km^2$, and Pecci and Smiraglia (2000, p. 75) give '…a total surface area of about 800 $km^2$'. These differences may be partly because of the complexity of the basin, partly the quality of maps and images. However, the main point and difference arises from being stricter about just where there is glacier ice. Mayer et al. (2006) recognise this too and give an estimate of 524 $km^2$ for the actual glacier.

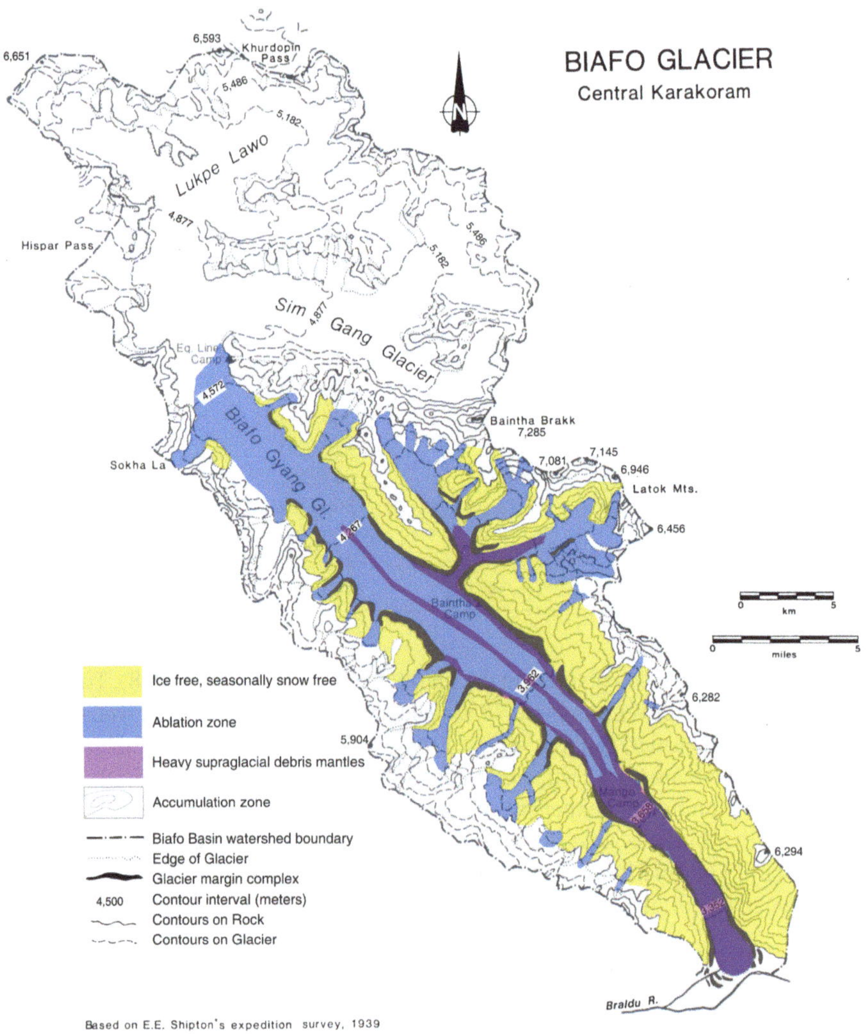


**Fig. 3.1** Biafo Glacier basin showing topography and some main terrain features (After Shipton/RGS archive 1939)

the same area as Biafo's, covering 72 % of the basin. However, the accumulation zone makes up less than a third of this and only 12 % of basin area, much less than Biafo or most valley glaciers described in the literature. Some small, disconnected tributaries on the south-facing north flank occupy less than 4 % of basin area. Dozens of minor ice masses occur high up, entirely within the frigid zone, difficult to delineate but with areas that probably add up to less than 1 % of the basin.

Rock walls comprise 61 % of the whole basin, two-thirds more than the glacier itself (Plate 3.2). In the high, perennially frozen zone, they may be thinly veneered with ice, snow from recent storms or avalanche flutings but are too steep to support a build-up of snow or glacier ice. In addition, steep rock walls make up nearly all of Baltoro's ice-free and seasonally snow-free areas.

**Plate 3.1** Biafo Glacier accumulation zone; a view looking northwards across the extensive firn basin known as Lukpe Lawo or 'Snow Lake' from and elevation of 5,200 m. Lukpe Lawo Brakk (6,593 m) is the peak in the centre background (Hewitt 1985)

**Plate 3.2** Rock walls on the south, Biacherdi Pk (6,781 m) flank of Baltoro Glacier, seen from glacier's irregular debris-covered surface about 30 km above the terminus (Hewitt 2005)

Substantial covers of heavy supraglacial debris are usually associated with lower ratios between ablation and accumulation zones (Benn and Evans 1998, p. 84). The heavy mantle on the lower Baltoro covers 85 $km^2$, almost three times more than Biafo's (Plate 3.3). That is, about one-third of the ablation zone, a considerable fraction but

**Plate 3.3** Baltoro ablation zone just above Urdukas (4,240 m) 15 km above the terminus and upwards for 10 km, showing the heavy debris mantle that obscures nearly all of the ice (Hewitt 2005)

showing that clean or thinly covered ice is twice as extensive. The debris cover has been widely interpreted as a major control over this and other glacier behaviour (Nakawo et al. 2000; Mayer et al. 2006; Smiraglia et al. 2008; Milhalcea et al. 2008). Since it comprises just 7 % of total glacier area this needs to be reconsidered (Chap. 6).

### *3.2.3 The Toltar–Baltar Glacier*

The third example is one of a series of extremely steep basins on the southwest flanks of Batura Mustagh in Hunza and introduces some characteristics even more widely observed in the Karakoram, but rarely described elsewhere. The maximum watershed point is 7,750 m and total fall 4,760 m, slightly less than Baltoro but 500 m more than Biafo. Given the shorter length, steepness is far greater. Terrain proportions are very different from Biafo and more extreme than Baltoro's (Fig. 3.2).

Half of the perennial snow and ice zone consists of rock walls, and a series of disconnected, small and minor ice masses high up are about 20 % of the glacier cover. Much of their surface is buried in windblown or avalanched snow, and they terminate in ice cliffs from which ice avalanches descend to the main glacier system. The total glacier cover is rather less than half the basin area. A novel feature is the near absence of a discernible 'accumulation zone', while the ablation zone comprises nearly all of the main ice mass. The ice streams of the main glacier all

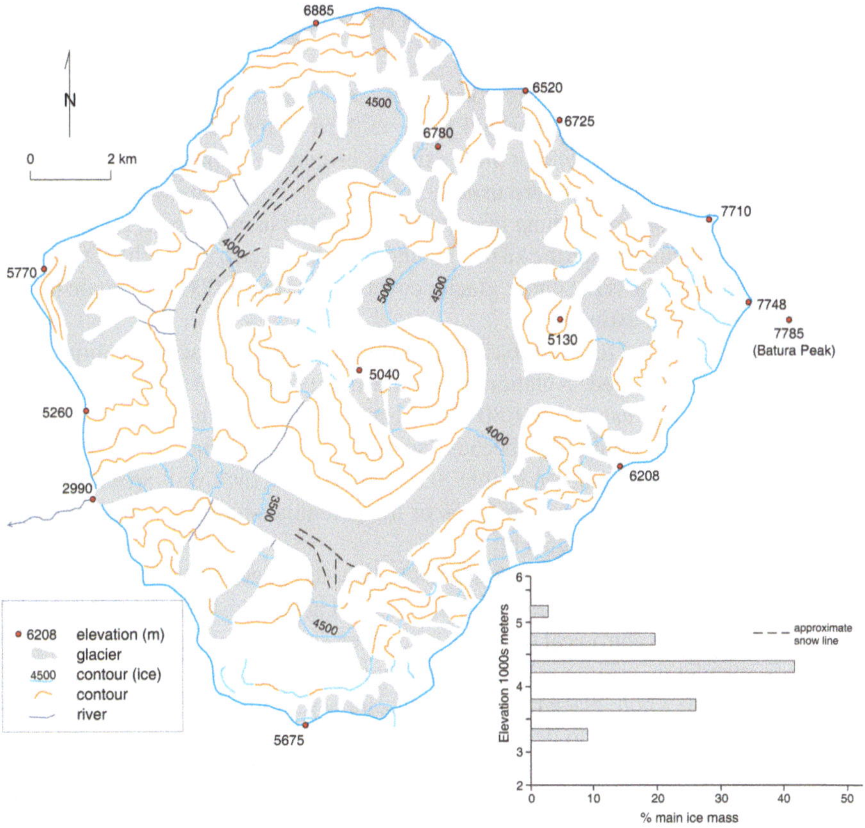


**Fig. 3.2** Topographic map and glacierised area of the Toltar–Baltar Glacier, Batura Range and Hunza River Basin (After Alpensvereinskartographie 1995)

commence at the base of rock walls more than 2,000 m below the higher parts of the watershed and are buried in avalanche cones fed from above. They quickly become heavily mantled in debris and in a greater proportion than lightly covered or clean ice. Even at summer's end, however, snow avalanche aprons and cones are almost continuous on the ice streams or their margins in the ablation zone as low down as 3,700 m beneath northerly slopes, and 4,200 m below southerly ones. They not only protect the ice from ablation but add to it.

Major differences between the three examples involve the extent of their accumulation and ablation zones, of rock walls in the source zones and of debris-covered ice. Their average gradients differ, with Toltar–Baltar's almost four times greater than Biafo's and three times Baltoro's. Nevertheless, as already noted in relation to their long profiles, most of their ice stream areas are concentrated in the same elevation band (see Fig. 2.3). Now it is necessary to ask how representative these examples are, compared with the set of 42 larger glacier basins.

## 3.3 Terrain Features of the Main Set of Glaciers

It is of particular note that the area of glacier ice in the set of basins is barely half the perennial snow and ice cover and just one-third of total area (Table 3.3). It is barely half the estimates found in most existing inventories, apparently because the perennial snow and ice cover are usually treated as synonymous with glacier area (see Table 1.1). This is something that seems reasonable in most areas, but not in the Asian high mountains. Since the 42 basins amount to three-quarters of the Karakoram Glacier zone, it suggests the entire glacier cover is less than 8,000 km$^2$, again barely half of most existing estimates.

Ablation zone areas turn out to comprise slightly more than accumulation zones as conventionally defined. Ablation zones are more than 60 % of the glacier ice cover in half of the cases, over 80 % in one quarter and, in some, over 90 % (Table 3.4). Without doubt, it is because so much of the perennially frigid zone is too steep to support glacier ice (Table 3.5). On average, rock wall areas comprise 65 % of glacier basins, and a huge 72 % of the perennial snow or source zones rising, in some cases, to more than 90 %.

The results may seem to raise doubts as to how the huge valley glaciers are sustained at all, if accumulation areas are so limited and ablation zones relatively so large and high up in these subtropical latitudes. It has been suggested the protection afforded by debris mantles is critical in limiting ablation. However, they are also relatively small parts of the whole glacier area and concentrated in the lowest parts. The main explanation will be shown to lie in the exceptional amounts of snow carried by snow and ice avalanches from the steep rock walls to feed the glaciers. These and some other aspects of terrain distributions lay the groundwork for looking at glacier mass balance and dynamics in the next three chapters. More immediately they raise questions about the kinds of ice masses that occur in the region and how best to characterise them.

## 3.4 Glacier Types

For a long time, German-language studies have recognised some distinctive types of glaciers in High Asia. Much attention was given to the *Turkestan* or 'Lawinen' (avalanche) type, referring to glaciers fed mainly or wholly by avalanches (von Klebelsberg 1925–1926, 1938). Elsewhere, they are called 'fall' glaciers but are small and have been treated as something of a curiosity (Benn and Evans 1998, p. 19). In most of the Himalaya and Inner Asia, they occur in large numbers and many in the Karakoram are of substantial size.

Östreich (1911–1912) seems to have been the first to refer to 'Mustagh-type' glaciers, based on Karakoram experience. These are largely avalanche nourished but have deeply incised, sometimes extensive, ice streams in the perennial snow zone. They include most of the largest valley glaciers such as Baltoro, Panmah and Batura.

In the past there were extended discussions of these classes, the criteria used to define them and lists of examples (Visser and Visser-Hooft 1938; von Wissmann 1959, pp. 29–47). Interest has fallen off more recently and only rarely appears in

**Table 3.3** Perennial snow and ice (PSI) and glacierised areas for the main set of Karakoram Glaciers

| Glacier | PSI (km$^2$) | PSI (% basin area) | Glacier cover (km$^2$) | Glacier cover (% basin area) |
|---|---|---|---|---|
| Siachen | 1,110 | 79.3 | 750 | 53.6 |
| Baltoro | 910 | 71.7 | 535 | 42.1 |
| Biafo | 645 | 75.4 | 540 | 63.2 |
| Hispar | 630 | 80.3 | 290 | 36.9 |
| Batura | 535 | 75.4 | 260 | 36.6 |
| Chogo Lungma | 495 | 71.7 | 150 | 21.7 |
| Panmah | 460 | 67.6 | 180 | 26.5 |
| Rimo | 480 | 78.4 | 400 | 65.4 |
| Skamri | 376 | 73.7 | 175 | 34.3 |
| Kondus–K. | 365 | 74.5 | 130 | 26.5 |
| Chiantar | 360 | 82.6 | 258 | 59.2 |
| Braldu | 335 | 77.9 | 195 | 45.3 |
| Khurdopin | 317 | 76.4 | 85 | 20.5 |
| Sarpo Laggo | 330 | 84.6 | 205 | 52.6 |
| Virjerab | 296 | 76.1 | 87 | 22.4 |
| Terong NS | 220 | 62.0 | 172 | 48.5 |
| Yashuk Y. | 160 | 48.5 | 79 | 23.9 |
| Hasanabad | 247 | 77.2 | 55 | 17.2 |
| Bilafond | 230 | 79.3 | 95 | 32.8 |
| N. Shukpa | 240 | 85.7 | 75 | 26.8 |
| Charakusa | 190 | 71.7 | 88 | 33.2 |
| Kukuar | 186 | 82.7 | 86 | 38.2 |
| Toltar–Bal. | 150 | 73.2 | 49 | 23.9 |
| Yazghil | 152 | 69.1 | 56 | 25.5 |
| Sosbun | 160 | 74.4 | 45 | 20.9 |
| Karambar | 155 | 75.6 | 62 | 30.2 |
| Barpu | 65 | 32.8 | 32 | 16.2 |
| N. Gasherbr | 152 | 80.0 | 58 | 30.5 |
| Sherpigang | 125 | 65.8 | 52 | 27.4 |
| Ghondo'ro | 138 | 76.7 | 42 | 23.3 |
| Momhil | 119 | 68.0 | 34 | 19.4 |
| Singhi | 145 | 85.3 | 63 | 37.1 |
| Malangutti | 105 | 61.8 | 50 | 29.4 |
| Urdok | 128 | 79.0 | 51 | 31.5 |
| Gyong | 105 | 65.6 | 40 | 25.0 |
| Chong Khum | 128 | 82.6 | 42 | 27.1 |
| Sokha Lungma | 105 | 70.0 | 35 | 23.3 |
| Bualtar | 62 | 44.3 | 34 | 24.3 |
| S. Shukpa | 95 | 70.4 | 62 | 45.9 |
| Gharesa | 108 | 80.0 | 28 | 20.7 |
| Kutiah | 105 | 80.8 | 31 | 23.8 |
| Kyagar | 108 | 83.1 | 35 | 26.9 |
| **Total** | **11,527** | – | **5791** | – |
| **Mean** | **274.5** | **73.1** | **137.9** | **32.4** |

**Table 3.4** Areas of main connected glaciers, their accumulation and ablation zones

| Glacier | Main glacier ($km^2$) | Accum. zone ($km^2$) | Ablation zone ($km^2$) | Abl. zone (% main glacier) |
|---|---|---|---|---|
| Siachen | 580 | 370 | 210 | 36.2 |
| Baltoro | 495 | 180 | 260 | 52.5 |
| Biafo | 460 | 330 | 130 | 28.3 |
| Hispar | 282 | 80 | 202 | 71.6 |
| Batura | 220 | 75 | 145 | 65.9 |
| Chogo Lungma | 115 | 45 | 110 | 95.7 |
| Panmah | 167 | 70 | 105 | 62.9 |
| Rimo | 370 | 300 | 70 | 18.9 |
| Skamri | 170 | 48 | 118 | 69.4 |
| Kondus–K. | 109 | 30 | 80 | 73.4 |
| Chiantar | 249 | 165 | 84 | 33.7 |
| Braldu | 165 | 75 | 80 | 48.5 |
| Khurdopin | 75 | 22 | 52 | 69.3 |
| Sarpo Laggo | 200 | 105 | 95 | 47.5 |
| Virjerab | 75 | 20 | 54 | 72.0 |
| Terong NS | 164 | 114 | 50 | 30.5 |
| Yashuk Y. | 47 | 7 | 40 | 85.1 |
| Hasanabad | 45 | 5 | 40 | 88.9 |
| Bilafond | 89 | 55 | 34 | 38.2 |
| N. Shukpa | 70 | 40 | 40 | 57.1 |
| Charakusa | 72 | 15 | 57 | 79.2 |
| Kukuar | 80 | 6 | 74 | 92.5 |
| Toltar–Bal. | 38 | 3 | 35 | 92.1 |
| Yazghil | 51 | 30 | 21 | 41.2 |
| Sosbun | 36 | 5 | 22 | 61.1 |
| Karambar | 36 | 7 | 29 | 80.6 |
| Barpu | 28 | 4 | 24 | 85.7 |
| N. Gasherbr | 52 | 25 | 27 | 51.9 |
| Sherpigang | 35 | 32 | 13 | 37.1 |
| Ghondo'ro | 36 | 7 | 29 | 80.6 |
| Momhil | 32 | 15 | 20 | 62.5 |
| Singhi | 60 | 15 | 45 | 75.0 |
| Malangutti | 44 | 28 | 16 | 36.4 |
| Urdok | 50 | 15 | 35 | 70.0 |
| Gyong | 36 | 20 | 16 | 44.4 |
| Chong Khum | 39 | 21 | 18 | 46.2 |
| Sokha Lungma | 25 | 4 | 21 | 84.0 |
| Bualtar | 32 | 7 | 25 | 78.1 |
| S. Shukpa | 60 | 45 | 15 | 25.0 |
| Gharesa | 27 | 11 | 16 | 59.3 |
| Kutiah | 28 | 8 | 20 | 71.4 |
| Kyagar | 30 | 18 | 13 | 43.3 |
| **Total** | **5,074** | **2,477** | **2,590** | – |
| **Mean** | **120.8** | **59.0** | **61.7** | **60.6** |

**Table 3.5** Rock walls in the main set glacier basins comparing total areas with percentage in the glacier source zones

| Glacier | Rock walls total ($km^2$) | Rock walls (% basin area) | Rock walls (% source zone) |
|---|---|---|---|
| Siachen | 670 | 47.9 | 27.7 |
| Baltoro | 780 | 61.4 | 67.2 |
| Biafo | 385 | 45.0 | 27.1 |
| Hispar | 495 | 63.1 | 80.7 |
| Batura | 450 | 63.4 | 74.6 |
| Chogo Lungma | 490 | 71.0 | 83.9 |
| Panmah | 470 | 69.1 | 76.5 |
| Rimo | 205 | 33.5 | 14.3 |
| Skamri | 323 | 63.3 | 62.3 |
| Kondus–K. | 355 | 72.4 | 82.1 |
| Chiantar | 180 | 41.3 | 36.5 |
| Braldu | 226 | 52.6 | 54.0 |
| Khurdopin | 310 | 74.7 | 89.5 |
| Sarpo Laggo | 191 | 49.0 | 53.2 |
| Virjerab | 243 | 62.5 | 85.7 |
| Terong NS | 160 | 45.1 | 23.4 |
| Yashuk Y. | 225 | 68.2 | 83.3 |
| Hasanabad | 260 | 81.3 | 97.2 |
| Bilafond | 174 | 60.0 | 57.7 |
| N. Shukpa | 202 | 72.1 | 80.2 |
| Charakusa | 170 | 64.2 | 84.2 |
| Kukuar | 135 | 60.0 | 93.9 |
| Toltar–Bal. | 141 | 68.8 | 96.5 |
| Yazghil | 145 | 65.9 | 70.0 |
| Sosbun | 160 | 74.4 | 94.7 |
| Karambar | 135 | 65.9 | 90.0 |
| Barpu | 158 | 79.8 | 85.7 |
| N. Gasherbr | 130 | 68.4 | 77.1 |
| Sherpigang | 140 | 73.7 | 62.4 |
| Ghondo'ro | 135 | 75.0 | 92.6 |
| Momhil | 110 | 62.9 | 73.8 |
| Singhi | 98 | 57.6 | 83.3 |
| Malangutti | 116 | 68.2 | 68.9 |
| Urdok | 109 | 67.3 | 82.6 |
| Gyong | 125 | 78.1 | 75.0 |
| Chong Khum | 112 | 72.3 | 79.2 |
| Sokha Lungma | 116 | 77.3 | 93.3 |
| Bualtar | 96 | 68.6 | 75.0 |
| S. Shukpa | 71 | 52.6 | 37.5 |
| Gharesa | 117 | 86.7 | 87.5 |
| Kutiah | 95 | 73.1 | 89.2 |
| Kyagar | 98 | 75.4 | 83.3 |
| **Total** | **9,506** | – | – |
| **Mean** | – | **65.1** | **72.2** |

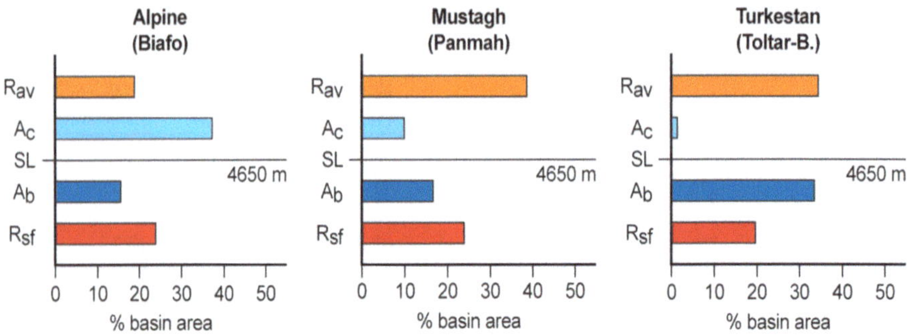


**Fig. 3.3** Terrain proportions that help distinguish glaciers of Alpine (Biafo), Mustagh (Panmah) and Turkestan (Toltar–Baltar) types. $A_c$ and $A_b$ designate conventional accumulation and ablation zone areas, respectively. $R_{av}$ and $R_{sf}$ refer to steep rock walls, respectively, in the upper frigid source zones and seasonally snow-free lower areas

English-language studies. The German nomenclature is sometimes mentioned (Mercer 1975). Goudie et al. (1984, p. 391) confirm that 'The incised reservoir and avalanche types are dominant in the Karakoram, with narrow, high gradient accumulation zones…'. They do not take the matter further.

English-language classifications tend to focus on size and morphology, sometimes climatic types (Hambrey 1994, p. 37; Benn and Evans 1998, pp. 18–19). In most cases Karakoram Glaciers, except the smallest, would fall into just one class, their 'valley glaciers'. The influential work of Ahlmann (1948, pp. 59–63) includes a 'Himalayan' class, but for smaller, steep ice masses.

### 3.4.1 *Classification by Nourishment*

Past efforts to define Karakoram Glaciers focused primarily on nourishment but were weakened by confusion over related morphological criteria (Hewitt 2011). When the problems are rectified, as detailed below, the classes can serve to reveal useful features and a clear relation to terrain types (Fig. 3.3). Four basic nourishment types can be identified:

1. *The Turkestan-type*: glaciers fed more or less entirely by avalanches of snow and ice from higher areas. Main ice streams commence below a perennial snow zone where there may be no basin areas gentle enough for snow build-up. There may be ice masses, usually small and minor, not connected to the main ice network (Plate 3.4).
2. *The Mustagh type*: glaciers also predominantly avalanche fed but with some ice streams that commence in the perennial snow zone (Plate 3.5). These may add up to quite extensive accumulation zones, but are smaller in area than surrounding rock walls.
3. *The Alpine type*: glaciers that are predominantly fed by direct snowfall in extensive accumulation zones (Plate 3.1). They are typical of the European Alps and most mountain glaciers described in the literature, but a lesser fraction in the Karakoram.

**Plate 3.4** A typical, small Turkestan-type glacier, a tributary of Panmah Glacier's Nobonde Sobande branch, fed by avalanches descending from the Skamri massif and with heavy debris mantle on the ablation zone (Hewitt 2005)

**Plate 3.5** Typical features of a Mustagh-type glacier; Chiring Glacier tributary of Panmah Glacier with prevailing rock wall terrain and deeply incised ice streams. Some, in the upper, frigid areas, like the left-of-centre tributary, have areas of conventional accumulation although avalanche nourishment is greater (Hewitt 2005)

4. *The Wind-fed type*: glaciers that depend largely or wholly on snow redistributed and carried to them by wind action. They include many of the high elevation 'cornice apron' and 'niche' glaciers (Dolgushin 1961; Kotlyakov et al. 2010) and can occur on lee slopes below the perennial snow zone.

For consistency the Turkestan-type class should only designate glaciers whose ice mass or main ice streams commence below, or not far into, the perennial snow zone. However, it is difficult to exactly define the upper limit of connected ice streams when avalanches bury them in huge cones and sheets of snow. Bergschrund-like crevasses that may otherwise signal the head of a glacier are commonly smothered. For convenience, the Turkestan-type will be restricted to ice masses where less than 20 % of their area is at and above the reported 'snowlines' (von Wissmann 1959). Adopting a 0 % cut-off might seem better but, apart from avalanche burial, there is great local variability according to aspect, ruggedness, shading, prevailing winds and wind action (Chap. 2). Toltar–Baltar Glacier, a typical Turkestan-type, has main ice streams that commence in extensive cones and aprons of avalanched material below, at, or spread somewhat across where a snow limit would appear on Alpine-type glaciers.

Turkestan-types do not occur among the largest of the main set of glaciers and are a modest share of intermediate sized valley glaciers. Their greatest numbers are of small size. Countless small and minor examples occur in the lesser ranges with interfluves close to the glaciation limit.

In the Mustagh type, well-defined ice streams do commence in basins in the perennial snow zone where snowfall can accumulate directly on them. They must comprise at least 20 % of the main glacier and, for classification purposes, with an upper cut-off of 50 %. Glacier ice generally covers a smaller fraction of the upper basin area than rock walls, and avalanches are responsible for the larger part of nourishment.

The Alpine type is defined by substantial accumulation areas and comprising over 50 % of the main connected glacier. Larger examples occur where extensive, high-altitude snow basins develop as in the Rimo, Siachen, Biafo and Chiantar Glaciers.

It is important to highlight the existence of wind-fed ice masses and the evidence of snow redistributed by wind. Windward slopes become stripped, while deposition in leeward areas is evident in massive cornice development, especially above the snowline. The higher peaks and arêtes are rarely without snow plumes driven by high winds, which explains the many narrow but laterally extensive glaciers on their lee flanks, usually disconnected from and high above the main ice streams. In such places it is likely that wind-borne snow far exceeds direct snowfall and may become the dominant support of ice masses. For the many small ice masses on lee flanks in lesser ranges windblown snow can be more important than orientation and shading from sunlight. Wind redistribution also intervenes in much of the snow ultimately feeding Turkestan- and Mustagh-type glaciers. It needs to be underlined as one of the most neglected but important aspects of the Karakoram cryosphere.

These considerations are a first step towards interpreting the features described at Biafo, Baltoro and Toltar–Baltar and some of the differences between them. Respectively, they fit the Alpine-, Mustagh-, and Turkestan-type profiles, differentiated by the predominant mode of nourishment.

**Table 3.6** Revised classification for Karakoram Glaciers based on traditional nourishment and morphology classes

| Nourishment type | 'Caldron' type | 'Ice stream' type, narrow, incised |
|---|---|---|
| Turkestan-type | Hinarche, Surgin, Masherbrum, Kukuar | Charakusa, Karambar, Hasanabad, Toltar |
| Mustagh type | Kutiah, Skamri, Khurdopin, Kondus | Baltoro, Batura, N. Shukpa, Panmah, Hispar, Chogo Lungma |
| Alpine type | Chiantar, Sarpo Laggo | Siachen, Biafo, Rimo |

### *3.4.2 Combined Nourishment and Morphological Classification*

In the past, there has been some confusion about placing glaciers in the nourishment typologies. Visser and Visser-Hooft (1935–1938) and von Wissmann (1959, pp. 1146–1147) came to different conclusions about a number of examples. Problems may have arisen because maps were generally poor, those who could visit the glaciers only saw parts of them, and authors like von Wissmann depended largely on second-hand accounts. Satellite coverage now provides a better basis, especially upper basin coverage. However, it cannot compensate for some limitations in the classifications themselves.

Mustagh-type glaciers have been divided into *Firnkessel* (= firn caldron) and *Firnstrom* (= firn stream) types (von Wissmann 1959). These morphological differences are of some interest, but usage of them has been misleading. Firstly, calling them 'firn' types infers a significant role of direct snowfall, when avalanches are far more important.[3]

It makes more sense to call Alpine types like Biafo or Siachen 'firn stream' glaciers. Von Wissmann (1959) does classify them as such, but then places them in the Mustagh category. It seems more logical to follow Visser and Visser-Hooft (1938) who called them Alpine types because direct snowfall nourishment is predominant. Further confusion arises when Baltoro is classed as *Firnkessel* type by von Wissmann (1959), but *Firnstrom* by Visser and Visser-Hooft (1938). If it is used at all, the 'firn' category should be reserved for Alpine-type glaciers. To reduce confusion over morphological criterion, the 'stream' type is taken to imply a length at least three times the width, and 'caldron' type where it is less (Table 3.6).

Secondly, using the two incised 'stream' or 'caldron' forms only for Mustagh-type glaciers is problematic. Siachen Glacier's accumulation zone, for example, includes tens of kilometres of ice streams that are deeply incised and, arguably, of firn stream type. Firn basins of Biafo and Chiantar are surrounded by high, steep-walled interfluves and resemble the caldron type. Most Turkestan-type glaciers have extensive, deeply incised ice streams or caldrons, if entirely in the ablation zone.

[3] 'Firn' usually refers to snowfall that survives in accumulation areas at the end of the summer (see Chap. 4).

Choosing a main set of largest glaciers, which are located in the highest, most humid parts of the Karakoram, increases the chances of extensive ice streams in the perennial snow zone. The proportion of Alpine-type glaciers is probably exaggerated and of Turkestan-type underestimated compared to the whole region. These and other relations of nourishment types will be more fully explored in the next several chapters.

Classification of glaciers by nourishment and morphology reinforces the importance of verticality. The rapid, downslope cascade of snow, ice and debris is underlined. One can argue for maintaining something of the neglected Karakoram 'tradition' with its nourishment focus. At the same time, there are other aspects or types of glacier systems that need to be explored.

A fairly elaborate morphological classification, developed for inventories of glaciers in the former Soviet Union, remains in use in Asia (Kotlyakov et al. 2010). It includes four types of 'valley glacier'.[4] A similar classification is employed in inventories of glaciers in China (Yao 2007). Table 3.7 summarises glacier classes from this literature and indicates where Karakoram types might be placed.

## 3.5 Valley Glacier Complexes

The largest glacier systems involve contiguous sets of topography-controlled valley glaciers. Wherever the interfluves rise above about 6,500 m, glaciers fan out in all directions around topographic highs. They generate a series of glacier complexes associated with the highest peaks or massifs, from Kampire Dior (7,168 m) in the west to Sasir Kangri (7,672 m) in the east, and across four degrees of longitude.

Just to the east of the highest K2 Massif is a complex that includes Siachen, Rimo, Baltoro and several intermediate glaciers or a *Transection* system, defined as 'web-like patterns with ice *diffluences* where ice flow splits, and *confluences* where ice flow from two or more tributaries converges…' (Benn and Evans 1998, p. 19). Near the heart of this complex is the only Karakoram example of a substantial 'summit' or ice cap form (Kotlyakov et al. 2010). The Teram Shehr dome drains to the Siachen, Rimo and North Terong Glaciers, one of the few places where glacier ice extends uninterrupted across the watersheds.

The Rimo is, perhaps, the least typical large Karakoram Glacier: an Alpine-type perched near the edge of the Tibetan Plateau with watersheds smothered in snow and a more limited role of avalanched slopes. This seems to explain the unusually small areas of supraglacial debris towards its three terminal lobes.[5] Diffluence in the Rimo system means it is partly a watershed of the Indus–Indian Ocean drainage and partly of the interior Yarkand–Tarim Basin.

---

[4]Their typology for ice masses on volcanoes has no application in the Karakoram.

[5]The Rimo was virtually off limits to researchers for almost eighty years. One must go back to early twentieth-century expedition reports to appreciate the glacial landscapes (Workman 1914; Dainelli 1924–1935). Originals of the magnificent photo collections are in the Italian Geographical Society's Photographic Archive in Rome.

**Table 3.7** Classification of ice masses roughly in order of size, with particular reference (in bold) to those of importance in the Karakoram or its literature and in nourishment and morphological classes (After Visser and Visser-Hooft 1935–1938; von Klebelsberg 1938; Pillewizer 1956; Schneider 1962–1963; Kick 1964; Hewitt 1968; Mercer 1975; Benn and Evans 1998)

| Morphological type | Morphology and nourishment types | Definitions | Karakoram examples |
|---|---|---|---|
| **A. Minor ice masses** | | | |
| 1. Niche, wall-sided, carapace and glacierets | | 1. Small masses where sufficient snowfall or avalanche deposits are trapped in the lee of ridges or peaks, on rock walls and in chutes | Countless |
| 2. Cirque | | 2. Filling armchair-like bedrock basins | Many in lesser ranges; rare on main range |
| 3. Avalanche apron | | 3. Perennial snow aprons or cones well below the snow and firn lines (extreme 'Turkestan' type?) | Many at elevations 3,000–4,000 m |
| **B. Valley glaciers** | | | |
| | **Avalanche fed** | | |
| | 1. Turkestan or ('Lawinenkessel') Avalanche caldron | 1. Valley glaciers lacking névé basins, main ice streams commencing below the firn limit and climatic snowlines, fed by snow and ice avalanches from the slopes | Shispar, Toltar–Baltar, Hassanabad, Yashkuk Y, Lupghar |
| | | 'Incised reservoir' | Malangutti, Momhil |
| | 2. Mustagh type | 2. Valley glaciers with limited ice streams confined in narrow, steep-sided troughs or basins that commence in the accumulation zone, but are a small part of its area, largely avalanche fed which obscures the firn limit | Hispar, Barpu, Chogo Lungma Baltoro, Batura |
| | I. Caldron type also called firn caldron ('Firnkessel') | | |
| | II. Stream type also called firn stream ('Firnstrom') | | |

(continued)

**Table 3.7** (continued)

| Morphological type | Morphology and nourishment types | Definitions | Karakoram examples |
|---|---|---|---|
| | 3. Alpine snow fed with firn and firn lines | 3. Valley glaciers with extensive ice streams in accumulation zone, with mainly direct snowfall nourishment | Chiantar, Braldu |
| | | 4. With extensive, relatively open accumulation basin(s) at the head, although high, steep-walled interfluves may surround them | Siachen, Biafo |
| | | 5. With elongated, sometimes walled-in but extensive streams in the accumulation zone | Sasir Kangri; K2-Baltoro Kangri; Distaghil Sar-Trivor: Rakaposhi |
| **C. Valley glacier complex** | | | |
| 1. Radial complex | | 1. Or 'radiating type' where multiple ice streams fan out from a common source area | Batura Mts system |
| 2. Transection | | 2. Web-like patterns, with ice diffluences, where ice flow splits and confluences, where ice flow from two or more tributaries converge…' | East Karakoram, Rimo–Siachen–Terong–Saltoro complex |
| **D. Ice caps (small)** | | | |
| Firn plateau ('Zentrale Firnhaube') | | Small (10–50 km diam.) snowfall-fed ice cap draping and smothering interfluve area | Teram Shehr ice cap, upper Siachen |
| **E. Ice sheets** | | | |
| Ice cap (large) | | | None since LMG |
| Continental I. S. | | | None |

## 3.6 Minor, Disconnected Ice Masses

At the other extreme are countless minor ice masses (<2 km length). Most occur either at the highest or the lowest levels of glaciation. The former include a whole suite of types disconnected from the main ice streams and generally suspended above or within rock wall areas, variously called 'hanging', 'slope', 'niche' and 'apron' glaciers, the smallest, *glacieret* types (Benn and Evans 1998; Kotlyakov et al. 2010). For them to exist, local topography must allow build-ups sufficient to generate some glacier ice. A combination of avalanches, wind transport and direct snowfall contribute to most of them. Unlike the Turkestan or Mustagh glaciers identified so far, the minor ice masses at high elevations have *no ablation zone*. Much as polar glaciers may calve into the sea, these terminate in ice cliffs from which ice avalanches serve the role of 'ablation' (Plate 3.6). By contrast, minor ice masses at lower elevations commonly lack a conventional 'accumulation zone'. Below interfluves close to the glaciation limit avalanches feed the many small and minor Turkestan-type ice masses.

Cirque glaciers are rare or absent from much of the high Karakoram. However, on offshoots of the Karakoram and surrounding, lesser ranges, there are large numbers, for example, in the Aghil, Hindu Kush, Kun Lun, Ladakh, Zanskar, Deosai and Gilgit Ranges (Burbank and Fort 1985; Schmidt and Nüsser 2012). They do seem to differ in various ways from the classic cirques and may involve a somewhat different evolution (see Chap. 7).

**Plate 3.6** Ice avalanche descending from a disconnected high elevation tributary glacier of Kaberi Glacier in the Kondus–Kaberi system in the east central Karakoram (Hewitt June 1998)

## 3.7 Climatic and Thermal Classes

Karakoram Glaciers also raise some singular questions with respect to classifications based on climatic or thermal conditions (Singh and Singh 2001, pp. 449–452). Their location some 1,500 km from the nearest ocean suggests they must be 'Continental' glaciers, and that does apply in some parts of the range. However, according to widely used criteria, precipitation exceeding 1,000 mm water equivalent is in a 'maritime' regime, and this applies in the Mustagh Karakoram and to the larger glaciers and greatest ice covers (Jijun et al. 1984; Yafeng et al. 2010, F130).

Thermal classes identify two basic types of glacier: those referred to as *warm* with most of the ice at pressure melting point or *cold* with all or most of the ice below the melting point. Warm glaciers are not frozen to the bed although their surface layers and ice near the terminus may be sub-zero in winter. Cold glaciers are typically frozen to the bed, but, in thick slow-moving ice, geothermal or frictional heating can cause melting at the bed and produce a thin layer of 'warm' ice. Others have a period of surface warming and ablation in summer. Mixed configurations are referred to as *polythermal* glaciers.

It has been common to name or attribute the various thermal configurations to latitudinal zones: polar and subpolar for cold and temperate or midlatitude for warm glaciers. However, while Karakoram glaciers are located in the subtropics, such measurements as exist show a complete range of thermal regimes. The extreme cold at high altitude means glaciers there have properties identified with polar or subpolar types (Benn and Evans 1998, p. 95). Smaller, south-oriented glaciers and the ice tongues reaching the lowest elevations have the warm ice and unfrozen beds of the temperate type.[6] The larger glaciers and many intermediate ones are of mixed thermal regimes, cold or sub-zero ice in the upper reaches and some or all possible types of polythermal configuration in ablation zones (Paterson 1994). The exceptional number of surge-type glaciers suggests that mixed and thermally unstable conditions at the bed are widely present (Chap. 11).

To further complicate the situation, some glaciers in the far eastern, upper Shyok Basin and the north slope draining to the Yarkand seem to combine polar and tropical attributes. Most of them appear to be cold glaciers, frozen to the bed, and with shearing structures at the termini that resemble those of polar glaciers. There are examples of the distinctive 'subpolar', lobate or piedmont type. On the other hand, ablation zones and some higher accumulation zones have large areas of highly developed *penitents*, also called *penitentes* or *nieves penitents*. These are forms developed in the intense sunlight and sub-zero ambient temperatures usually associated with high-altitude tropical glaciers (Smiraglia 2011).

It seems that Karakoram Glaciers have some properties in, and some between those, ascribed to maritime and continental, tropical, temperate, polar and/or subpolar classes. Their climatic and thermal characteristics place them in intermediate, compound or

[6] The Karakoram adds force to Hambrey's (1994, p. 43) view that the thermal categories of 'polar' and 'subpolar' are 'misleading and best avoided'. For similar reasons, the same can be said of 'temperate', 'tropical' and 'subtropical' – geographical designations for thermal phenomena that are not confined to these or any particular region. *All* are present in the Karakoram!

complex regimes. One result of the extreme elevation range is to make classes devised in and for other contexts less appropriate.

It has been proposed that the regional climate is responsible for the presence of Mustagh- and Turkestan-type glaciers (Pillewizer 1956; Kick 1964). However, actual patterns of precipitation and temperature and the importance of avalanche and wind nourishment mainly reflect the orography and steep terrain. Their vertical organisation radically modifies regional climate. In the predominant Mustagh- and Turkestan-type glaciers, the larger part of nourishment on all timescales comes not from precipitation directly but processes *beyond and above* the glacier surface. These, in particular avalanches and wind, more or less drastically intervene in and modify snow distribution, density, thermal and other properties.

## 3.8 Ice Depths and Volumes

The terrain data and glacier classifications are derived mainly from surface conditions. Some sense of the volumes of ice involved is also needed. Transformation of snow to active ice usually occurs when it is tens of metres thick and hence, with many years of accumulated snowfall, longer in cold climates and shorter in warmer ones. A major complication in the Karakoram is that an overwhelming proportion of snow is delivered to glaciers by wind and avalanche. It arrives at the glacier surface already much denser than direct snowfall. Most of it is emplaced at lower elevations than where it arrives as precipitation. This should speed up the process of conversion to glacier ice. Data are lacking to say by how much, but with the terrain and thermal diversity noted conversion to, and depths of, active glacier ice seem likely to be very variable (Chap. 4). Of course, the mass of the glacier itself is transferred, and its thicknesses determined by ice movement. In the larger ones thickness is the product of many centuries of inputs and gravitational adjustment (Chap. 7).

Biafo is one of the few large glaciers with fairly reliable ice depth profiles measured using ice-penetrating radar and over a range of elevations (Fig. 3.4). The greatest depths were found at about 45 km above the terminus, near the equilibrium line and reach a maximum of about 1,400 m (Hewitt et al. 1989). The ablation zone ice thins downstream to about 550 m at the Baintha profile, 25 km from the terminus, due not just to ablation but to spreading which doubles the width of the glacier. By 10 km from the terminus depths decline to about 300 m, but at a location where the glacier has narrowed to one-third the width at Biantha. Above the equilibrium line profile in the accumulation zone, a profile 5 km for the Sim Gang tributary has a greatest thickness of about 500 m. The depths along the larger Lukpe Lawo tributary, of which Sim Gang is probably a hanging glacier, should be greater but were not measured. The total volume of ice in the basin could be of the order of 100 $km^3$.

Radio-echo soundings along a midline profile of Hispar Glacier recorded ice depths between 450 m at 16 km above the terminus and 590 m at 25 km (Dong et al. 1984). The results are comparable to Biafo measurements at similar elevations and distances into the ablation zone. On the Miar tributary of Barpu Glacier at 3,400–3,500 m elevation, radar ice depths of up to 560 m were measured near the

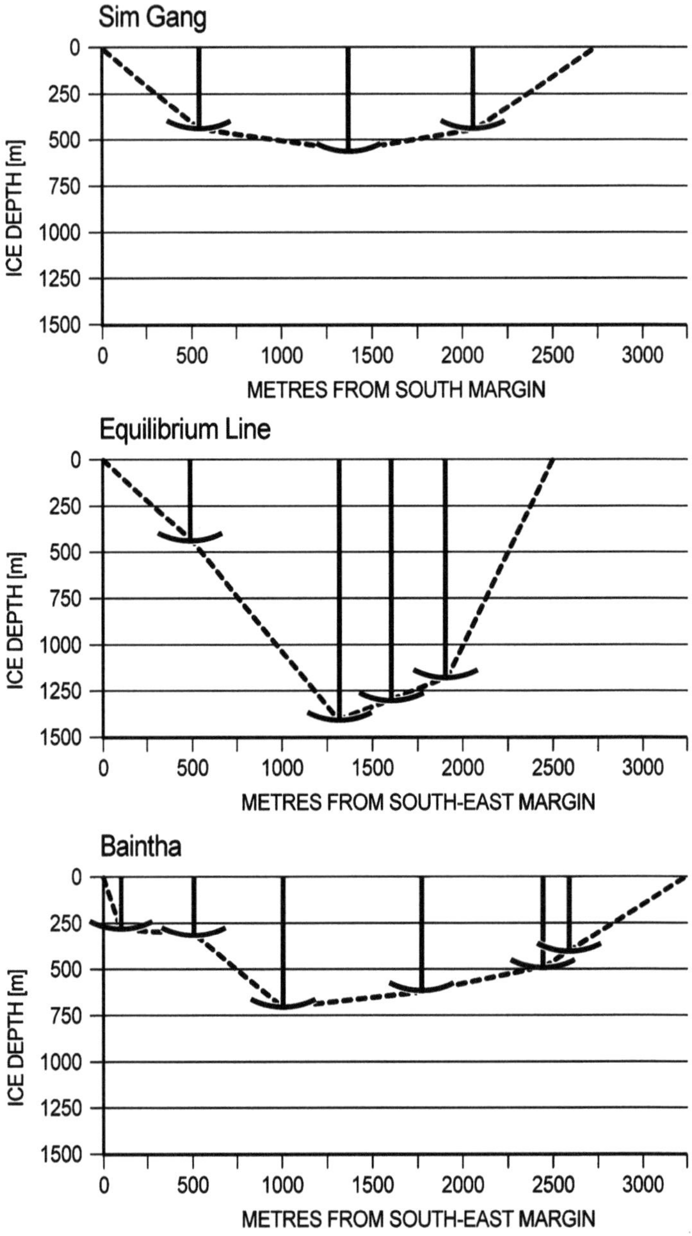


**Fig. 3.4** Biafo Glacier ice depths determined from radar sounding at three locations, Sim Gang in the accumulation zone, equilibrium line (4,650 m), and Baintha (4,080 m) in the ablation zone (After Hewitt et al. 1989)

mid-glacier point (Young and Schmok 1989). This is at the base of a high icefall where conditions suggest an over-deepened basin. Radio-echo soundings along the much smaller Ghulkin Glacier in Hunza showed depths rising from between 110 and 140 m about 2 km above the terminus to as much as 425 m in the mid-ablation zone, about 9 km from the terminus (Francis et al. 1984). If these few depth measurements available are representative, they indicate that the Karakoram Glaciers should comprise a total ice volume of not less than 2,500 $km^3$.

Valley glacier thicknesses also involve down valley movement of ice affected by the topography. These help determine the morphology of ice masses and where the greatest depths occur, how they vary along a glacier, and in relation to the other environmental conditions (Chap. 7). Generally, the data on ice thicknesses reinforce the area–altitude profiles and show glacier mass to be even more concentrated between about 4,000 and 5,000 m where ice stream surface areas as well as depths are greatest. The centre of mass should rise by some hundreds of metres towards the eastern and northern flanks. It seems likely to be lowest around the deeply incised Hunza Basin that cuts through the Karakoram main axis.

## References

Ahlmann HW (1948) Glaciological research on the North Atlantic coasts, Research series 1. Roy Geogr Soc, London

Alpensvereinskartographie (1995) Karakoram orographical sketch map 1/250,000. Swiss Foundation for Alpine Research, Zurich, 2 sheets

Benn DI, Evans DJA (1998) Glaciers and glaciation. Hodder Arnold Publication, London

Bishop MP, Colby JD (2011) Topographic normalization of multispectral satellite imagery. In: Singh VP, Singh P, Haritashya UK (eds) Encyclopedia of snow, ice and glaciers. Springer, Dordrecht, pp 1187–1197

Burbank DW, Fort M (1985) Bedrock control of glacial limits: examples from the Ladakh and Zanskar Rages, northwestern Himalaya, India. J Glaciol 31(108):143–149

Dainelli G (1924–1935) Relazioni Scientifiche della Spedizione Italiana De Filippi nell'Himalaia, Caracorum e Turchestan Chinese (1913–1914). Series II, 10 vols. Zanichelli, Bologna

Dolgushin ID (1961) Main features of the modern glaciation of the Urals. Int Assoc Sci Hydrol 54:335–347

Dong ZB, Ferrari RL, Francis MR, Musil G, Oswald GKA, Zhang X (1984) Impulse radar ice-depth sounding on the Hispar glacier. In: Miller KJ (ed) The International Karakoram Project, vol 2. Cambridge University Press, Cambridge

Dyhrenfurth GO, Ertl H, Roch A (1939) Baltoro: Ein Himalaya-Buch. B. Schwabe, Basel

Francis MR, Miller KJ, Dong ZB (1984) Impulse radar ice-depth sounding of the Ghulkin Glacier. In: Miller KJ (ed) The International Karakoram Project, vol 2. Cambridge University Press, Cambridge, pp 111–123

Goudie AS et al (1984) The geomorphology of the Hunza Valley, Karakoram Mountains, Pakistan. In: Miller KJ (ed) International Karakoram project, vol 2. Cambridge University Press, Cambridge, pp 359–410

Hambrey MJ (1994) Glacial environments. University College London Press, London

Hewitt K (1968) Geomorphology of the mountain regions of the Upper Indus Basin, vol 2. University of London, Dissertation

Hewitt K (2011) Glacier change, concentration and elevation effects in the Karakoram Himalaya, Upper Indus Basin. Mt Res Dev 31(3):1–13, http://dx.doi.org/10.1659/MRD-JOURNAL-D-11-00020.1

Hewitt K, Wake CP, Young GJ, David C (1989) Hydrological investigations at Biafo Glacier, Karakoram Himalaya, an important source of water for the Indus River. Ann Glaciol 13:103–108

Jijun L, Derbyshire E, Xu S (1984) Glacial and paraglacial sediments of the Hunza Valley, North West Karakoram, Pakistan: a preliminary analysis. In: Miller KJ (ed) The International Karakoram Project, vol II. Cambridge University Press, Cambridge, pp 496–535

Kick W (1964) Der Chogo-Lungma Gletscher im Karakorum II. Z Gletschkd Glazialgeol 5(1):1–59
Kotlyakov VM, Dyakova AM, Koryakin VS, Kravtsova VI, Osipova GB, Varnakova GM et al (2010) Glaciers of the former Soviet Union. In: Williams RS, Ferrigno JG Jr (eds) Satellite image atlas of glaciers: ASIA, vol 1386-F, United States Geological Survey, Denver, professional paper. U.S. GPO, Washington, pp 1–93
Mayer C, Lambrecht A, Belo M, Smiraglia C, Diolaiuti G (2006) Glaciological characteristics of the ablation zone of Baltoro glacier, Karakorum, Pakistan. Ann Glaciol 43:123–131
Mercer JH (1975) Glaciers of the Karakoram. In: Field WO (ed) Mountain glaciers of the Northern Hemisphere. United States Army Cold Regions Research and Engineering Laboratory, Hanover, pp 371–409
Mihalcea C, Mayer C, D'Agata C, Diolaiuti G, Lambrecht A, Smiraglia C, Vuillermoz E, Tartari G (2008) Spatial distribution of debris thickness and melting from remote-sensing and meteorological data, at debris-covered Baltoro glacier, Karakoram, Pakistan. Ann Glaciol 48(1):49–57
Nakawo M, Raymond CF, Fountain A (2000) Debris-covered glaciers, vol 264, International Association of Hydrological Sciences (IAHS) publication. Oxfordshire, Wallingford
Östreich K (1911–1912) Der Tschogletscher in Baltistan. Zeitschrift für Gletscherkunde 6:1–30
Paterson WSB (1994) The physics of glaciers, 3rd edn. Pergamon, New York
Pecci M, Smiraglia C (2000) Advance and retreat phases of the Karakorum glaciers during the 20th century: case studies in Braldo Valley (Pakistan). Geogr Fı´s Din Quat 23(1):73–85
Pillewizer W (1956) Der Rakhiot-Gletscher am Nanga Parbat im Jahre 1954. Z Gletschkd Glazialgeol 3(2):181–194
Schmidt S, Nüsser M (2012) Changes of High Altitude Glaciers from 1969 to 2010 in the Trans-Himalayan Kang Yatze Massif, Ladakh, Northwest India. Arct Antarct Alp Res 44(1):107–121
Schneider HJ (1962–1963) Die Gletschertypen, Versuch im Sinne einer einheitlichen Terminologie. Geogr Taschenbuch 276–283
SIHP (Snow and Ice Hydrology Project) (1990) Final report, vol 1. Cold Regions Research Centre, Wilfrid Laurier University, Waterloo
Singh P, Singh VP (2001) Snow and glacier hydrology. Kluwer, Dordrecht
Smiraglia C (2011) Epiglacial morphology. In: Singh VP, Singh P, Haritashaya UK (eds) Encyclopaedia of snow, ice and glaciers. Springer, Dordrecht, pp 262–268
Smiraglia C, Mayer C, Mihalcea C, Diolaiuti G, Belo M, Vassena G (2008) Himalayan-Karakoram glaciers: results and problems in the study of recent variations of major non-polar glaciers. In: Bonardi L (ed) Terra glacialis, special issue: mountain glaciers and climate changes in the last century. Servizio Glaciologico Lombardo, Milan, pp 149–164
Sugden DE, John BS (1976) Glaciers and landscapes. Edward Arnold, London
Visser PhC, Visser-Hooft J (1935–1938) Wissenschaftliche Ergebnisse der niederländischen Expeditionen in den Karakorum und die angrenzenden Gebiete in den Jahren 1922, 1925 und 1929–30. E.J. Brill, Leiden
von Klebelsberg R (1925/1926) Der Turkestanische Gletschertypus. Z Gletschkd Glazialgeol 14:93–209
von Klebelsberg R (1938) Die Zusammensetzung des Talgletscher. Z Gletschkd Glazialgeol 26:22–43
Wissmann, H von (1959) Die heutige Vergletscherung und Schneegrenze in Hochasien mit Hinweisen auf die Vergletscherung der letzten Eiszeit. Akademie der Wissenschaften und der Literatur in Mainz. Abhandlungen der mathematisch-naturwissenschaftlichen Klasse 14:1103–1431
Workman FB (1914) The exploration of the Siachen or rose glacier. Eastern Karakorum, London
Yafeng S, Mi D, Yao T, Zeng Q, Liu C (2010) Glaciers of China. In: Williams RS Jr, Ferrigno JG (eds) Satellite image atlas of glaciers: ASIA, vol 1386-F, United States Geological Survey, Denver, professional paper. U.S. GPO, Washington, pp 127–166
Yao T (2007) Map of the glaciers and lakes on the Tibetan Plateau and adjoining regions. Xi'an Cartographic Publishing House, Xi'an
Yoshizawa I, Miyamori T (1988) The Baltoro Glacier, 1:100,000 map. The Mountaineering Journal. Yama-Kei Publishers, Tokyo
Young GJ, Schmok JP (1989) Ice loss in the ablation area of a Himalayan Glacier: studies on Miar Glacier, Karakoram Mountains, Pakistan. Ann Glaciol 13:289–294

# Chapter 4
# Glacier Mass Balance I: Snowfall and Glacier Nourishment

**Abstract** The inputs to glacier mass balance and phenomena that influence them in the Karakoram are considered in this chapter. Snowfall in the source areas of the glaciers is the first concern. Observations at Biafo Glacier, an Alpine type, offer a rare, relatively detailed quantification of high-altitude snowfall in the accumulation zone. The data show that maximum precipitation in the Central Karakoram occurs in glacier source zones. Between 4,800 and 5,800 m snowfall exceeds 1,000 mm water equivalent annually, rising to as much as 2,500 mm in some years and locations. This is much more than at valley weather stations, and whereas the latter have a dominant winter season precipitation, an 'all-year' input regime is indicated for the glacier basins in the Mustagh Karakoram. Summer snowfall is partly from Arabian Sea moisture but the summer monsoon proves significant in most years. Establishing the presence and quantities of firn and firn limits is problematic. In Biafo's accumulation zone, extensive, changeable surface and near-surface subzones are described in cold, dry and wet snow, summer horizons with dirt and ice layers, and complicated relations to the ablation zone below. Thick, clean snow layers from individual, mainly winter, storms are observed and also depth hoar. Such information, from the accessible high elevation areas of an Alpine-type glacier, is a basis to consider what happens on the much greater numbers of less accessible Mustagh- and Turkestan-type glaciers where the intervening role of avalanching is critical. Avalanches dominate snow inputs and continue to add mass far into what otherwise appear as ablation zones. Thus, a significant component of avalanche nourishment occurs below where firn lines and equilibrium line altitudes are conventionally placed. At higher elevations and around interfluves, the evidence of wind action is particularly important in feeding small and minor ice masses and as an influence on avalanching.

**Keywords** Glacier accumulation • High-altitude snowfall • Biafo Glacier • Snow pits • Snow chemistry • Precipitation gradients • Precipitation sources • Firn • Firn limits • Avalanche nourishment • AAR • Cornices • Cornice glaciers

K. Hewitt, *Glaciers of the Karakoram Himalaya: Glacial Environments, Processes, Hazards and Resources*, Advances in Asian Human-Environmental Research,
DOI 10.1007/978-94-007-6311-1_4, 

## 4.1 Glacier Mass Balance

The preferred modern approach to glacier maintenance, hydrology and responses to climate is through *mass balance.* This depends upon the relation between inputs or gains to the glacier and outputs or losses from it. Active glaciers are open systems with mass continually being added and removed. Typically, thickening and advance or thinning and retreat of glacier tongues are in response to increase or decrease in mass balance, respectively. Mass balance is generally measured or estimated over a calendar or *budget* year (Paterson 1994). These terms relate to standard practices for measurement and reporting. The main consideration in valley glaciers is usually the ratio between inputs of snowfall or *accumulation* on the upper glacier and loss of ice to surface melting or *ablation* on the lower glacier. The two are linked and moderated by movement of ice between upper and lower areas.

Saying 'usually' implies that inputs and outputs of Karakoram glaciers are *un*usual in various respects. They require a careful working through of the relevant phenomena and such measurements as are available. It is more than can be accomplished in a single chapter. Firstly, therefore, inputs or glacier nourishment are considered; outputs in the following chapter and Chap. 6 will draw together the results.

The dominant inputs to Karakoram glaciers, already evident from terrain relations, are from avalanches and wind action. They intervene between snowfall and its delivery to glacier surfaces. However, while it has long been recognised that avalanche nourishment will alter the terms of mass balance (Meier 1962; Kasser 1967), this has received very little attention anywhere, likewise wind redistribution. As a rule, mass balance studies have avoided glaciers with any or, at least, significant avalanche or wind nourishment. It is partly due to the difficulties and dangers involved, partly because glacier monitoring practice has been advanced mainly in regions where direct snowfall prevails (WGMS 2008).

With rare exceptions, measurements relevant for mass balance of Karakoram glaciers have been restricted to ablation, mainly in debris-mantled ice. Mass balance estimates are most often based on indirect approaches and deduce inputs from weather station data and river discharge records, sometimes remotely sensed snow cover and surface change (Ferguson et al. 1984; Archer and Fowler 2004; Bolch et al. 2012; Gardelle et al. 2012). These are useful approaches but what happens in glacier basins remains a matter of assumption, a 'black box'. Yet mass balance has developed as a measurement-based or accounting approach. Its value comes from an insistence upon quantitative results. An assessment for the Karakoram can hardly be credible without giving priority to such actual inputs as have been measured.

## 4.2 High Elevation Snowfall and Accumulation Zones

> … perpendicular rock faces have to be eliminated from the problem, and the solution sought on even, regular slopes, where the depth of snow represents the average snowfall….
> (Workman and Workman 1905, p. 261)

Adequate mass balance estimates for the Karakoram have been limited above all by lack of information about high-altitude snowfall. The extreme conditions involved are certainly a factor but also the difficulties of direct instrumental measurement in high mountains. Snow gauges and snow pillows, even where they can be installed and maintained, are notoriously unreliable (Goodison and Metcalfe 1992; Hansen and Davies 2002). Much of the perennial snow zone on glaciers lies above elevations where winter snow courses are practicable. Conditions on avalanche-fed glaciers largely preclude measurements.

In the Karakoram a partial solution can be found in those few Alpine-type glaciers with extensive, conventional *accumulation zones* where mass is added largely by direct snowfall. Although a minority of Karakoram glaciers, they nevertheless offer unique opportunities to measure high-altitude snowfall sequences in snow pits and cores. There are still problems, including the need to avoid areas of disturbance and redistribution of snow by avalanches, wind action and crevasses. However, with careful site selections, snow build-up in such areas can provide good indications of high-altitude snowfalls and at similar elevation to the majority of basins where measurement may not be possible. The seasonality and air mass provenance of snow can be investigated and is as important as knowing annual amounts.

### 4.2.1 The Biafo Accumulation Zone

Between 4,700 and 6,000 m elevation, Biafo Glacier has vast accumulation basins located in the heart of the Mustagh Karakoram and suitable for investigating snowfall (Fig. 3.1). In the 1980s, snow accumulation was observed there over a number of years and in some more limited observations in upper basin areas of the Mustagh-type Hispar and Khurdopin Glaciers. Such work has not been repeated. If somewhat dated, the observations provide the *only* relatively comprehensive profiles yet available to reconstruct high-altitude snowfall and its contribution to mass balance. They identify sources of precipitation, seasonal, elevation- and storm-related variables, and provide insights into year-to-year fluctuations of key variables. Concurrent meteorological and ablation rate measurement programmes help to estimate mass balance input and output variables in subsequent chapters (Hewitt et al. 1989; Wake 1989).[1]

Snowfall was measured between 4,800 and 5,800 m, an elevation zone that comprises more than half the entire Biafo Basin. The sites were in the 'the Snow Lake' or Lukpe Lawo area (Shipton 1938). Methods employed snow pits and drill cores to establish vertical profiles. Samples were retrieved for snow density, chemistry and isotope analyses (Plates 4.1 and 4.2a, b). The pits revealed the importance of major snowfall events, intermittent freeze–thaw, ice lenses, dirt layers and depth hoar

[1]The main observations of snow chemistry and isotope work in 1986 were by Cameron Wake (Wake 1987; SIHP 1990).

**Plate 4.1** Snow measurements at Lukpe Lawo, Biafo Glacier testing the snow corer (Hewitt 1985)

(Table 4.1). The example given, a trial pit 10 km into the major accumulation basin of Biafo, reveals highly variable properties. There is an alternation of clean, major snowstorm layers in all seasons and relatively uniform, clean winter snow. Summer horizons include many ice lenses and dirty snow lawyers reflecting surface modifications by wind-borne dust, free–thaw and percolation down to freezing layers. There were no layers indicative of wind or avalanche activity, but some ice lenses did occur in uppermost winter horizons as well. Percolation and refreezing seem to explain the uniform 0 °C at this elevation although in winter the snow falls at lower temperatures. The pit suggests a 'warm' or 'temperate' glacier at this elevation. Higher and more exposed sites had uniform sub-zero 'polar' or 'cold' temperatures at depth, if higher than mean air temperatures. When compiled for an entire profile, estimates can be made of annual or budget year accumulations. In the example given, the full budget year identified had a total input of about 1.7 m water equivalent. Although the winter horizons seem much thicker, summer is deceptive because warmer snow is more readily packed down and so much of it is melted and refrozen in higher-density ice layers. Chemistry and isotope work were also able to bracket seasonal contributions and variability and help identify precipitation sources.

**Plate 4.2** Biafo Glacier trial excavation for snow pit studies. (**a**) At Snow lake Camp (5,300 m). (**b**) At Hispar Pass (5,150 m) (Hewitt 1985)

A sense of the environments and constraints that must be dealt with is given by considering site selection procedures. As far as possible, areas were avoided with signs of wind redistribution. Cornices indicate wind movement across interfluves around the basin, and there are areas of 'snow dunes' within the Lukpe Lawo where there is persistent funnelling of air from adjacent slopes. No significant wind-deposited or packed snow layers were observed at chosen sites, although some removal by wind was indicated at some of the more exposed, higher sites. Likewise, slopes with avalanche activity and places where crevasses occur were avoided. Profiles revealed in crevasse walls were investigated, but while annual layers are well defined, the snow is unduly modified by intermittent thawing, percolation and refreezing of moisture (Plate 4.3).

**Table 4.1** Summary findings in a trial snow pit at Lukpe Lawo ('Snow Lake') Camp (5,210 m) August 6–7, 1985. Weather conditions: 10/10 cloud cover, site in cloud layer ('fog'), snowed steadily, no wind, air temp −1.0–1.1 °C. Snow densities from weighing of samples from snow tube (3.64 cm, radius) pushed into pit walls (Observations by Gordon Young and Kenneth Hewitt; cf. Wake 1987, pp. 51–52)

| Layer (cm) | Verbal | Hardness estimate | Density ($kg/m^{-3}$) | Temp (°C) |
|---|---|---|---|---|
| 0–4 | New snow | V. soft | 250 | 0.0 |
| 4–14 | Dirty snow | Soft | 400 | |
| 14–22 | Cleaner snow | Soft | 430 | |
| 22–36 | Mixed rotten snow, thin ice lenses | Medium | 450 | |
| 36–38 | White snow (single storm layer) | Medium | 450 | |
| 38–41 | Many thin ice layers/lenses + 1 cm basal layer | Hard | – | |
| 41–47 | White snow (single storm layer) | Medium | 500 | |
| 47–50 | Mixed ice lenses and snow | Hard | – | |
| 50–67 | White snow (single storm) | Medium | 550 | |
| 67–68 | Ice lens | Ice | – | |
| 68–79 | White snow (single storm) | Hard | 550 | 0.0 |
| 79–80 | Ice lens | Ice | – | |
| 80–133 | Uniform white snow (spring 1984–1985) | Hard | 600 | |
| 133–137 | Major ice lens, variable thickness | Ice | – | |
| 137–219 | Uniform white snow (winter 1984–1985) | Hard | 600 | |
| 219–235 | Depth hoar | Soft | – | |
| 235–239 | Depth hoar | V. soft | – | |
| 239–308 | V. uniform hard white snow | Hard | 620 | |
| 308–328 | Looser, more granular snow | Medium | 540 | |
| 328–356 | Hard, granular and lumpy snow (previous summer horizon) | Hard | 650 | |
| 356–358 | Ice layers | Ice | – | |
| 358–370 | Dense, clean, granular snow (winter 1983–84) | V. hard | 650 | 0.0 |

Elevation and site-specific conditions affected the nature and usefulness of results. Horizons with ice lenses and dirt layers were present to some extent in all the profiles. Summer melting of surface snow is greatest in sites below about 5,000 m and decreased up to the highest site at 5,660 m. Chemical and isotope signatures were spoiled or less reliable at the lower sites, but generally well preserved and clear above 5,100 m (Wake 1987).

The perennial questions of differential over- and under-catch of snow had to be considered. The windiness of high elevations is notorious. Yet, a surprising observation at Lukpe Lawo was the remarkable stillness of the air during the storms that bring most of the snowfall. I have observed this in summer and winter. Snowfall mostly occurs *within the cloud layer*, usually in a steady vertical descent of snowflakes. There is a striking contrast to the windiness observed at both the higher elevations around the ridge tops and down the glacier below the cloud layer. Perhaps there is slowing or stalling of storm movement by the severe topography

**Plate 4.3** Crevasse walls at Lukpe Lawo (4,900 m) reveal the scope of seasonal and annual layers in firn (Hewitt 1985)

surrounding the open accumulation basins. If so, the absence of wind effects may involve another form of 'over-catch' as snow, swept into the air above the open basins, settles out to add to direct precipitation. However, this seems the norm rather than an anomaly.

### *4.2.2 Snow Accumulation at Biafo, 1983–1988*

The measurements confirm the relatively heavy snowfall at these elevations in the Central Karakoram (Fig. 4.1). Averages at all sites exceeded 1,000 mm water equivalent (w.e.) in a budget year, an order of magnitude greater than precipitation at valley weather stations below 3,500 m. Yearly estimates ranged from a low of 850 mm (w.e.) at one site to more than 2,300 mm at another.

The results are similar to Chinese estimates for Batura Glacier (BIG 1979). These were based on ablation and flow rates measured on the lower glacier, and they concluded that snowfall in upper glacier areas had to be around 1,000–2,000 mm. At a single site at 4,840 m in a single year, they measured accumulation of 1,034 mm (ibid. 961). For K2 Glacier on the Karakoram north slope, Ding (1990) reported single-year measurements of 800–900 mm (w.e.) at about 4,900 m and 1,500–2,000 mm at higher elevations below the peak of K2. For the latter, over-catch on the lee slopes of the massif is likely.

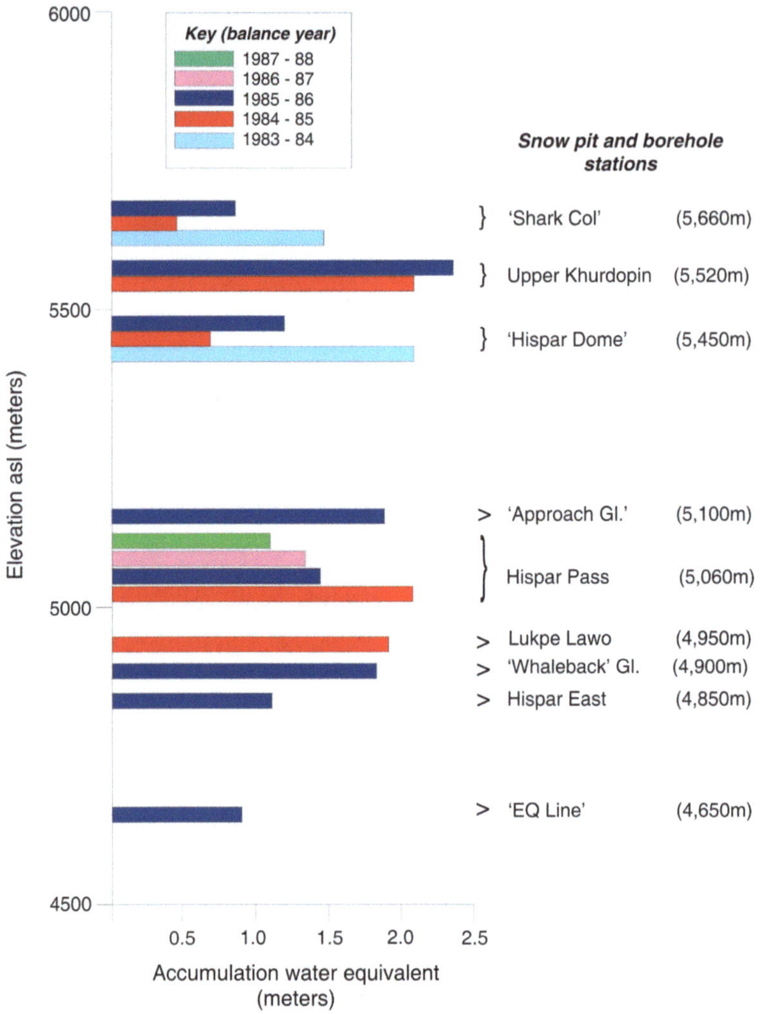


**Fig. 4.1** Snow accumulation in the 1980s snow pits by year, elevation and depth, upper Biafo Glacier (After Wake 1987; Hewitt et al. 1989; Hewitt 2011)

### 4.2.3 *Moisture Sources*

Sources of precipitation help to reveal how glacier nourishment and health relate to climate systems and may be affected by global climate change. Chemical signatures in snow at the three higher Biafo sites showed that winter snowfall comes from westerly sources; the Atlantic Ocean, Mediterranean, Black and Caspian Seas (Chap. 1). Wake (1987, p. 109) found low sodium and chloride concentrations, notably as compared to summer Indian Ocean sources.

In late spring and early summer, Arabian Sea moisture may be drawn into the Karakoram before snowfall from monsoonal systems, sometimes as a part of them. In every year, some summer snow samples included moisture identified with the monsoon, as indicated by isotope or chemical signatures coming from the vast lowland agricultural areas of the subcontinent (ibid.). However, details were complicated. On July 27, 1987, a heavy fall of snow at Hispar Dome (5,045 m) came from the west, one of several indications that westerly sources can remain important even through summer. Considerable variability is indicated from year to year in the contributions of winter and summer precipitation according to source and at different elevations and sites. Variations in monsoonal activity seem to be critical, much as for the Ladakh Himalaya to the southeast (Mayewski et al. 1980).

### *4.2.4 Temporal and Spatial Variability*

> … this knot of lofty mountains [the K2-Gasherburm massif] attracts the great mass of snow-clouds, and gets the share which ought to fall on the Karakoram, while [those] of less height, attract the clouds to a less degree, and are in consequence almost bare of snow….
> (Younghusband 1904, p. 244)

The zone of maximum precipitation is of special interest. Typically, mountain precipitation increases to a certain elevation then declines again. The Biafo data suggest maxima in the Central Karakoram occur above 4,900 m and, hence, in glacier accumulation zones. The broadest elevation range of concurrent data from snow pits was in 1984–1986. From these Wake (1987, p. 111) placed the maximum between 4,900 and 5,100 m. There was some decline above and below this band. Other years show a somewhat different picture and a distinct possibility that maximum precipitation occurs at or up to 5,600 mm, perhaps even higher. However, the highest values of more than 2,000 mm (w.e.), recorded on the upper Khurdopin Glacier at 5,520 m, may represent 'over-catch'. Khurdopin commences on the lee slope for prevailing westerly and southerly airflows. It could also mean that inputs high up in the adjacent Hispar and Biafo Basins are diminished by wind action rather than reduced precipitation. The lower average falls at Shark Col site (5,660 m) could be due to wind stripping. Wake (1987, p. 71) pointed this out but noted how measurements over 3 years of record were in the same ratio to the nearby, more sheltered Hispar Dome site (5,060 m). Mean accumulation there ranged between a high of 1,830 mm (w.e.) in 1984–1985 and low of 1,160 mm in 1987–1988. There were signs, however, in the snow core below 12.2 m depth that parts of the record were lost. This suggests the possibility of more extreme windstorms in some years, or episodes of melting and runoff in unusually warm or sunny summers, that can remove an entire annual layer (Wake 1987, p. 108). This is 300 m above the apparent firn limit and equilibrium line altitude (ELA) deduced from the 1980s observations.

Our records showed a vertical increase in precipitation for the whole basin leading to highest snowfalls in the accumulation zone. Wake (1989) felt there were indications of a reversal of precipitation at about 5,200 m, but the data are not wholly convincing. Other researchers assume a continued increase upwards, although without providing actual measurements to support the view (Mayer et al. 2006).

As already noted (Chap. 2), Karakoram ice cover and glacier lengths continue to increase up to the highest watershed elevations, suggesting a net increase in inputs to the glaciers. As suggested below, this probably relates more to concentration of snow by avalanches down precipitous slopes than to greater snowfall amounts high up. It was also observed that the spread and lowest elevations of glaciers coming from the highest watersheds do not differ significantly with orientation. If inputs are affected by preferential wind redistribution towards the north and east flanks, other factors, possibly thermal, may intervene to maintain similar outward spread of ice streams.

Seasonal incidence of snowfall is another critical concern. At Biafo a little over half the average high-altitude snow accumulation occurred in winter – but *almost half* in summer. The data are well constrained at the highest sites by stratigraphy, chemistry and isotope signals where, on average, 45 % of accumulation came in May–September (Wake 1989). This is, again, in contrast to valley weather stations lower down (see Chap. 1). Similarly, while the summer contribution varied from 30 % of annual precipitation to as much as 78 %, at the weather stations, it varied between 20 % and 45 % (Fowler and Archer 2006). Evidently, however, year-to-year variations are considerable and may turn out to be among the most sensitive to global climate change.

For the moment, a key finding at the Biafo profiles is that the input regime is essentially an *all-year* one. This is of major importance for the mass balance regime in the Mustagh Karakoram glaciers, at least (Chap. 6). It places the snowfall regime as intermediate between the ‘summer accumulation types’ of the Greater Himalaya and dominant ‘winter accumulation’ of the Caucasus and European Alps to the west (Ageta and Fujita 1996). It seems likely that the monsoon contribution increases southeastward towards Ladakh and declines towards the north and northwest (Mayewski et al. 1980).

Seasonal incidence is important but also how most snowfall occurs in large storm events, evident in snow pits from the incidence of thick, uniform bands of snow. Winter snowfall, in particular, shows well-defined, uniform snow layers from a small number of large, individual storms. Large contributions from individual summer storms were also observed. Moreover, heavy snowfalls were recorded at Snow Lake when little or none appeared even in records at a base camp station at 4,050 m. Between June 15 and 25, 1986, 1,100 mm of snow (300 mm, w.e.) were measured at 4,650 m, but only a trace at the Baintha station 600 m lower down (Wake 1987, p. 51). It is hardly surprising that precipitation records at the much lower permanent weather stations hardly reflect what is happening in glacier source areas (Winiger et al. 2005).

## 4.3 Problems with Firn

*Firn* is a generally accepted term for the intermediate state of snow in transition from snowfall to glacier ice. It has to survive beyond one summer. In mountain glaciers it is commonly treated as synonymous with accumulation zone snow. The origin and early use of word was in climates where there is some surface melting and modification by wetting and refreezing. According to Paterson (1994, p. 9), firn should be restricted to such 'wetted snow'. He questions its use for the sub-zero 'dry snow zone' that prevails in polar glaciers and at higher altitudes in mountain glaciers. However, this creates more problems than it solves in the Karakoram.

On the one hand, at Biafo the extent of dry versus wetted snow and the elevations where they occur vary greatly from year to year. A spell of clear weather can mean intense sunshine and daytime surface melting over huge areas where freezing temperatures otherwise prevail. Short episodes of superficial melting may occur as high as 6,000 m elevation. Conversely, in the snow pits at Biafo, thick winter horizons were observed to remain below freezing throughout the summer and beneath most points where temporary melting occurred at the surface. At elevations above about 5,200 m, most of the snow at depth was below the freezing point, even the dirtier summer snow and ice lenses from earlier years. Summer snowfall is warmer but also brings recurring episodes of high albedo helping to compensate for intense solar radiation, which is also offset by high nocturnal heat loss.

For most of the year, however, the whole Biafo accumulation zone has Paterson's (1994) cold, dry snow at the surface, and perennially so above about 5,500 m on most south-facing slopes, and 4,800 m on north facing. Thus, in the high Karakoram context, accumulation zones have both extensive dry snow areas and wet snow areas with percolation of meltwater. Each has large seasonal and year-to-year fluctuations in extent. In any case, as Paterson points out, it is hardly obvious just when snowfall becomes firn. The latter is commonly described as having a density of 400–500 kg $m^{-3}$ but there are intermediate states involving wind, frost and sun exposure that occur between this and new, dry snow of 30–70 kg $m^{-3}$. Calling accumulation zone snow firn only if and when there is some melting is overly complicated and difficult or impossible to track. In addition, problems arise in recognising and restricting firn to *year-old* snow (Paterson 1994). The all-year precipitation regime in the Karakoram, and frequency of summer snowstorms, mean that year-old snow is rarely observed at the surface, if ever. In this context the distinction to emphasise is between snowfall as a direct contributor to glacier ice and the much larger contributions of avalanched and wind-borne snow discussed below.

If the term is to be used in the Karakoram, it seems best to refer to firn as all snowfall that contributes *directly* to accumulation, whether it remains dry, is wetted, wind packed or refrozen. It does mean lumping together all the diverse forms and qualities of frozen materials as 'firn', but it may be impossible to effectively separate them.

Conversion of snow to dense, polycrystalline glacier ice, when nearly all the air is forced out, can take some decades, less in more humid, warmer areas and longer in colder, drier ones (Obbard et al. 2011). In accumulation areas like those of Biafo, annual firn is first deposited in layers some metres thick, usually of cold snow, sometimes in summer and lower parts, at or close to melting point. It is then increasingly compacted with depth. Crevasse wall profiles suggest significant melting and refreezing at the surface in most years and below about 5,200 m. Transformation to ice is not reached at depths less than 40 m (Plate 4.3). Temperatures in pit profiles for several years at and above 4,800 m were sub-zero although, up to 5,200 m, they included horizons with some summer thaw, percolation and refreezing. Evidently, transformation of snow to ice is spatially and temporally complex here. Extensive accumulation areas at these elevations display a series of annual ‘sandwiches’ of thermally and physically varying, layered firn. Existing observations suggest transformation from cold, dry snow is predominant in higher and north-facing areas and in uniformly high accumulation zones as at Rimo Glacier.

These investigations also raised problems with the notion of the *firn limit*, defined as the lowest firn to survive on the glacier at the end of the ablation season. Typically it is said to coincide fairly well with the highest elevation(s) at which ablation exposes glacier ice. It is commonly treated as equivalent to the climatic or annual snowline on the glacier (Mercer 1975). It is widely inferred to approximate the *equilibrium line altitude* (ELA), the interface or ‘hinge’ between the areas of net annual gains and losses. In mountain glaciers this is an elevation as well as a budget term (Benn and Evans 1998, p. 67). It is an important notion in relation to mass balance changes associated with year-to-year and climatic changes.

In the Karakoram, however, firn limits and ELAs raise similar or even more complex challenges than those of snowlines discussed earlier (Chap. 3). The greatest problems arise with avalanche-nourished Mustagh- and Turkestan-type glaciers. Firn limits can be hard or impossible to identify on the former. The latter lack them altogether. The conventional ELA hardly applies to either. Since relations of inputs and losses are involved, the implications for ELAs must await consideration in Chap. 7. Even at Biafo Glacier, where an approximate ELA can be determined from mass balance measurements (Chap. 6), complex and changeable sets of the conditions make it difficult to place the firn limit.

## 4.4 Subzones in Accumulation Areas

> … conditions vary from one point of a glacier to another; very few glaciers can be fitted into a single category. Thus, to speak about different zones in a glacier is better than trying to classify entire glaciers…. (Paterson 1994, p. 9)

In those Karakoram glaciers with conventional accumulation zones, a firn limit may be discerned in late summer. Typically, however, it is in a complicated band of snow and slush that tend to be refrozen overnight. There may be superimposed ice, where meltwater has been refrozen onto the former glacier ice surface (Fig. 4.2). Elsewhere, such phenomena have received a good deal of attention each being identified with subzones of the accumulation zone (Müller 1962; Paterson 1994,

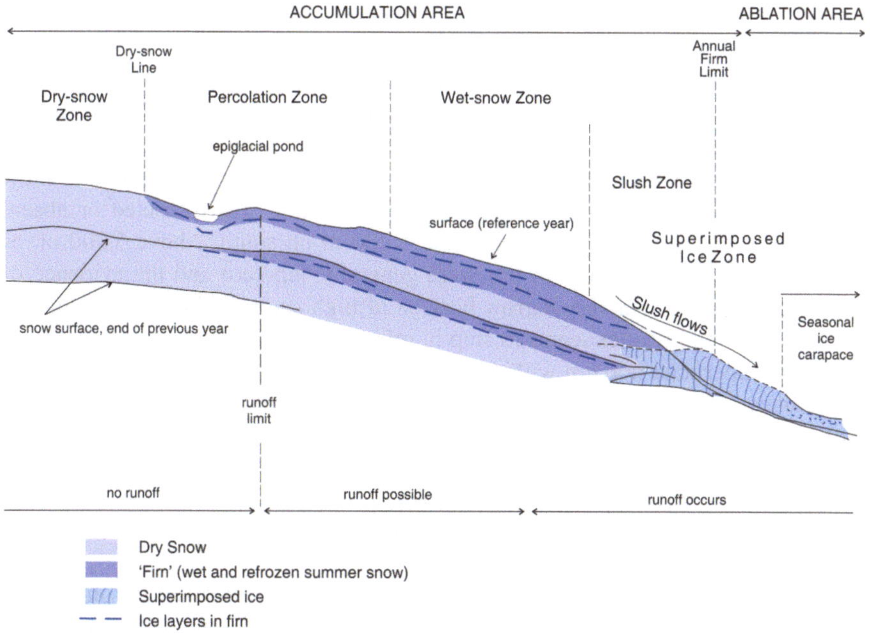


**Fig. 4.2** Subzones in lower accumulation zone areas based on observations at Biafo Glacier and terminology after Müller (1962), Paterson (1994) and Menzies (1995). Note that the set of zones may stretch over more than 10 km and can migrate even further in a given year

p. 10; Hambrey 1994, p. 53; Menzies 1995, p. 116). These descriptions from other regions do convey an impression that the subzones comprise relatively narrow fringes of accumulation zones, more or less closely related to the firn limit. Perhaps this does apply in many midlatitude and subpolar valley glaciers. On the larger Karakoram glaciers, however, in the accumulation areas, a broad belt of transitions or subzones may extend over 10–15 km and more than 1,000 m of elevation. Menzies (ibid) reports that 'most mountain glaciers' lack dry snow and percolation subzones. On Karakoram glaciers, however, both are extensively developed. Dry, cold snow prevails at higher elevations or where shaded by high peaks. Below this, in clear weather at the height of summer, subzones of wetted and nocturnally refrozen snow may be very extensive. There are large inter-annual changes and week to week in the summer.

A well-known division into wet snow versus percolation subzone can be found in the Karakoram: the percolation zones 'B' and 'A', respectively, of Müller (1962, p. 305). They do have distinctive patterns. Only near the firn limit and in late summer was wet snow found to comprise the whole snow pack down to the previous year's surface. Here, and below the firn limit, superimposed ice may be found. In other places, wet snow and percolation are confined to firn in the upper part of the year in question, usually summer or late spring layers. It is rarely deeper than about 60 cm. Even when very extensive in area, this is a relatively narrow horizon above deeper old firn. Most of the winter snow remains intact, perhaps with a layer of depth hoar at the interface with the previous year's firn.

In late summer the dry snow zone usually prevails above about 5,800 m but can vary by some hundreds of metres according to orientation and topographic shading, summer weather and the lateness of the season. A percolation zone was observed near the surface as high as 5,700 m at Biafo. However, the 'ripe' isothermal snow pack, widely described in association with percolation, was restricted or absent above 4,800 on north-facing surfaces and 5,200 m on south. Above 5,200 m, as noted, temperatures at depth in firn were generally sub-zero and the evidence of percolation limited to a very narrow horizon of high summer (Wake 1987, p. 65). Where it occurs the latter seems to stop in refrozen layers at variable depths, but a relatively superficial horizon within the annual snow pack. The vertical ice glands and drainage tubes seen in such firn elsewhere were not found (Paterson 1994, p. 10). At lower elevations but many kilometres into the accumulation zone, saturated and even slushy snow could form the upper few tens of centimetres in sunny weather. Typically it is a daytime phenomenon in clear weather and refreezes over night. The deeper layers in the highest snow pits were unaffected by such diurnal melting or percolation.

Superimposed ice is normally described as meltwater refrozen on glacier ice below the previous year's firn (Sugden and John 1984, p. 17). In this strict sense it was observed at Biafo as high as 4,600 m, near the apparent firn limit. Over most of the accumulation zone, however, percolation and refreezing of moisture is not superimposed on glacier ice. As such it presents a rather different picture from that usually described. Most ice in snow pits at Biafo consisted of more or less continuous layers in the summer firn of the year in question. These layers were 'superimposed', but on much thicker sections of firn with few or no ice layers, the winter and spring snow from early in *the same budget year*.

In the snow pits at Biafo, Wake (1987, p. 63) reported the ice layers comprised 6–14 % of the profiles, greatest around 4,800 m and declining towards higher elevations and in such thinner firn as occurs down to about 4,600 m. More crucially, ice layers are associated mainly with summer accumulation. This makes the summer contribution appear thinner and proportionately less than winter layers. When the densities or water equivalents are determined, however, the summer contribution to mass balance emerges as roughly equal, sometimes greater.

These observations are important in relation to years when surface snow has been reported to almost disappear from accumulation zones, even as high as 6,000 m, leaving an icy surface. Shipton (1938) observed this at Biafo's 'Snow Lake' in 1937. It was reported from satellite imagery in 1977 and 1979 (Shroder and Bishop 2010, F225) and, on the ground, in August 2009 (SP. Hasan Karrar, pers. comm.). However, while these events record significant summer extremes, they are deceptive. It may seem that firn has disappeared and glacier ice is exposed, but neither happened. Rather, superficial firn melts and washes away to expose a more or less continuous crust of ice layers that had formed below the surface but *within* the firn. They are not glacier ice, and a little excavation will show more layers of firn beneath them. Only a fraction of accumulation of the year involved is lost, and the real extent of firn does not change dramatically. Indeed, the ice layers in the firn may increase prior to such events. They are associated with clear sky conditions and alternating intense daytime sunshine and large nocturnal heat loss. Since this encourages conversion of superficial firn to the harder-to-melt ice layers, it may actually reduce net ablation losses.

In recent years, something like the implied 'apocalyptic' destruction of firn seems to have been occurring on glaciers elsewhere in the Himalaya and, if true, is alarming (Chap. 12). Possibly the processes described above are critical in how and how quickly climate change can affect the glaciers. However, there is no evidence that such destruction of firn is happening in the high Karakoram.

The subzones of 'possible' and 'no runoff' in accumulation areas described by Menzies (1995, p. 116) can be widely developed in the larger Alpine- and Mustagh-type glaciers of the Karakoram. Surface seepage networks arise over sloping firn in the lower parts of accumulation areas. In warm, clear weather networks of tiny rills cover the surface, but are not channels, and it is moot whether this should be called 'runoff' or 'interflow'. It seems that percolating moisture is induced to move laterally over impermeable layers, the ice layers observed in snow pits. However, the 'rills' indicate that seepage develops linear rather than sheet flow in the upper firn. The moisture may drain to the ablation zone, contributing to meltwater yields that short circuit the usual pattern of glacier input–output relations. Others remain trapped in the accumulation zone where they may contribute to surface ponds, water-filled crevasses, but will eventually freeze. A highly variable 'runoff limit' is involved (Menzies op cit).

A related phenomenon involves the ponds of water in depressions far into the Biafo and other accumulation zones, and as high as 5,400 m. They persist for many years. These 'epiglacial ponds' (Smiraglia and Dioliauti 2011) are fed by melting of near-surface snow and seepage along the networks of superficial drainage lines just mentioned. Presumably, sub-zero snow and ice at depth seal the ponds. In winter they freeze over and the ice cover insulates them. I have also observed water-filled crevasses up to similar elevations, evidently collected from surface melt and seepage. They are far from the englacial drainage system, suggesting eventual refreezing and contributions to glacier ice.

### 4.4.1 Slush Zone and Slush Flows

Müller's (1962) notion of a *slush zone* is relevant to Karakoram glaciers of the Alpine and Mustagh type. In summer, a subzone develops at the lower limits of the wet snow zone where saturated snow and small slush flows occur. Slush flows, also called 'slush avalanches' when large, will suddenly flow across and out of this subzone, even down small gradients. Underlying ice layers in firn may again be important in forming impermeable substrates in what are, otherwise, deep, permeable snow layers. Slush flows may commence well below, at or even above where an annual firn limit occurs. Most summer slush flows are small and local, but large ones may occur and comprise major events following heavy spring or summer snowfalls and in extreme warm weather. In this regard it is important to note that

**Plate 4.4** Slush flow in heavy snow pack at 4,900 m on Hispar Glacier, late June (Hewitt 2002)

slush flows are just as likely to occur in heavy snow on the ablation zone and their location can follow the upward migration of seasonal snow before it reaches the firn limit.

Slush flows on an exceptional scale were observed on the Hispar and Biafo Glaciers around 4,500–4,800 m in late June of 1999 and at Panmah Glacier around 4,500 m in late May 2005 (Plates 4.4 and 4.5). They occurred during bright, sunny days following heavy, late spring snowstorms. In each case, the snow pack appeared intact until the first break out of a slush flow, many of which grew down glacier by incorporating further wet snow. On Hispar and Biafo some flows travelled as much as 6 km down the glaciers, generally along pre-existing channels or depressions in the underlying ice. Some were large and dangerous slush avalanches.

Where there was little relief on the glacier, slush flows developed within surrounding snow pack or in what Müller (1962, p. 306) referred to as a 'slush lake'. Those at Panmah emerged within such a 'lake' and were relatively slow moving. Initially they accelerated to push a slightly raised tongue of floating, partly drained slush through the snow pack. After a few tens of metres of movement, they tended to deviate into the surrounding wet snowpack, forming a sinuous tongue and losing momentum until stalled.

At Biafo in 1999, recurring large slush flows gave rise to a 'slush swamp' (Benn and Evans 1998, p. 106). It covered about 3 km$^2$ on the upper ablation zone at about 4,550 m elevation (Plate 4.6). Whereas slush lakes form in pre-existing snowpack, slush swamps occur where slush flows become stalled in large collecting areas or depressions lower down. In the Biafo example, the complex systems of levees left by

**Plate 4.5** Slush flow in a slush lake at 4,700 m on Panmah Glacier, late June (Hewitt 2005)

**Plate 4.6** Slush swamp at 4,700 m on Biafo Glacier below Pamshe Pk (6,000 m) in late June (Hewitt 2002)

the flows were the only ground firm enough to make a way through the treacherous swamp. Channels and ponds were filled with slush and water to depths of at least a metre or two.

The literature creates an impression that slush flows are confined to higher latitudes (Hestnes 1998; Gude and Scherer 1998). However, they are another phenomenon widely observed on Karakoram glaciers but never investigated there. Unlike higher latitudes, rain is not known to be involved. For the larger slush flows, a run of clear weather and soaking of the upper snow pack are preconditions. It might be felt that the slush flows belong with glacier ablation and losses, which in part they do and so must be returned to later (Chap. 5). However, their treatment has commonly been in relation to the subzones discussed above.

## 4.5 Alternative Sources of Glacier Nourishment

> The most distinctive characteristic of valley glaciers is the presence of ice-free slopes overlooking the glacier surface. Such slopes are important sources of snow and ice accumulation in the form of avalanches…. (Benn and Evans 1998, p. 19)

Useful as the Biafo situation is for measuring high elevation snowfall and developments in accumulation zones, it is fairly exceptional. Glacier ice derived from direct snowfall is much less significant than that derived from avalanches. Indeed, firn as normally understood, or whether wet or dry, cold or warm, is absent as an intermediate stage in creating the larger fraction of Karakoram glacier ice. In part this is because more of the glacier source areas consist of rock walls than of conventional accumulation zone. As a result, avalanche processes intervene to control the pattern, intensity and qualities of material feeding most of the glaciers. They may be dirty or clean, and their thermal properties and run-out distances would seem critical to these inputs.

As the quote above indicates, this is not saying anything very new, yet research into avalanche nourishment of glaciers is practically non-existent for *any* region or glacier. The problem is partly one of the difficulties studying such a dynamic and dangerous condition. Yet, in other contexts, notably where avalanches are indeed major hazards, there is a great deal of work on them. In the Karakoram they seem important enough that ways need to be found to establish and document their roles.

### *4.5.1 Avalanche Nourishment: A Terrain Ruggedness Effect*

> The difficulty here of course is that there is a constant transfer by avalanches, especially in Alpine glaciers [sic],[2] from one area to the other. (Meier 1962, p. 262)

Due to the extent and steepness of rock walls, high-altitude snowfall observed at Biafo does not generally remain in place. Its rapid funnelling to the glaciers entails

[2]Meier is using the term 'Alpine' to refer to rugged mountain contexts rather than a type of glacier.

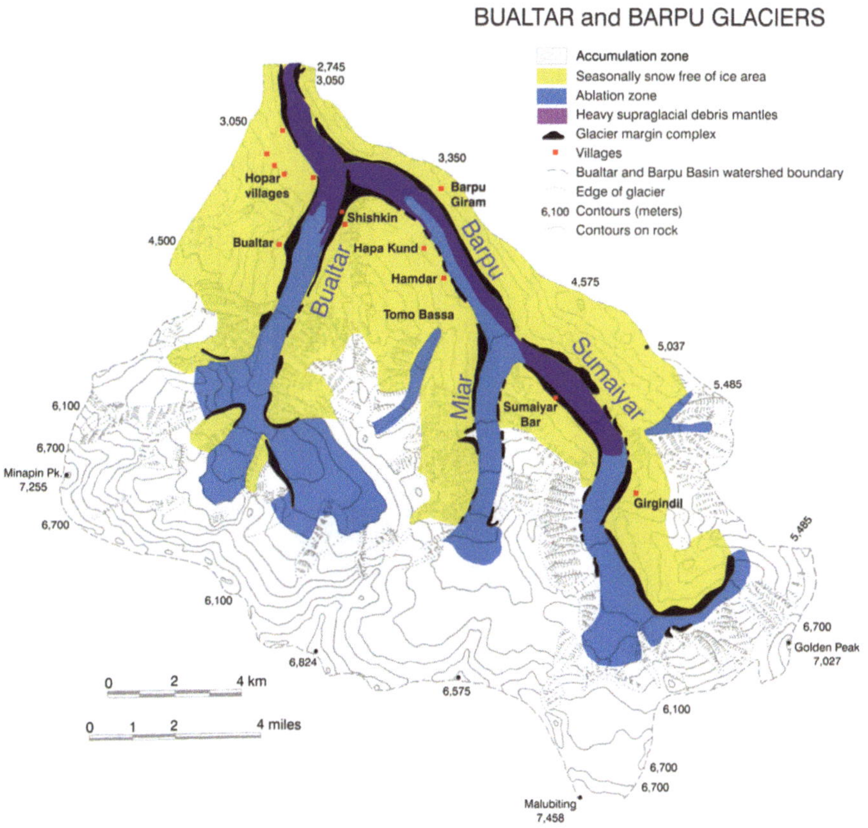


**Fig. 4.3** Bualtar–Barpu Glacier system, where Bualtar and the Sumaiyar tributary are Turkestan-type, almost wholly avalanche nourished, and the Miar tributary is Mustagh type. The wholly avalanche-fed Gannissh Chissh tributary is in the lower right corner (see Plates 4.7 and 4.8)

a different notion of accumulation (Benn and Evans 1998, p. 79). In fact, since they imply direct snowfall and areas of firn, accumulation zones should be treated just one form of glacier source area and direct snowfall just one type of input.

Snow avalanched onto glacier surfaces is associated with a whole range of hitherto un-researched developments and subzones, more extensive in the Karakoram than its Alpine-type glaciers suggest, and at least as complex and important as what was described from them. Cones and tongues of avalanche snow also go through processes of development on the way to becoming glacier ice. In avalanche-fed glaciers, a distinctive concern in mass balance inputs is their exceptional scale *below* where snowlines or ELAs are typically placed. With Mustagh- and Turkestan-type glaciers, inputs take place in what is normally regarded as the 'ablation zone', a major challenge in relation to conventional views and for determining mass balance.

Turkestan-type glaciers were defined as commencing below the accumulation zone and firn limit as normally understood. The Sumaiyar Bar tributary of the Barpu–Bualtar Glacier system illustrates the point (Fig. 4.3). Sumaiyar Bar derives its nourishment mainly from the precipitous, north-facing slopes of the Malubiting (7,453 m) and

**Plate 4.7** Avalanches from the flanks of Spantik (7,030 m) descending to the upper Sumaiyar Bar Glacier, the main branch of the Barpu Glacier (Hewitt August 1987)

Gannish Chhish or Spantik (7,027 m) massifs.[3] The importance of avalanches is clear from the way ice streams commence from large cones of avalanched snow tending to create uneven long and cross-profiles. Because of large and far-travelled avalanches, access to the upper basin is dangerous or impossible for 8–10 months of the year and the glacier surface at risk from them all year (Plate 4.7). The main ice stream is built from a series of avalanche cones, some steep and narrow, some as much as a kilometre across the base (Plate 4.8). Meanwhile, cones of avalanche snow survive from year to year as low down as 3,700 m along north and northeast-facing slopes and hence are 'inputs'.

In 1987, an avalanching episode on the east-facing rock walls opposite Girgindil lasted from mid-May to mid-June. Day and night there was an almost continuous roar of avalanches, although they had been rare in the weeks just before and ceased altogether afterwards. This depositional zone of the avalanches stretched from about 3,700 m upwards along the glacier for some 4 km. The interfluve above is between 4,800 and 5,300 m, which explains the timing of the avalanches and the fact that they ended early in summer. This is just one indication of the diversity of spatial and vertical patterns of avalanche inputs.

---

[3]The latter is also known as 'Golden Peak' or 'Yengutz Har' on the Shipton 1939 survey and, on Chogo Lungma side, as Spantik.

**Plate 4.8** Complex developments where snow and ice avalanches descend from steep walls and disconnected tributary glaciers to nourish the upper Sumaiyar Bar Glacier on a southwest-facing slope at about 5,000 (Hewitt 1987)

### *4.5.2 Subzones in Avalanche-Nourished Areas*

As already described, subzones in accumulation basins have received much scrutiny; not so, in avalanche-nourished areas. Yet, Turkestan-type glaciers and Mustagh types are also subject to snowfall, surface melting, freeze–thaw, percolation and superimposed ice at cold interfaces and to rilling by surface runoff, slush and slush flows in warmer weather (Plate 4.9). More importantly, other complicated developments arise on and around avalanche deposits, partly dependent on whether the material is wet or dry, and the occurrence of freeze–thaw cycles. The much higher densities of avalanched snow restrict percolation but some occur and, depending on the thermal state of avalanched material and seasonal temperatures, refreezing at depth. Irregular water tables develop in warm weather and springs may bubble out of the packed snow, an aspect of englacial hydrology not yet explored.

Avalanching onto ablation zone areas means that complicated patterns arise as dry and wet avalanche snow covers some parts, while glacier ice is exposed in others. Cleaner avalanches follow storms and dirty ones, sunny weather. Superimposed ice and (avalanche) snow survive from one year to the next beside places where ablation and runoff are large and seasonal snowfall is removed. At higher elevations there is so much interaction between wind and avalanche processes, and their responses to snowfall events, together affecting patterns of nourishment.

**Plate 4.9** The east slopes of Hispar Pass (5,150 m) where downslope winds have deposited layers of packed snow and 'Workman Peak (5,885), in the middle highest interfluve, is crowned by large cornices above slopes swept by avalanches and continually receiving lee-slope wind-borne snow (Hewitt 2002)

### *4.5.3 Estimating Avalanche Inputs*

> Climatic factors influence accumulation on avalanche-fed glaciers only in so far as the size and frequency of avalanches is climatically controlled. (Paterson 1994, p. 55)

Snowfall amounts and elevations at Biafo offer a means to estimate those where snow is avalanched. Some simple assumptions and calculations suggest the orders of magnitude involved. In the Toltar–Baltar Basin, for example, all but a tiny fraction of the source zone area of about 75 $km^2$ is above 4,500 m and below 6,000 m. Suppose the annual snowfall is similar to Biafo's in the same elevation band or at least 1,000 mm w.e. over about 70 $km^2$. If all of it is avalanched to the glacier, it would be about 70,000 $m^3$ of water or equivalent to 3.5 m of ice over the 30 $km^2$ of the main ablation zone. If the precipitation, or precipitation plus over-catch, is closer to 1,500 mm, it would exceed a 5 m ice equivalent over the ablation zone. It is fairly crude, but these estimates are of the same order as known average ablation rates in the Central Karakoram at the elevations involved and where over half of the ice is heavily debris mantled (Chap. 5). There is no sense of massively greater or diminished inputs compared to snowfalls, only a much greater concentration by elevation and ice mass area compared to Alpine-type glaciers.

Various qualifications must be noted. It is not known how far snowfall, where the surfaces are steep rock walls, compares with direct snowfall in open accumulation areas. During major snowstorms the rock walls are within the cloud layer as the snow falls or drifts into them. Close to the interfluves, the conditions suggest major wind stripping on windward slopes, heavy blow-over and excess deposition to leeward slopes. The avalanche system provides more efficient and rapid vertical transfer and spatial concentration from source zone to the main glacier.

### 4.5.4 Wind Redistribution and Nourishment: A Topoclimatic and Terrain Effect

An unknown but large fraction of Karakoram snow is redistributed within and between glacier basins by wind action. In places, deposition of wind-borne snow far exceeds direct snowfall and becomes the dominant form of nourishment. Satellite imagery shows this is especially so for lesser but rugged ranges in the region, including the Chip Chap, Aksai Chin and Chang Chenmo glaciers at the eastern limits of the Karakoram.

Evidence of snow redistribution is recorded by slopes stripped bare of snow, wind-packed and streamlined surfaces on windward slopes, in snow drifts, cornices and snow spilling down lee slopes from their crests (Plate 4.9). There is massive cornice development on leeward sides of mountain ridges and peaks permanently above the snowline and, temporarily, at lower elevations in winter (Kalvoda and Goudie 2007, photos 4–6).

Such ubiquitous features show how wind assumes a major role in the distribution and depth of snow. Wind directly feeds innumerable minor ice masses at the highest and lowest elevations of glacierisation. It exercises nearly total control of snow on the ground around interfluves, over lee-slope deposits including cornices and cornice glaciers, commonplace features and significant primers of avalanche activity.

Visible features suggest that as elevation and steepness increase, relatively more of snowfall, or where it comes to rest, is wind controlled. Wind strength, incidence and direction are major factors, but terrain and elevation are an immediate influence over patterns of snow deposition or removal.

The vast complex of peaks and interfluves in the high Karakoram preclude simple relations between slope orientation and wind nourishment. Any given glacier may lose inputs from its windward flanks or tributaries to wind stripping or blow-over of snow, but benefit from the same on its leeward flanks. There is no systematic evidence of high elevation wind fields, but highly variable and asymmetrical inputs of snowfall and wind redistribution must be anticipated. And these are permutations in gain and *loss* in what is usually treated as an 'input' to mass balance – a large but, as yet, indeterminate part.

The 'wind drift accumulation of snow' is thought to be highest at higher altitudes (Menzies 1995, p. 107). Some Soviet and post-Soviet glaciological work in the Central Asian high mountains recognised its importance for glacier formation and made some attempts to assess it. At an elevation of 3,750 m on Mount Elbrus (5,642 m) in the Caucasus Mountains, it was estimated that wind transported about 100,000 t of snow per kilometre 'normal to the direction of transport' (Krenke et al. 1970). It is not clear how to extrapolate this to larger areas or other mountains but it sounds like a great deal of snow. Wind speeds and snowfalls in the high Karakoram are likely to be as great or greater than on Elbrus. Evidence of wind control is overwhelming on interfluves above 5,500 m, meaning in 70 % or more of source zones. If the sorts of snowfalls delivered are comparable to those of Biafo's gentler accumulation zone, greater mass is likely wind redistributed than, perhaps, the total amounts feeding glaciers of the European Alps, or the Caucasus and other less extensive high mountains in Asia.

At the highest elevations, wind may be more important in feeding ice masses than either avalanching or direct snowfall. Locally minor and even small glaciers are sustained by snow dropping out of fierce winds across interfluves and coming to rest where there are minor shelf-, depression-, and niche-like basins, and suspended amid extreme slopes. Individual masses are small but, in total, they are sizeable reservoirs within large glacier basins. These snow and ice masses may be a quite conservative element, being 'cold', slow moving and, perhaps, less sensitive to precipitation changes. Inputs by wind-borne snow and outputs by ice avalanches have been totally ignored in mass balance studies.

In the lesser ranges wind assumes importance close to the glaciation limit (see Chap. 2). The mainly small glaciers tend to have a single dominant orientation, usually countless small ice masses on lee flanks, where windblown snow can be more important than shading from sunlight. It partly explains why the lowest glaciation thresholds are on northeasterly slopes, reflecting the prevailing west and southwest winds (cf. Evans 2006. 182).

Physical modifications of snow by wind action are probably second only to those of avalanching. Wind action serves to break up snow crystals, produce solid grains and redistribute and compact snow in snowdrifts (Benn and Evans 1998, p. 68). It remains a matter of casual observation in the high mountains, but windblown and modified snow is encountered throughout. It is denser, accelerating transformation towards glacier ice. Data may be lacking for the Karakoram, but it seems valid to suggest that more snow may be contributed to glacier ice that is wind redistributed and compacted than direct snowfall. It is second only to avalanched snow over which it has a major influence.[4]

## 4.6 Snowfall and 'Subzones' Below the Snowline

Mass balance studies not only ignore avalanched snow but tend to disregard snowfall below the perennial snow or accumulation zones. However, the amounts, interannual and spatial variability of this snow are important in the Karakoram context. Seasonal snow cover is, at once, an element in water discharge from glacier basins, especially the spring–early summer component, in albedo values over the glacier, in helping to regulate the length of the ablation season and for the length of covering and exposure of ice.

[4]For further discussion of the topic of wind regimes, (see Sect. 7.8).

Again, the glacier surface subzones described above are usually presented as applying only above the firn limit and in accumulation zones. Yet, similar or identical processes occur in the seasonal snow cover and ice surfaces of the ablation zones, in particular as the transient snowline moves up-glacier. Subzones of wet snow, percolation, ice layers and superimposed ice are observed and influence the course and timing of exposure of ablation zone ice to renewed melting losses. They exist in areas below the firn limit over more weeks of the year than at or close to it.

The role of snowfall in the ablation zone of Karakoram glaciers reflects the interrelations of three factors: the orographic effect that gives increased snowfall with elevation; the vertical gradient which gives declining temperatures, especially the shorter ablation season with elevation; and the impact of valley winds in redistributing snow on the glacier surface.

At Biafo, in 1986, transient snowlines and snow depths on the glacier were observed by Wake from May 29 (1987, pp. 50–56). On the first date, he reported the seasonal limit of snow at about 4,600 m elevation on the southwest-facing slopes beside the glacier and about 4,000 m on the northeast-facing slopes of its other flank. The winter snow cover on the glacier was extremely variable, with many bare patches and areas of superimposed ice exposed, while much deeper and denser snow lay in depressions and along the margins of medial moraines due to wind redistribution through the winter. Some sense of the orographic effect on precipitation can be gained from snow depths measured below 4,700 m in mid-June 1986 (Wake 1987, p. 656). Snow had virtually disappeared from the glacier below 4,000 m, but increased to a 90 % cover from 4,300 upwards. Between 4,200 and 4,650 m, snow depths showed a fairly consistent increase with elevation, amounting to a roughly tenfold increase. Remnants of the winter snowfall were largely confined to depressions where they had been deposited by wind movement.

Between June 5 and 15, when only trace precipitation was recorded at the terminus weather station, three storms deposited heavy snowfall on the mid- and upper parts of the glacier. At 4,650 m near where the firn limit would be identified later in the year, about 1.1 m of snow fell, or 300 mm (w.e.). On June 18 a profile of snow cover on the glacier itself was measured over a distance of 14 km from 4,200 to 4,620 m elevation (Table 4.2.). Over the lower 3 or 4 km, the new snow cover was already patchy and variable in depth. At about 4,300 m 60–90 % of the glacier remained covered, and above this, depths were determined along a continuous profile. The profile reflects a general increase in snowfall with elevation. The irregularities may reflect wind conditions during and following the storms, but also uneven residual winter snow.

The data provide an example of the spectrum of snow cover conditions on the glacier, influenced by season and particular weather conditions. These will affect the timing of transient snowlines and when glacier ice is exposed to expand the ablation zone to given elevations. However, along with wind redistribution, freeze–thaw means that apparent snow depths will not accurately represent the winter, nor even the recent storm contributions. Through the winter, but especially in fall and spring, freeze–thaw gives rise to superimposed ice, icy layers and refrozen snow that build a carapace over the ablation zone. It seems more extensive in tropical conditions. Since

**Table 4.2** Snow depth profile for Biafo ablation zone, 1986 (After Wake 1987)

| Elevation (m) | Distance from EQ line site (km) | Snow depth[a] (m) |
|---|---|---|
| 4,650 | 0 | 2.3[b] |
| 4,590[c] | 1.5 | 2.0 |
| 4,550 | 2.0 | 1.0 |
| 4,540 | 2.4 | 0.9 |
| 4,530 | 2.8 | 1.0 |
| 4,520 | 3.2 | 1.1 |
| 4,510 | 3.6 | 1.45 |
| 4,500 | 4.0 | 1.2 |
| 4,480 | 4.6 | 0.75 |
| 4,465 | 5.2 | 0.6 |
| 4,450 | 5.8 | 1.2 |
| 4,435 | 6.4 | 0.5 |
| 4,420 | 7.0 | 0.45 |
| 4,400 | 7.6 | 0.5 |
| 4,380 | 8.2 | 0.05 |
| 4,360 | 8.8 | 0.35 |
| 4,340 | 9.4 | 0.7 |
| 4,320 | 10 | 0.6 |
| 4,300 | 11 | 0.2–0.9 |
| 4,200 | 14 | 0–0.3 |

[a]Each depth is the average of four soundings
[b]Snow density measured at this site on June 15, 1986, was 390 kg/m$^3$
[c]Between this and the next observation site is a moderate icefall section with crevasses

it is a characteristic of the ablation zone, it is left for the next chapter (Sect. 5.10). The point to emphasise here is how processes normally described for the subzones of the accumulation area are important in regulating ablation zone developments as well. As with avalanche inputs, this complicates the conventional views and separation of ablation and accumulation zones.

## 4.7 Inputs Summary

The main inputs to Karakoram glaciers, in likely order of mass contributed, are as follows:

(i) *Snow avalanches*: overwhelmingly important for Turkestan-type and predominantly for Mustagh-type glaciers. They contribute significantly to most Alpine types or their tributaries.
(ii) *Wind-redistributed snow*: unknown but possibly larger than the direct snowfall contribution, critical at high elevations and around interfluves, for small wind-fed glaciers of lesser ranges and in priming much of the avalanche contribution.

(iii) *Direct snowfall*: dominant only on Alpine types and an important fraction of inputs on many Mustagh types, although there has been no actual monitoring to test these views.
(iv) *Ice avalanches*: from disconnected, steep tributaries high up and on steep icefalls.

There are four distinct paths to glacier ice, five if one accepts that many avalanches begin with wind-redistributed snow. Here is another reason why Ahlmann's (1948) classification of glaciers into *temperate, subpolar* and *high polar* is difficult to apply in High Asia. Similar difficulties arise with *tropical* and *subtropical* classes of Kaser and Osmaston (2002). The snowfall/firn temperature and ablation conditions that were invoked as separating these broad regional and 'geophysical' classes are *all* found in this one region. Most or all apply to the larger Karakoram glaciers. It is another way in which verticality intervenes to reconfigure glacial conditions.

Finally, in the High Asian subtropical environment, a great deal of energy and moisture is absorbed in freeze–thaw over glacier surfaces and meltwater percolating down to sub-zero ice or snow layers, where it refreezes. It was shown how this applies to lower parts of accumulation or source zones, at least up to 5,500 m. In most of the ablation zones, temporary, seasonal 'inputs' further complicate matters. Some have recognised its importance of these phenomena elsewhere in High Asia (Müller 1958), but they have been largely neglected in the Karakoram.

## 4.8 Accumulation–Area Ratios (AARs)

The inputs summary has implications for a widely employed measure related to glacier nourishment, the *accumulation–area ratio* (AAR), or the accumulation zone divided by total ice area (Paterson 1994, p. 36; Benn and Evans 1998, p. 83). It may be given as a percentage or a ratio and is generally used in relation to mass balance to differentiate glaciers according to climatic regime and for year-to-year changes (Dyurgerov and Meier 2000).

In polar glaciers the AAR can reach or approach 1.0. Temperate ones are usually shown to have AARs of 0.6–0.7 (Paterson 1994, p. 36). Kaser and Osmaston (2002, p. 48) found that the ideal tropical glacier has an AAR > 0.8. However, glaciers undergoing rapid retreat in recent years have AARs less than 0.2 as the accumulation zone of temperate, tropical and subpolar glaciers partly or entirely disappears (WGMS 2008, Table C, pp. 155–164).

In the Karakoram, these uses of the AAR are problematic. The main set of valley glaciers has AARs representative of *all possible 'climatic' types* in any and all years (Table 4.3). If all AARs are found in the same region, they are hardly useful for assigning climatic regime there, or perhaps generally, and their responses to climate change will be ambiguous. Some large glaciers like Biafo and Chiantar have AARs around 0.6–0.7, so, possibly, year-to-year variations in the AAR might track

**Table 4.3** Accumulation–area ratios (AARs) for the main set of largest valley glaciers

| Glacier | AAR | Glacier type |
|---|---|---|
| Rimo | 0.8 | Alpine |
| Biafo | 0.7 | '.' |
| Chiantar | 0.66 | |
| S. Shukpa | 0.61 | |
| Terong N+S | 0.6 | |
| Siachen | 0.56 | |
| Sarpo Laggo | 0.53 | |
| N. Shukpa | 0.49 | Mustagh |
| N. Gasherb'm | 0.48 | '.' |
| Baltoro | 0.47 | |
| Gharesa | 0.4 | |
| Panmah | 0.37 | |
| Batura | 0.34 | |
| Khurdopin | 0.32 | |
| Hispar | 0.29 | |
| Chogo Lungma | 0.29 | |
| Kutiah | 0.29 | |
| Skamri | 0.29 | |
| Kondus–Kab. | 0.29 | |
| Virjerab | 0.28 | |
| Ghondoghoro | 0.24 | |
| Bualtar | 0.22 | |
| Charakusa | 0.21 | |
| Braldu (Sh.) | 0.21 | |
| Barpu | 0.2 | |
| Karambar | 0.19 | Turkestan |
| Yashkuk Yaz | 0.15 | '.' |
| Hinarche | 0.15 | |
| Aling | 0.12 | |
| Surgin | 0.12 | |
| Masherbrum | 0.11 | |
| Hasanabad | 0.1 | |
| Minapin | 0.095 | |
| Kukuar | 0.07 | |
| Toltar–Baltar | 0.04 | |

climatic responses. However, as described above, there are serious problems deciding when a change in the *appearance* of the accumulation zone or firn would actually signal a change in AAR or mass balance. Most Karakoram valley glaciers have ratios below 0.5. For a great many, like Toltar–Baltar, they are close to zero. And these are the more typical cases, the Turkestan-type glaciers. Also of note are the thousands of small and minor ice masses above the snowline and disconnected from a main glaciers with AARs of 1.0.

On the other hand, AARs can be used to differentiate the glaciers according to nourishment type. In fact, the different nourishment classes imply different accumulation–area ratios, suggesting they can be used in the classifying glaciers found in the Karakoram. Discussion of the broader significance of this finding will be resumed when looking at mass balance as a whole and in relation to glacier dynamics (see Chap. 7).

## References


Ageta Y, Fujita K (1996) Characteristics of mass balance of summer-accumulation type glaciers in the Himalayas and Tibetan Plateau. Z Gletschkd Glazialgeol 32:61–65

Ahlmann HW (1948) Glaciological research on the North Atlantic coasts, research series 1. Royal Geographical Society, London

Archer DR, Fowler HJ (2004) Spatial and temporal variations in precipitation in the Upper Indus Basin, global teleconnections and hydrological implications. Hydrol Earth Sys Sci 8(1):47–61

Benn DI, Evans DJA (1998) Glaciers and glaciation. Hodder Arnold Publication, London

BIG (Batura Investigations Group) (1979) The Batura Glacier in the Karakoram Mountains and its variations. Sci Sin 22(8):958–974

Bolch T, Kulkarni AV, Kääb A, Huggel C, Paul F, Cogley JG, Frey H, Kargel JS, Fujita K, Scheel M, Bajracharya SR, Stoffel M (2012) The state and fate of Himalayan glaciers. Science 336:310–314

Ding Y (1990) Nourishment condition of existing glacier on the north side of Karakoram Mountains. J Glaciol Geocryol 12(4):237–334

Dyurgerov MB, Meier MF (2000) Twentieth century climate change: evidence from small glaciers. Proc Nat Acad Science (PNAS) 97(4):1406–1411

Evans IS (2006) Local aspect asymmetry of mountain glaciation: a global survey of consistency of favoured directions for glacier numbers and altitudes. Geomorphology 73:166–184

Ferguson RI, Collins DN, Whalley WB (1984) Techniques for investigating meltwater runoff and erosion. In: Miller KJ (ed) International Karakoram Project. 2 vols. Cambridge University Press, Cambridge, UK

Fowler HJ, Archer DR (2006) Conflicting signals of climatic change in the Upper Indus Basin. J Clim 19(17):4276–4293

Gardelle J, Berthier E, Arnaud Y (2012) Slight mass gain of Karakoram glaciers in the early twenty-first century Nat Geosci online 15 April, doi:10.1038/ngeo1450

Goodison BE, Metcalfe Jr (1992) The WMO solid precipitation measurement intercomparison: Canadian assessment. WMO Technical Conference on Instruments and Methods of Observation (TECO-92). Vienna, 11–15 May 1992, pp 221–225

Gude M, Scherer D (1998) Snowmelt and slushflows: hydrological and hazard implications. Ann Glaciol 26:381–384

Hambrey MJ (1994) Glacial environments. University College London Press, London

Hansen S, Davies NA (2002) Windshields for precipitation gauges and improved measurement techniques for snowfall, US Department of Agriculture, Forest Service, Watershed, Soil and Air pubs, www.fs.fed.us/t-d/pubs/htmlpubs/htm02252325/index.htm

Hestnes E (1998) A contribution to the prediction of slush avalanches. Ann Glaciol 6:1–4

Hewitt K (2011) Glacier change, concentration and elevation effects in the Karakoram Himalaya, Upper Indus Basin. Mountain Research and Development 31(3):1–13, http://dx.doi.org/10.1659/MRD-JOURNAL-D-11-00020.1

Hewitt K, Wake CP, Young GJ, David C (1989) Hydrological investigations at Biafo Glacier, Karakoram Himalaya, an important source of water for the Indus River. Ann Glaciol 13:103–108

Kalvoda J, Goudie AS (2007) Landform evolution in the Nagar region. In: Kalvoda J, Goudie AS (eds) Geomorphological variations. Nakladatelstvi, Prague, pp 87–125

Kaser G, Osmaston H (2002) Tropical glaciers. Cambridge University Press, Cambridge, UK
Kasser P (1967) Fluctuations of glaciers, 1959–1969. International Association of Hydrological sciences (IAHS) Publication, Brussels
Krenke AN, Borovik ZS, Rototayev KP (1970) Snow accumulation on the glaciers of the Caucasus. NIGMI 45, Gidrometeoizdat, Leningrad
Mayer C, Lambrecht A, Belo M, Smiraglia C, Diolaiuti G (2006) Glaciological characteristics of the ablation zone of Baltoro Glacier, Karakorum, Pakistan. Ann Glaciol 43:123–131
Mayewski PA, Pergent GP, Jeschke PA, Ahemad N (1980) Himalayan and Trans-Himalayan glacier fluctuations and the South Asian monsoon record. Arc Alp Res 12:171–182
Meier MF (1962) Proposed definitions for glacier mass budget terms. J Glaciol 4:252–261
Menzies J (1995) Glaciers and Ice Sheets. In: Menzies J (ed) Modern glacial environments: processes, dynamics and sediments, vol 1, Butterworth-Heinemann, vol 4, Oxford. Chap., pp 101–138
Mercer JH (1975) Glaciers of the Karakoram. In: Field WO (ed) Mountain glaciers of the northern hemisphere. United States Army Cold Regions Research and Engineering Laboratory, Hanover, pp 371–409
Müller F (1958) Eight months of glacier and soil research in the Everest region. In: The mountain world 1958/59. Swiss Foundation for Alpine research, Harper, New York, pp 191–208
Müller F (1962) Zonation in the accumulation area of the glaciers of Axel Heiberg Island, NWT, Canada. J Glaciol 4:302–313
Obbard RW, Baker I, Lomonaco RW (2011) Firn. In: Singh VP, Singh P, Haritashya UK (eds) Encyclopedia of snow, ice and glaciers. Springer, Dordrecht, pp 290–293
Paterson WSB (1994) The physics of glaciers, 3rd edn. Pergamon, New York
Shipton E (1938) Blank on the map. Hodder and Stoughton, London
Shroder J, Bishop M (2010) Glaciers of Afghanistan. In: Williams RS, Ferrigno JG Jr (eds) Satellite Image Atlas of Glaciers: ASIA, United States Geological Survey, Denver, Professional Paper 1386-F, Washington, F-3, pp 167–199
SIHP (Snow and Ice Hydrology Project) (1990) Annual report, 1989. Cold Regions Research Centre, Wilfrid Laurier University, Waterloo
Smiraglia C, Dioliauti G (2011) Epiglacial morphology. In: Singh VP, Singh P, Haritashya UK (eds) Encyclopedia of snow, ice and glaciers. Springer, Dordrecht, pp 262–268
Sugden DE, John BS (1984) Glaciers and landscape. Edward Arnold, New York
Wake CP (1987) Spatial and temporal variation of snow accumulation in the Central Karakoram, northern Pakistan. Master's thesis, Department of Geography, Wilfrid Laurier University, Waterloo
Wake CP (1989) Glaciochemical investigations as a tool to determine the spatial variation of snow accumulation in the Central Karakoram, northern Pakistan. Ann Glaciol 13:279–284
WGMS (2008) Fluctuations of glaciers 2000–2005 IX. World Glacier Monitoring Service, Zurich
Winiger M, Gumpert M, Yamout H (2005) Karakorum – Hindukush – western Himalaya: assessing high-altitude water resources. Hydrol Process 19:2329–2338
Workman FB, Workman WH (1905) From Srinagar to the sources of the Chogo Lungma Glaciers. Geogr J 25:245–268
Younghusband FE (1904) The heart of a continent: a narrative of travels in Manchuria, across the Gobi Desert, through the Himalayas, and Chitral, 1884–1894. John Murray, London

# Chapter 5
# Glacier Mass Balance II: Ablation Losses

**Abstract** The negative or loss factors in mass balance are examined, mainly processes, controls and rates of ablation on glacier surfaces and some other contributors to water yields from glacier basins. Limited but actual ablation measurements exist for three of the large glaciers, Baltoro, Biafo and Batura. They are described and interpreted in relation to differing nourishment and debris covers. Ablation is largely due to solar radiation but regulated by surface conditions in ablation zones. The length of the ablation season is important, as are the extent of exposed ice and of debris covers. These vary strongly with elevation, leading to verticality effects. Quantities and duration of seasonal snow cover also increase with elevation. A further control is the 'carapace' of refrozen meltwater, snow and icy layers superimposed on ablation zone ice in the fall and spring shoulder seasons. It also migrates vertically. The interactions of elevation and seasonal snow cover, debris mantles and ice surface topography are keys to specific and net annual ablation losses. When snow and the icy carapace are removed, when debris covers are thin or absent and when weather conditions are favourable, ablation rates are high and similar over a broad range of elevations. Hence, ablation season length and differences in surface conditions are critical for specific and total losses. On lower ablation zone areas, heavy debris mantles protect ice against the longer ablation seasons and higher air temperatures. However, areas of clean-to-dusty or dirty ice, thin or scattered debris, are more extensive and much more critical for net and total ablation losses and water yields. Thinner debris covers, dust, dirt, cold-tolerant algae and so-called cryoconite fragments can drive ablation rates much higher than for clean ice, in places more than double. These are found predominantly in the mid- and upper ablation zones. They are more sensitive to weather and climatic variability than where there are heavy debris mantles. However, geomorphological events in glacier basins, notably rock falls and massive rock slope failures, can suddenly alter and increase the protective role of debris covers for some decades as they are transported down the glacier. They are recurrent phenomena in the Karakoram which, along with motion instabilities and glacier surges, may alter mass balance or glacier advance and retreat.

K. Hewitt, *Glaciers of the Karakoram Himalaya: Glacial Environments, Processes, Hazards and Resources*, Advances in Asian Human-Environmental Research,
DOI 10.1007/978-94-007-6311-1_5, 

**Keywords** Ablation rates • Albedo • Radiation fluxes • Verticality • Supraglacial debris • Ice facets • Penitents • Cryoconite • Freeze–thaw

## 5.1 Introduction

The term *ablation* applies to all the ways in which glaciers lose mass. However, melting at and over the glacier surface is overwhelmingly important, except for glaciers in areas of volcanic activity or in high latitude glaci-marine environments and perennially frigid regions. With valley glaciers most ablation occurs in the lower warmest areas comprising *ablation zones*, defined as the area with an annual net loss of ice. This is the negative part of mass balance which, as noted, mainly concerns its relation to snow accumulation and other inputs. Ablation is also sufficiently complex in the Karakoram to warrant initial treatment on its own.

The processes driving ablation involve energy sources and fluxes, especially specific types and transformations of energy (Barry 1992, p. 209; Paterson 1994, p. 57; Armstrong and Brun 2008). The radiative fluxes are of overriding significance. In high mountain environments, received solar or short-wave radiation is the most critical. When there is cloud cover, diffuse short-wave radiation scattered from the atmosphere can become an important source. In some regions, long-wave radiation from the atmosphere and land surfaces is significant but only assumes importance locally in the Karakoram, mainly along glacier margins, on or beside prominent medial moraines and for ice facets within debris-mantled areas as described below. It has a fairly insignificant net role and likewise rainfall in these environments. Available measurements point to short-wave solar radiation as responsible for 90–95 % of ablation on Karakoram glaciers (Untersteiner 1957; BIG 1979; SIHP 1990).

Actual ablation or ablation rates are controlled by the partitioning of heat at the surface (Copeland 2011). This depends largely upon how much of the received radiation is absorbed and how much of the heat derived actually reaches ice at the melting point. Given the dominant role of solar energy, a decisive factor is the reflectivity of the surface, determining how much short-wave energy is absorbed. The usual measure employed is *albedo*, the fraction of received solar radiation that is reflected back or the reflecting power of a surface (Grenfell 2011). It may be expressed as a decimal, with 1.0 for a perfect ('white') reflector and 0.0 for total or 'black body' absorption, or as a percentage from 100 % to zero, respectively. The albedo of fresh snowfall can be close to 0.9, while old snow and ice may be as low as 0.4–0.3 but, in cleaner environments, as high as 0.7. Soot or black carbon can approach zero albedo. Most debris surfaces are in the range 0.1–0.15. For fine-grained sediment, albedo varies with moisture content. The main differences on ice surfaces arise from impurities and debris, especially debris mantles that melt out from the snow and ice or dust that arrives as wind-borne material.

Albedo distributions over space and time are of critical importance and highly variable on the glaciers of interest (Untersteiner 1957). The reflectivity of snow and ice over the ablation zones varies sharply from site to site at any given time and is changeable on time scales from minutes to years. In this, cloudiness and the presence and extent of snow, dust, dirt and debris are key factors. In the ablation season, debris is constantly shifted around by on-ice processes, and the extent of thin debris covers or atmospheric dust changes with weather conditions (Kirkbride 2011). Albedo is also influenced by micro-roughness or pitting and 'weathering' of snow or ice surfaces. Hence, ablation is itself a source of changing albedo as the texture of snow and ice surfaces is changed, and as surface dust and dirt build up, or are washed away. Reflectivity also depends upon surface slope and orientation, important mainly with respect to the solar angle. Shading and reflection from surrounding terrain are locally important in the mountains. In general, all of these are dynamic variables.

*Net short-wave radiation*, the share of the radiation absorbed per unit area of the surface, is an important measure connecting albedo to energy conditions (Van den Broeke et al. 2011). It can be derived where instruments measure incoming and outgoing radiation per unit area, for example, watts per square metre. This separates the radiative energy absorbed from that reflected. It provides both a *net* value from the ratio of reflected to incoming short-wave radiation and an absolute measure of the amount of energy absorbed and, therefore, available for ablation. In many ways this is a more satisfactory approach since, according to the intensity of sunshine, there may be huge differences in the amount of energy actually absorbed or reflected back for any given albedo. Strong sunlight may support lesser rates of ablation where albedo is high than weak sunlight where it is low. For such reasons also, net radiation can vary enormously over ablation zones through the day. Moreover, the albedo effect and whether or how much of absorbed radiation actually causes ablation depend strongly on other environmental factors, notably dust and debris covers, freeze–thaw, wind redistribution of snowfall, on-ice relief and micro-topography.

Short-term on-ice experiments at Chogo Lungma Glacier by Untersteiner (1957) gave some useful insights into Karakoram conditions. He measured dramatic differences in the energy absorbed according to albedo and large differences in ablation rates under identical weather conditions but differing albedo. He showed how changing weather conditions were responsible for constantly changing albedo on ice and snow surfaces. He also found that cloud cover had a major impact on ablation rates (ibid, p. 20). Cloud cover increases with elevation and is much greater for mid- and upper ablation zones than lower ones. However, on cloudy days the total or global radiation appeared to be relatively high in the Karakoram, and ablation was higher than observed in, say, the European Alps for comparable cloud covers.

Mass balance assessments must take account of these complex conditions affecting received energy at the surface, its albedo, including snow cover and heat transfer through or insulation by varying debris mantles (Plate 5.1). Sub-zero temperatures in snow and ice, seasonal and diurnal, mean they must be warmed for ablation to begin. Such considerations help to explain why rates of ablation and net losses on Karakoram glaciers vary widely. They involve diverse local patterns and large differences and even reversals with elevation. It is not always easy or wise to extrapolate from measurements at specific sites to others, especially at other elevations, between on- and off-ice areas, and where the type or rates of glacier surface processes differ.

**Plate 5.1** Diversity of surfaces in snow, ice and debris covers that influence albedo and ablation rates, Chiantar Glacier mid-ablation zone, in early summer (Hewitt 1993)

## 5.2 Verticality Relations

Ablation amounts and duration, or contributions from different parts of the glaciers, vary with elevation and not only in one direction. The length of the ice ablation season declines upwards, mostly as a function of seasonal temperatures. However, ablation is also influenced by the intensity of sunshine and cloudiness, both of which tend to increase with elevation but with opposite consequences. Seasonally, ablation is partly regulated by snowfall on the glacier, which increases in amounts and duration upwards, but is subject to more complicated patterns in its albedo, redistribution by wind, and how long it takes to melt off. Heavy debris mantles usually increase towards lowest ablation zone areas, where their influence is greatest. Conversely the importance of thin and scattered material increases up-glacier and is greatest in the extensive mid- and upper ablation zone areas.

The depletion of solar radiation by absorption and back scattering in the atmosphere is inversely related to elevation. The highest intensities of received sunlight tend to occur at higher elevations. The solar beam is likely to be reduced in lower ablation zones by greater dustiness. In the Karakoram above 4,500 m on clear days, it is not unusual for received radiation to approach the 'solar constant': the amount of sunlight arriving at the Earth's distance from the sun, and unaffected by the atmosphere, or roughly 1.368 kW/m$^2$ (Barry 1992, p. 29). On some days, in the upper ablation zone and firn basins at Biafo Glacier, we measured values higher than this. It is thought to be due to greater downward atmospheric reflectance and reflections from steep snow-covered terrain. These effects are offset by greater incidence of cloud covers at high elevations. Nevertheless, received radiation often seems unusually high even under cloud cover, presumably also due to the multiplication of reflections and scattering between the surface and cloud (Untersteiner 1957).

Although solar radiation powers ablation, it is most effective where air temperature remains above zero, where the glacier ice surface is exposed – not buried under debris, snow or icy layers – and has been raised to zero. In some regions and conditions *sublimation,* direct conversion of ice to vapour can be a factor and also the reverse, *rime or* direct conversion from vapour to ice. I have not found any evidence that they play a significant role in the Karakoram, and they appeared negligible in the measurement programs at Biafo.

The vertical air temperature gradient is important primarily as it regulates the length of the melting season and the duration of daily above-freezing conditions. Other things equal, ablation losses should be highest towards the termini of the glaciers because the season is so much longer. Near the lowest glacier termini in the Hunza valley at around 2,300 m, ablation can occur over almost the whole year (BIG 1979). In various years of observations at Biafo Glacier, it was found that the ablation season lasts about 8 months at 3,000 m, the terminus (Hewitt 1968), and three months at about 3,800 and declines to a couple of weeks at 4,600 m (SIHP 1990). Around the firn limit at about 4,800 m, melting occurs over no more than days or a series of part days with nocturnal freezing and may not be sufficient to expose glacier ice. At given elevations temperatures rise above zero on many more days or months but are offset by equivalent or stronger nocturnal freezing (see Fig. 2.6). There are large year-to-year variations in the length and timing of the ablation season and critical factors such as debris and seasonal snow cover intervene everywhere to control the extent of exposed ice. An important consequence is that because of heavy debris, maximum ablation losses rarely occur on the lower parts of glaciers where conventional mass balance curves say that it should (Benn and Evans 1998, pp. 78–79). They do not occur where the strongest sunshine is measured because of the shortness of the season. The mid-ablation zone areas are overwhelmingly important.

There are, however, only a few Karakoram glaciers and expeditions to them that have conducted detailed observations of any of these variables. Even fewer offer a basis to determine specific ablation throughout the ablation zone or ablation regimes. Nevertheless, they must guide the analysis.

## 5.3 Ablation at Batura Glacier

The Batura Glacier Investigation Group (BIG 1979) made detailed measurements of ablation over the lower 20 km of Batura Glacier where it has a relatively low gradient. Ice was exposed at the surface over most of the upper half of the section. The lower part becomes increasingly buried under heavy debris and completely so in the last 4 km. Results reflect complicated interactions between elevation, ablation season length, debris covers and topography on the ice.

For similar surface conditions of exposed ice, the highest annual ablation losses were recorded towards the terminus and declined up-glacier. Near the terminus at 2,500 m, ablation was observed on as many as 315 days of a year. Annual losses of 18,400 mm (w.e.) were recorded at one site at 2,610 m. However, such high values were confined to small areas where debris was thin and ice showed through. Despite the long ablation season and exceptional potential for ablation, the extent of thick debris mantles at the same elevations meant that net losses were much smaller than in higher parts of the ablation zone (Yafeng and Wenying 1980). Where there are considerable debris-mantled areas around 2,600 m, annual ablation averaged less than 2,000 mm (w.e.) and declined to less than 200 mm near the terminus with the heaviest debris covers. Between 2,900 and 3,500 m, it ranged from 3,000 mm to over 10,000 mm in quite extensive areas with mostly thin and scattered debris. In some sections it exceeded 14,000 mm annually.

A conspicuous feature was how ablation rate patterns varied more strongly in a longitudinal direction than by elevation, except for the terminal lobe. The longitudinal pattern seems to involve a combination of debris covers dominated by medial moraines, local ice surface morphology, also predominantly longitudinal, and apparently complex and irregular surface velocity distribution (BIG 1979, p. 966). Such longitudinal patterns are widely present in the morphology and debris bands of most mid-ablation zone ice streams. They may well reflect similar patterns of ablation, but there are no comparable measurements for other glaciers.

## 5.4 Ablation at Biafo Glacier

In the 1980s, ablation was measured at cross-glacier profiles on Biafo Glacier between 3,700 and 5,000 m (Hewitt et al. 1989). Ablation stakes at the on-ice meteorological stations were observed twice daily, at first light and sunset (Plate 5.2; Table 5.1). Other stakes across the glacier were visited every 5–7 days.

At 4,080 m ablation rates from mid-July to mid-August, the peak season, averaged 60 mm (w.e.) daily on clean-to-dusty ice (Hewitt et al. 1989; Hewitt 2005). Ice loss varied from 50 to 90 mm daily for clear or slightly cloudy conditions. Maxima could exceed 90 mm on slightly dirty ice and in continuous clear weather. On cloudy, cool days losses declined to around 1.6 mm. The total annual ablation for the network varied from 5,000 to 7,000 mm (w.e.).

Comparing ablation at stations and profiles between 3,800 and 5,000 m, two contrasting results appear. Under similar weather and ice surface conditions, daily ablation rates were about the same throughout. However, there were large differences in total annual losses over this elevation range due to differences in ablation season length, weather, snowfall and debris covers, related to the broad elevation gradients in these elements. There was very little difference between losses at and between ablation stakes at 3,730 m (Mango) and 4,080 (Baintha), partly because debris cover was less at the latter and the glacier is much wider and more open (Plate 5.3). Between 4,080 m and a station at 4,650 m near the firn limit, there was a large decrease,

**Plate 5.2** Setting of the mid-glacier ablation zone station at Baintha (4,080 m), Biafo Glacier (Hewitt August 1985)

**Table 5.1** Ablation rates from stake readings at meteorological stations on Biafo Glacier and related to cloud cover and noting when precipitation occurs (From unpublished SIHP records)

| *Station*[a] date (1985) | Ablation[b] (mm) (w.e.) | Cloud cover[c] tenths (min–max) | Precip. (mm) |
|---|---|---|---|
| *Mango (3,730 m)* | | | |
| July | | | |
| 21 | 39 | 1 (0–2) | 0 |
| 21/22 | 06 | | |
| 22 (day) | 57 | 0 (0–2) | 0 |
| 22/23 (night) | 11 | | |
| 23 | 63 | 1 (0–1) | 0 |
| 23/24 | 15 | | |
| 24 | 73 | 3 (1–4) | Trace |
| 24/25 | 18 | | |
| 25 | 15 | 10 (10) | 04 |
| 25/26 | 07 | | |
| 26 | 62 | 8 (1–9) | 0 |
| 26/27 | 09 | | |
| 27 | 51 | 7 (2–10) | 0 |
| *Baintha (4,010 m)* | | | |
| July | | | |
| 20 | 65 | 1 (0–2) | 0 |

(continued)

**Table 5.1** (continued)

| *Station*[a] date (1985) | Ablation[b] (mm) (w.e.) | Cloud cover[c] tenths (min–max) | Precip. (mm) |
|---|---|---|---|
| 20/21 | 14 | | |
| 21 | 65 | 1 (0–2) | 0 |
| 21/22 | 14 | | |
| 22 | 66 | 1 (0–3) | 0 |
| 22/23 | 11 | | |
| 23 | 45 | 1 (0–2) | 0 |
| 23/24 | 10 | | |
| 24 | 46 | 3 (2–4) | 0 |
| 24/25 | 05 | | |
| 25 | 41 | 10 (9–10) | 51 |
| 25/26 | 07 | | |
| 26 | 61 | 6 (1–8) | 0 |
| 26/27 | 07 | | |
| 28 | 68 | 1 (0–3) | 0 |
| 28/29 | 01 | | |
| 29 | 64 | 0 (0) | 0 |
| 29/30 | 01 | | |
| 30 | 62 | 1 (0–1) | 0 |
| 30/31 | 06 | | |
| 31 | 57 | 5 (0–8) | 0 |
| 31/1 | 02 | | |
| August | | | |
| 1 | 55 | 6 (0–10) | 0 |
| ½ | 04 | | |
| 2 | 42 | 9 (7–10) | 0 |
| 2/3 | 10 | | |
| 3 | 13 | 10 (9–10) | 0 |
| ¾ | 01 | | |
| 4 | 32 | 7 (4–10) | 0 |
| 4/5 | 25 | | |
| 5 | 57 | 9 (8–10) | 0 |
| 5/6 | 07 | | |
| 6 | 25 | 10 (10) | 09 |
| 6/7 | 15 | | |
| 7 | 08 | 10 (10) | 18 |
| 7/8 | 25 | | |
| 8 | 31 | 8 (7–9) | 0 |
| 8/9 | 09 | | |
| 9 | 40 | 3 (1–6) | 0 |
| 9/10 | 08 | | |
| 10 | 50 | 3 (1–6) | 0 |
| 10/11 | 10 | | |
| 11 | 50 | 7 (1–9) | 0 |
| 11/12 | 08 | | |
| 12 | 44 | 8 (3–10) | 0 |

(continued)

**Table 5.1** (continued)

| *Station*[a] date (1985) | Ablation[b] (mm) (w.e.) | Cloud cover[c] tenths (min–max) | Precip. (mm) |
|---|---|---|---|
| *Equilibrium line (4,650 m)* | | | |
| 26 July | 46 | 4 (2–6) | 0 |
| 26/27 (night) | 05 | | |
| 27 | 46 | 5 (1–9) | 0 |
| 27/28 | 06 | | |
| 28 | 51 | 2 (0–3) | 0 |
| 28/29 | 06 | | |
| 29 | 50 | 0 | 0 |
| 29/30 | 05 | | |
| 30 | 50 | 1 (0–3) | 0 |
| 30/31 | 02 | | |
| 31 | 48 | 3 (0–6) | 0 |

[a]Readings based on 'Ablation Squares' of nine stakes at mid-glacier Meteorological stations, where ice was exposed with some dirt and dust and scattered small clasts

[b]Water equivalents (w.e.) based on conversion of ice loss calculated from average of 'top-to-ice' stake readings at 08:00 am and 17:00 pm

[c]Average of hourly readings, 08:00–17:00 h

**Plate 5.3** View of the Biafo ablation zone in late July from above Baintha down glacier to the area of the Mango station (Hewitt 1985)

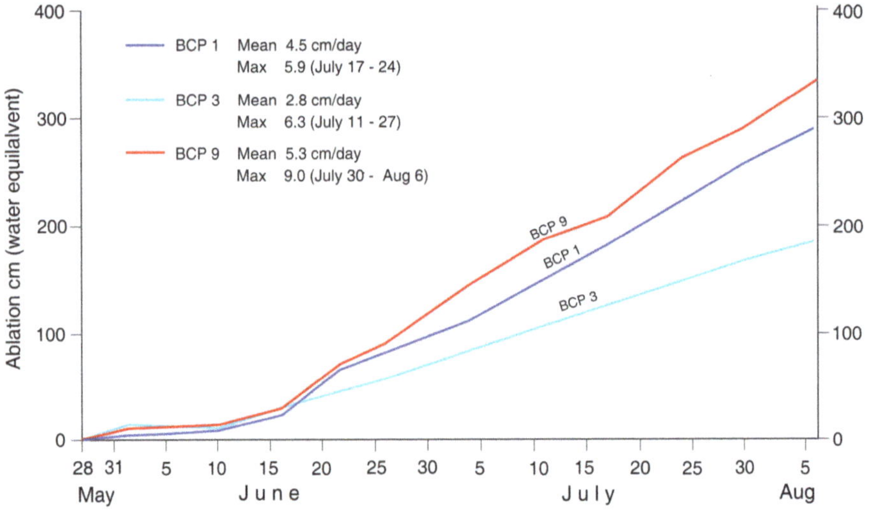


**Fig. 5.1** Ablation at three stakes on Biafo Glacier at the Baintha (4,050 m) cross section, May to early August 1986 (After SIHP unpubl. Data; see text)

mainly due to reduced ablation season length. Whereas ablation occurred over three months at 4,000 m, it was less than 1 month at the higher location.

Rates also differed according to surface conditions at each stake. For most profiles stakes were placed where ice was exposed at the surface, but local debris cover was altered enough over the season to increase rates compared to clean ice in some cases and suppress them at others. Seasonal ablation at three of the nine stakes along the Baintha Cross Profile (BCP) in 1986 illustrates some of the variables (Fig. 5.1). BCP 1 was in clean ice with only minor dust and occasional small pebbles. BCP 3 was on a major medial moraine with a 90 % cover of thin silt, sand and pebbles and 50 % scattered clasts from cobble to boulder sizes. BCP 9 was about 0.5 km from the glacier margin where the surface was considerably dirtier and changed less rapidly than the main ice streams.

The importance of solar radiation is revealed in comparing nocturnal and daytime ablation at on-ice meteorological stations where stakes were read at sunrise and sunset (Tables 5.2 and 5.3). These data from the middle of the ablation season show that the total night-time ablation was slightly less than 8 % of daily totals at the higher elevation site and almost 13 % at the lower one. Greater nocturnal heat loss at the higher site and dirtier ice at the lower one are likely factors controlling the differences. The considerable variation in nocturnal ablation at both sites relates to weather. Higher nocturnal losses coincided with days of heavy cloud but, therefore, also with reduced daytime losses. The data show overnight ablation is not negligible and of the same order, for example, as quantities for seasonal snowmelt. Sometimes, in cloudy stormy weather, they equalled daytime amounts. Here is an obvious factor in year-to-year differences and, potentially, where climate change could have an impact. Nonetheless, the data confirm the overwhelming role of daytime ablation and, again, of received solar radiation.

**Table 5.2** Nocturnal (N) versus daytime (D) ablation readings at Biafo Glacier I: 'Mango' weather station (3,730 m) stake array, July 22–28, 1985. Daytime readings are for ice loss (mm) between approx. 08.00 and 16:30, night-time 16:30 to 08:00 on the following morning. These are examples of individual stakes in four arrays and mean for 24 stakes in total (After SIHP unpubl. data)

| Stake date | | Night | | N | | N | | N | D | N | | |
|---|---|---|---|---|---|---|---|---|---|---|---|---|
| (July) | Day 22 | 26–23 | D 23 | 23–24 | D 24 | 24–25 | D 25 | 25–26 | 26 | 26–27 | D 28 | Total |
| R1C1 | 54 | 12 | 89 | 11 | 75 | 23 | 20 | 5 | 61 | 15 | 57 | 422 |
| R2C1 | 53 | 12 | 86 | 16 | 70 | 19 | 10 | 5 | 75 | 7 | 73 | 426 |
| R3C2 | 65 | 8 | 30 | 26 | 58 | 10 | 6 | 0 | 52 | 22 | 48 | 325 |
| R4C1 | 65 | 3 | 58 | 14 | 80 | 36 | 17 | 6 | 41 | 9 | 62 | 391 |
| Mean (*n*=24) | 63.2 | 11.5 | 68.8 | 1.7 | 80.2 | 20.4 | 16.3 | 8.2 | 68 | 9.7 | 56 | 404 |

**Table 5.3** Nocturnal (N) versus daytime (D) ablation readings at Biafo Glacier II 'Equilibrium Line' weather station (4,650 m) stake array, July 26–31, 1985. Daytime readings are for ice loss (mm) between 07.00 and 18:00, night-time 18:00 to 07:00 on the following morning; these are examples of individual stakes in four arrays and average for 24 stakes in total (After SIHP unpubl. data)

| | | N | | N | | N | | N | | | | |
|---|---|---|---|---|---|---|---|---|---|---|---|---|
| Stake | D 26 | 26–27 | D 27 | 27–28 | D 28 | 28–29 | D 29 | 29–30 | D 30 | N 3031 | D 31 | Total |
| R1C1 | 50 | 6 | 47 | 1 | 53 | 9 | 57 | 13 | 54 | 0 | 58 | 347 |
| R2C1 | 45 | 8 | 33 | 6 | 42 | 6 | 56 | 1 | 56 | 3 | 81 | 337 |
| R3C1 | 35 | 0 | 35 | 7 | 42 | 6 | 61 | 0 | 55 | 11 | 56 | 308 |
| R4C1 | 45 | 9 | 30 | 0 | 57 | 5 | 45 | 26 | 54 | 0 | 38 | 309 |
| Mean (*n*=24) | 49.9 | 4.9 | 49.8 | 6.5 | 55.6 | 6.7 | 54.6 | 5.9 | 55.3 | 1.9 | 52.8 | 344 |

Weather effects are complicated since heavy cloud often precedes or coincides with precipitation events. These not only bring low or absent ablation but changes in surface conditions that persist for hours or days after a storm. The impact is greatest in the case of snowfall because of its high albedo. Afterwards there is a secondary melting cycle of the snow cover, before ice will again be ablated.

These observations indicate why, even at the height of the season, weather conditions and disturbances give rise to large day-to-day and week-to-week fluctuations in ablation. Summer storms may shut down, melting altogether over much of the ablation zone. For example, immediately following the observations shown in Table 5.1, a storm moved into the basin. For several days at stations as low as 3,500 m, ablation declined to 10–15% of the values shown. It ceased altogether at two stations above 4,200 m where several centimetres of snow fell. It took nearly a week for the snow to disappear, partly because cloudy conditions persisted.

Such details lie behind large year-to-year fluctuations in water yields and river flows and suggest that climate change impacts on summer storminess or sunshine hours could have as great, or greater, consequences as changing temperature or precipitation averages. Recall that at most of these elevations, where the greatest expanses of exposed ice occur on Biafo, the ablation season is just a few weeks in length. The impact of summer weather on mass balance is clear.

## 5.5 On- and Off-Ice Observations in Glacier Basins

Experiments at Biafo Glacier showed some marked differences between weather stations on the ice and off the ice along glacier margins. A detailed snapshot of these relations at the same elevation (4,050 m) gives a sense of their complexity (Fig. 5.2). Except near the beginning and end of the sequence, both series' rise and fall together fairly consistently. Yet, temperature range is generally 10–15 °C greater at the off-ice station, and the mean daily temperatures are higher. Unexpectedly perhaps, the minima are 2–3 °C lower than the on-ice station, leading to more and larger freeze–thaw cycles off-ice and giving greater sensitivity to cloud and precipitation events.

Two features not apparent in this data are the water from melting around the on-ice station, whose greater heat capacity and evaporative cooling can help dampen temperature variations compared to the drier off-ice site. There were also significant differences in windiness and wind speeds, the latter are generally greater on the glacier from a recurring down-glacier wind through much of the daytime. Also unexpectedly, perhaps, this katabatic glacier wind seemed responsible for maintaining steadier temperatures day and night. It involved just the lower few metres of the atmosphere, so that the effect is diminished or absent a short way above the glacier margins and at the off-ice weather station.

Unfortunately, observations at on-ice stations have been few and far between and most of those available of short duration. The difficulties of installing and maintaining them are considerable. High up, they get buried under successive snowfalls and

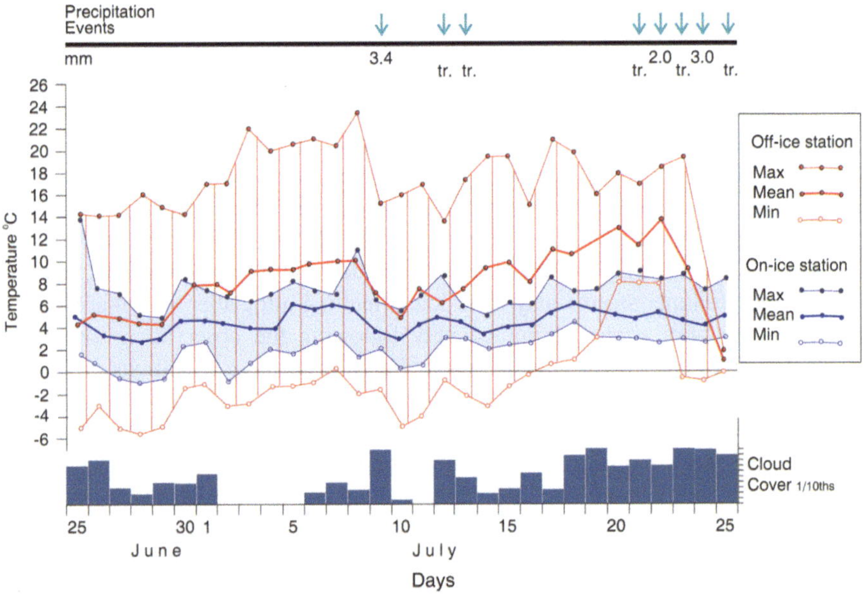


**Fig. 5.2** On- and off-glacier weather station readings at the same elevation, Baintha (4,050 m) cross section. June–July 1986 (After SIHP unpubl. Data; see text)

icings. In the ablation zone, they must be continually reset in summer to keep pace with surface melting. In the case of the Baintha station, for example, some 4–6 m of ice were ablated at the weather station during each of the two summer observation periods, requiring frequent redrilling and adjustment of instrument housings.

## 5.6 Supraglacial Debris and Ablation

The Himalayan ranges have been identified especially with 'debris-covered glaciers' (Nakawo et al. 2000). Different types and thicknesses of debris exercise a major control over ablation rates and patterns in the Karakoram. The importance of supraglacial debris follows, on one hand, from the importance of solar radiation and, on the other, from the enormous amounts of debris delivered to the glaciers by avalanches, rockfalls and wind-borne dust. Most studies have focused on the heaviest debris covers. This mainly serves to protect the ice in those glacier tongues that reach into the lower elevations, as seen at Batura. Debris mantles here can reduce or largely suppress ablation where, otherwise, a longer ablation season and hot summer conditions should lead to the largest net losses. To a lesser and more complicated degree differential insulation by heavier debris affects areas of medial and on-ice lateral moraine throughout ablation zones (Plate 5.4).

**Plate 5.4** Diversity of debris covers on Biafo Glacier near experimental site and Baintha (4,050 m) profile (Hewitt 1985)

**Table 5.4** Heavy debris covers as a percentage of ablation zones for 25 large and intermediate Karakoram glaciers (see Chap. 3, 3.2)

| Glacier | Proportion (%) |
|---|---|
| Siachen | 17 |
| Baltoro | 33 |
| Biafo | 23 |
| Hispar | 45 |
| Batura | 48 |
| Panmah | 42 |
| Chogo Lungma | 47 |
| Rimo | 11 |
| Khurdopin | 48 |
| Virjerab | 29 |
| Charakusa | 26 |
| Kukuar | 14 |
| Toltar–Baltar | 57 |
| Braldu (Shaks.) | 27 |
| Skamri | 25 |
| Chiantar | 12 |
| Sarpo Laggo | 14 |
| Kondus–Kaberi | 38 |
| Bilafond | 18 |
| Sherpigang | 15 |
| Hasanabad | 70 |
| Singhi | 8 |
| Bualtar | 32 |
| Gharesa | 12 |
| Gondogoro | 24 |

Heavy debris mantles cover as much as one-third of the ablation zones of the main set of larger Karakoram glaciers (Table 5.4). There is generally much less on glaciers of the far eastern upper Shyok Basin, but in total they probably exceed 1,200 $km^2$ for ablation zones of the whole region – more than half the total glacier cover of the European Alps. The protective role makes these areas relatively smaller contributors to water yields and, as indicated below, generally less sensitive to climatic factors that can alter ablation losses. However, conditions in apparent heavy covers are quite diverse, leading to significant variations in ablation rates that are difficult or impossible to differentiate without ground-based observations. Meanwhile, areas of thin or scattered debris that enhance ablation rates are twice as extensive as those of the heavy covers.

Observations on Karakoram and other Himalayan glaciers identify several critical thicknesses of debris as it affects ablation (Suzuki 2011). These follow from the extent to which the higher heat absorption of the debris is balanced by lower conductivity through to the ice surface. At a certain thickness ablation rates are about the same as for clean ice, usually in the range 3–10 cm varying with debris properties and ambient conditions. For progressively thinner covers ablation tends to increase to a second critical thickness, usually measured in millimetres, at which the highest ablation rates occur. Typically these are one-and-a half to two times greater than for clean ice. More rarely they may be up to four times greater, notably with dark volcanic ash or soot from combustion (Adhikary et al. 2000). As more of the ice is exposed, ablation rates again decline below this critical thickness. However, compared to clean ice, any scattered debris will tend to increase ablation, and rates vary somewhat with surface roughness.

For thicker debris, ablation rates decline up to a fourth critical thickness where penetration of warmth approaches zero and little or no ablation can occur, usually for covers averaging at least 1.5–2.0 m. At such depths ice is insulated from diurnal and seasonal penetration of surface warmth, and temperatures remain close to zero (Khan 1989, pp. 118–119). Conversely, the effect of ice temperatures on overlying debris ceases at about 40 cm. It is noticeable as a gradient of increased cooling as debris thickness declines (Mihalcea et al. 2008), an indirect measure of the heat required for ablation.

It has been noted that heavily mantled glacier tongues usually thin and retreat relatively slowly. It has no effect on advances, especially rapid ones, which are driven by the dynamics of the up-valley ice stream. With significant negative mass balance, terminal areas are likely to stagnate and leave masses of dead ice. These evolve slowly through *thermokarst* processes, melting again regulated by the debris cover but developing differently from active ice (Menzies 1995, p. 195).

In the eastern Karakoram, at Terong Glacier, Osmaston reported a daily ablation rate where debris was 3–4 cm thick the same as clean ice (Kaser and Osmaston 2002, p. 154). The rate under 5 cm was half the clean ice rate, while under 'grit' about 1 cm in thickness, it was double that of clean ice. The Terong ablation zone is relatively high, above a terminus at about 4,000 m.

Roughly similar results were found at an experimental site on Biafo Glacier at 4,100 m elevation in late July 1985 (Fig. 5.3). The highest rates recorded were at 0.5 cm and, at this thickness, the highest of all was for wet mud (silt and fine sand) and twice the maximum for clean ice. Debris between 1.5 and 3 cm thick had roughly the same rate as clean ice, but also areas with a 70 % cover of small stones and finer material. Some differences shown in the experiment evidently arise from the calibre, shape and colour or lithology of debris. Differences between average and maximum daily rates followed variations in cloud cover and clarity of the air, as indicated by changes in incoming short-wave radiation measured during the observation period. It was also observed that in thinner and scattered debris regimes, there are ablation-related cycles in which small-scale surface features develop and fade out, notably debris-covered cones.

Studies on Rakhiot Glacier at about 3,050 m in the nearby Nanga Parbat Range support the Karakoram data. Critical thickness for protection versus enhancement of ablation was 3–5 cm (Mattson and Gardner 1989).

Estimating ablation rates for a given glacier or year, let alone the whole region, encounters major problems with determining actual depths of debris, their extent and the variability of continuous covers. From a distance or remotely, it is almost impossible to differentiate in the critical regions between, on the one hand, 1–10 cm and, on the other, 10 cm and 1 m depth. Even when visiting a site, it is usually necessary to excavate to be sure of the depths. The thickest debris mantles nearly always

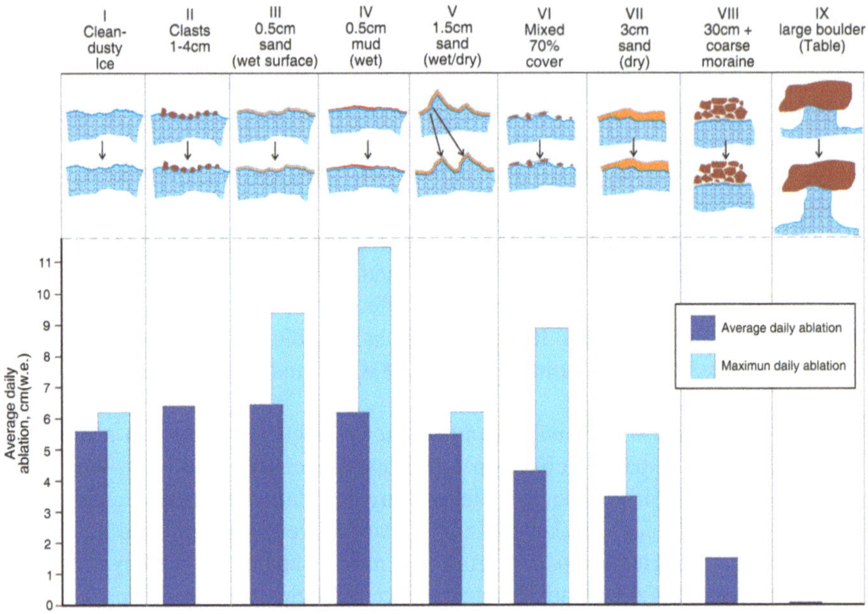


**Fig. 5.3** Rates of ablation at monitored plots on Biafo Glacier at 4,100 m over a ten-day period in 1986 to illustrate the importance of debris cover (After Hewitt 2005). Weather was clear to moderately cloudy, with cloud cover less than 4/10ths. Types classified as I through VIII, each represents conditions over tens of square kilometres in the ablation zone as a whole. Thin debris promotes higher rates and shows larger fluctuations according to weather, wetting and drying and sensitivity to small-scale patterns of build-up and disturbance (compare III, V and VII)

contain a large fraction of coarse clasts. The larger boulders complicate the picture by obscuring the ice surface and the presence or absence of other underlying debris. Isolated boulders go through cycles as ‘glacier tables’ with zero ablation beneath, but higher than average ablation around the margins due to an enhanced long-wave or thermal radiation (Plate 5.5).

These questions were addressed by Khan (1989) on Sumaiyar Bar Glacier, the largest tributary of Barpu Glacier, at 3,500 m elevation (Plate 5.6). The debris cover here is almost continuous and appears typical of heavier mantles. When measured at 100 m intervals for a total of 110 points, it had an average thickness of 12.4 cm (ibid, pp. 86–89). However, over half of the 1.1 km$^2$ area had a cover of less than 10 cm, and 70 % 15 cm or less. The thickest debris, exceeding 55 cm, covered barely 2 % and only near the glacier margins. To estimate the relation of debris thickness to ablation rates, measurements were made at 45 sites between May 26 and July 19 (see 5.8). The ‘most effective thickness for ablation’ was around 1 cm, with measured rates double those of clean ice (Fig. 5.4). The net rate for all sites was 23 mm of ice per day. The extremes were up to 90 mm with thin debris and less than 10 mm where debris exceeded 35 cm thickness. The critical boundary between enhanced and reduced ablation was a debris thickness around 3 cm (ibid, pp. 91–92).

**Plate 5.5** Glacier table at about 3,900 m on Biafo Glacier between Baintha and Mango. The difference between the table and general glacier surface is less than half the average annual ablation on clean-to-thinly dirt-covered ice in this area (Hewitt 1998)

**Plate 5.6** Debris-covered ice with facets on the ablation zone of Sumaiyar Bar, a tributary of Barpu Glacier, where experiments reported in the text (Khan 1989) were conducted. It is a wholly avalanche-fed,Turkestan-type glacier (Hewitt 1993)

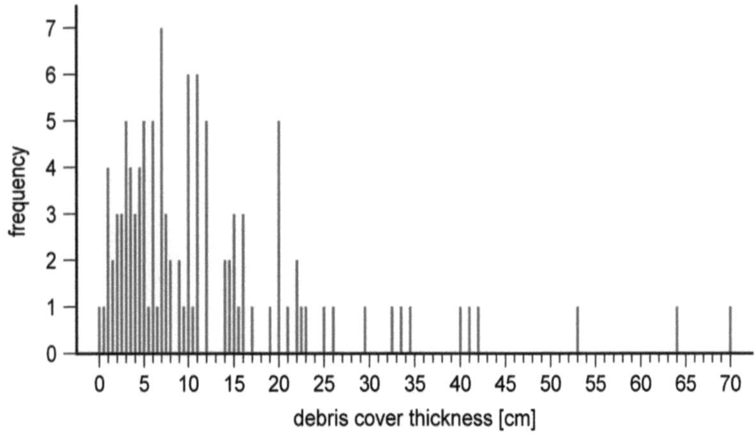


**Fig. 5.4** Distribution of debris cover thickness on Sumaiyar Bar tributary of Barpu Glacier (After Khan 1989)

This set of measurements was a basis for comparison with ablation of ice exposed in facets (see below).

Over the 3-month period, constant shifts were observed in debris thickness and morphology at most of the sites. Complications arose from ice movement, crevassing, on-ice topography, supraglacial drainage and ponds. This testifies to the dynamic nature of these covers on active glaciers, in contrast to an assumption that supraglacial debris mainly undergoes or represents 'passive' transport (Shroder et al. 1993, p. 146; Hewitt 1999, p. 232).

## 5.7 Ablation on Debris-Mantled Areas of Baltoro Glacier

Baltoro Glacier is notorious for an almost complete debris cover over the lower 40 km of the main ice stream (see Plates 3.2 and 3.3). It was first described by Godwin-Austen (1864). However, not until 2004 was there a concerted effort to measure ablation on the glacier (Mihalcea et al. 2006). This was carried out on the main debris-mantled ice stream between Urdukas (4,100 m) and Concordia (4,600 m). Stakes were arranged in a longitudinal profile that continued up the two arms above Concordia, with the highest observations at K2 Base Camp (5,033 m).

The role of debris thickness was the focus. Readings were taken over a ten-day period in late July and early August. Results broadly confirmed what has been described above in relation to debris thicknesses. Between 4,000 and 5,000 m ablation was found to vary much more as a function of the debris cover than by elevation. However, a second discovery was that in most areas observed local surface topography, specifically slope angle and orientation, influenced net ablation as much as debris thickness (Mihalcea et al. 2006, p. 296). The results also confirmed those at Barpu, showing debris covers to be highly variable and dynamic. Areas of much-reduced ablation alternated with relatively high values. Indeed, some of the highest losses were in areas that observations from a distance or remotely sensed would likely assign to heavily mantle zones and, hence, as low ablation areas. This may relate to meltwater movement over ice surfaces beneath the debris as well as constant relief inversion and debris redistribution.

Baltoro is one of many Mustagh-type glaciers where heavy debris can be found relatively far up the glacier. This reflects the great extent and extreme elevations of rock walls and avalanching in the upper Basin, including the greatest concentration of peaks over 8,000 m. It must be emphasised, however, that even at Baltoro the heavy mantles tend to be concentrated relatively low down. In total they apply to barely one-third of the whole ablation zone. Extensive areas of bare or thinly covered ice exist on many relatively steep tributaries down to where they join the main, debris-mantled ice stream.

## 5.8 Ablation-Enhancing Dusty and Dirty Conditions

Heavy debris mantles have received the greatest attention in the Himalaya, but in the Karakoram at least, areas of thin and scattered debris are more important for glacier hydrology and mass balance and, most probably, for climate response. They comprise the more extensive parts of ablation zones: on average, two-thirds for the 42 largest glaciers. In this case *debris-enhanced ablation rates* are involved. This differs from the focus of most research on 'debris-covered glaciers' and reduced or suppressed ablation (Nakawo et al. 2000; Han et al. 2006).

As a rule, what appears to be 'clean' ice on Karakoram glaciers has a fair amount of dust, dirt, rock particles, algae and impurities at the surface. A dusty atmosphere and avalanche and wind nourishment keep snow and ice relatively dirty. Summer conditions in these subtropical latitudes encourage algal communities and some other biota relatively high up that absorb radiation and pass the warmth to adjacent ice. The various experiments described above show marked increases in ablation if debris is thinner than about 3–5 cm and thicker than about 0.5 cm. Exact limiting values vary with elevation, debris composition, albedo and on-ice topography.

Differential ablation due to thin and scattered debris is difficult or impossible to pick up using conventional methods such as widely spaced ablation stakes, partly because the most diverse and dynamic parts of ice surfaces are involved. A host of positive and negative feedbacks arise through debris build-up and redistribution by collapses or flushing and depositing of finer material by surface meltwater. There are constant inversions of the local micro-relief on ice surfaces, incision and migration of on-ice meltwater streams, formation and drainage of superficial ponds. For thin debris, relative patchiness or continuity of the covers leads to significant variability. The detailed investigations at Biafo showed rates are affected by the specific character of the debris (Fig. 5.4). On the other hand, the sheer extent of any of the various covers need to be kept in mind, any one of the thickness classes mentioned adding up to hundreds of square kilometres.

Even specks of dust and isolated small clasts cause increased ablation compared to clean ice. In sunny weather, the particles tend to sink or, in effect, 'drill' themselves into the ice and to depths governed by the strength, duration and penetration of the sun's rays. The process affects surface micro-relief, which also tends to increase absorption of solar radiation and creates what are called 'water-filled cryoconite holes' (Benn and Evans 1998, p. 228). The ubiquitous presence in summer of small holes, pits and basins, from millimetres to tens of centimetres deep and wide was remarked upon by early modern visitors to Karakoram glaciers (Workman 1913/14).

A recent revival of interest in cryoconite materials concerns possible effects of global warming and other human activities (Jenkins 2010). Warming may increase water-filled cryoconite holes that support life forms and local ecosystems, and algal deposits and growth on snow and ice that absorb solar energy for photosynthesis (Margesin et al. 2002). In recent years, concern over human influences has increased. They include land use changes affecting dust in the atmosphere and its transparency on-ice deposition of artificially generated particulates and contaminants. Of special concern is enhanced ablation due to soot or black carbon derived from industrial areas, agricultural and domestic fires, from enhanced wildfire incidence or transportation corridors in the mountains (Pecci 2008). These can involve much darker particulates than most sediment eroded from rocks and, as noted above, enhance ablation accordingly. There are no data on this from the Karakoram, but observations have shown significant impacts on the Tibetan Plateau and parts of the eastern Himalaya (Ming et al. 2009).

## 5.9 Ice Facets and Relief Inversion in Heavy Debris Mantles

*Ice facets* are sloping surfaces that tend to be planar as well as steep, hence the name 'facets'. Most are close to 45°, some steeper where, for example, they are undercut by meltwater streams and the supraglacial ponds that often form below them. Facets characterise the more active debris-mantled ice in ablation zones, becoming rare or absent where movement is sluggish and debris build-up greatest. They seem to originate from processes that disturb ice surfaces, including crevasses and local surface steepening, or where supraglacial streams and supraglacial ponds undercut slopes and disturb debris covers, or as a random product of irregular debris covers and differential ablation. Once in existence, their high ablation rates and rapid recession lead to complicated evolution and inversions of ice surface relief. If facets rarely comprise more than 10 % of a glacier surface, in total their higher ablation rates directly affect 100 s km$^2$ of low- to mid-ablation zone areas.

At Khumbu Glacier in Nepal, Inoue and Yoshida (1980) were among the first to report much higher ablation rates on facets than surrounding relatively flat, debris-mantled ice. Compared to a daily average of 15 mm for the latter, facets averaged 23 mm, with a maximum of 45 mm. A doubling of local ablation rates was involved but those for facets seem low compared to experience in the Karakoram.

The experiment at Barpu Glacier at 3,500 m and from late May through early July showed an average ablation rate on facets of 57 mm of ice per day, more than

**Plate 5.7** Examples of ice facets in heavily debris covered ice on Baltoro Glacier near the junction with Liligo tributary (Hewitt 2005)

twice the average for all sites (Khan 1989). Ablation correlated closely with received short-wave radiation and varied significantly with orientation of the facets (Fig. 5.5). The highest rates, reaching 95 mm per day, were observed in clear weather on SW-facing slopes. Three factors seem to enhance the impact of facets on local ablation rates: thin veneers of fine debris washed down them from above, steeply angled surfaces and enhanced long-wave or thermal radiation from surrounding debris.

The experiments demonstrated the roles and complexities of differential debris covers and in relation to ice surface topography. Evidently, the local clustering of facets, and their variable distribution by elevation and on different glaciers, further complicates ablation and its relations to elevation and season. In many places facets coincide with or combine with ablation zone areas where avalanched snow is added to glacier surfaces below the perennial snow zone. There is, as yet, insufficient data to determine their total distribution in debris-covered areas or, therefore, their full role in mass balance. Satellite imagery may prove helpful in this regard.

## 5.10 The Annual Cycle on Ablation Zone Areas

Given the importance of ice loss and water yields, it is understandable that studies focus on the ablation season and define glacier areas involved in terms of ablation and ablation rates. However, it is a mistake to conclude that little or nothing happens in the same places and during the many months, when there is zero net ablation. Indeed, although there are few records or monitoring at these times anyone who does visit then is aware of constant changes. They include some more or less

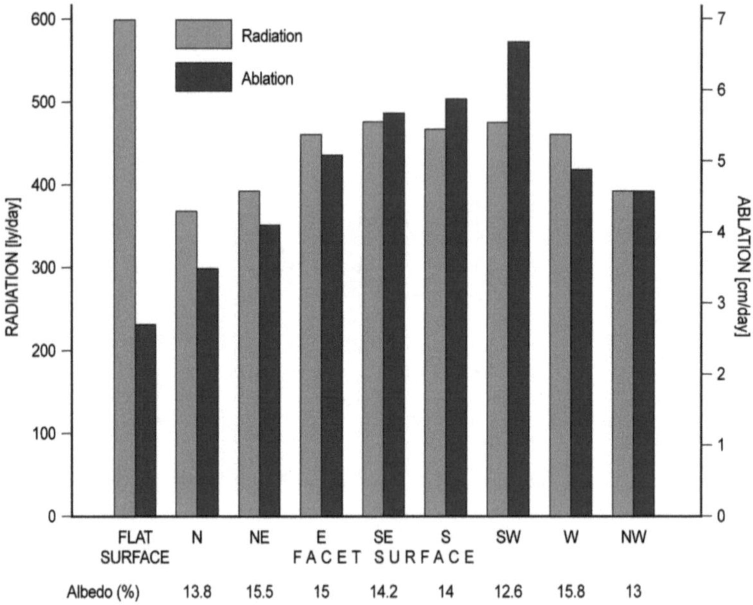


**Fig. 5.5** A comparison of incoming short-wave radiation and ablation of ice facets with different orientations for a 62-day period at 3,600 m on Sumaiyar Bar tributary of Barpu Glacier. Rates of ablation varied with orientation and closely related to received radiation. The flat surface was debris covered explaining the much lower ablation. Received radiation was higher even than south-oriented facets, apparently due to greater reflection from surrounding debris covered relief at this site (After Khan 1989)

substantial, if temporary, *additions* to mass by snowfall and avalanches. These occur over *most* ablation zone areas for much the greater part of each year.

From time spent on Karakoram glaciers outside the main ablation season, my experience suggests a cycle of surface changes that can be identified in terms of, at least, the usual four seasons, intersecting cycles that vary and migrate by elevation and other controls. The 'ablation zone' is normally defined by summer ice loss. In the other three seasons, however, there are not just no ablation losses, but also mass is added to the surface and redistributed over it. And it is not just a matter of the winter snowfall and freeze-up. Of special interest are the almost wholly neglected transitions between winter and the main ablation season, two transitional seasons distinguished by the roles of freeze–thaw cycles and redistribution of snow by strong winds.

Freeze–thaw cycles generate melting, percolation and refreezing that build an icy carapace over ablation zone surfaces, mainly an effect of fall and spring conditions. At these times, processes and surface subzones develop similar to those already identified with accumulation subzones (Chap. 4). A migrating patchwork of dry snow and wet snow is found, zones with percolation and superimposed ice, wind-redistributed and refrozen snow and what Müller (1962) referred to as 'icy layers'. They occur throughout the ablation zone and, in mid- to upper ablation areas, over the greater part of the year.

These transitional periods and winter proper, extend from a month or two near the termini to 11 months or more at the head of the ablation zones. And their influence is felt in the subsequent ablation season. The significance of the three cold seasons for mass balance involves more than snow in the accumulation zone. How the seasons develop and how large variation in their intensity and vertical migrations influence conditions from year to year are as important for the ablation season as summer weather. They are likely to be important in just how climate responses play out. To date, there are only a very few studies that even recognised this (Zichu et al. 1999). They are especially important in relation to glacier landscapes and landform developments and will be returned to in Chap. 8.

## 5.11 Concluding Remarks

In summary, given the dominant role of energy from received short-wave radiation in these high mountain glaciers, ablation rates and net ablation depend primarily on four conditions:

(i) Surface albedo: which is sensitive and changeable in response to weather, dust and debris and to ablation processes themselves
(ii) Debris thickness: the critical thicknesses for ablation rates and their distributions which need to be established for each region and elevation
(iii) Summer weather: especially the number of cloudy versus sunny days and incidence of snowfall on (parts of) the ablation zone
(iv) Spring and fall shoulder seasons: especially the build-up of a carapace of ice, icy layers, frozen and wind-packed snow over the ablation zones and how long it takes to be removed

The effect of these is moderated by the following:

(v) Surface slope and orientation, which affects solar angle and debris cover
(vi) Whether debris is wet or dry, especially thinner dirt layers
(vii) The ice 'weathering' layer, formed in above-zero sunny conditions, which disappears in cloudy weather and takes time to re-establish afterwards. It affects ablation rates mainly through surface albedo and roughness

In the Himalaya and other mountain areas of High Asia, most attention has been given to the role of thicker debris covers on ablation. The impression may even be gained that debris covers thick enough to suppress ablation are the rule. Rather, the observations assembled here suggest their importance has been exaggerated in relation to glacier hydrology and water yields and in responses to climate change. A case is made for much greater attention to the areas of 'thin' debris and relatively clean ice, which are larger parts of ablation zones and the main sources of water yields.

Conditions that vary with elevation turn out to have their greater effects through *net* ablation, or specific seasonal losses at given elevations, rather than potential ablation rates. The latter can be similar over a broad range of elevations for similar surface conditions. The positive effect of increasing received radiation with elevation is offset by declining duration of above-zero temperatures, by greater cloudiness and by the amounts and duration of the seasonal snow cover and of the icy carapace. Low down the pattern is reversed by increasing extent and thickness of supraglacial debris.

## References

Adhikary S, Masayoshi N, Katsumoto S, Shakya B (2000) Dust influence on the melting process of glacier ice experimental results from Lirung Glacier, Nepal, Himalayas. In: Nakawo M, Raymond CF, Fountain A (eds) Debris-covered glaciers. International Association of Hydrological Sciences (IAHS) Publication 264, Wallingford, Oxfordshire, pp 43–52

Armstrong RL, Brun E (2008) Snow and climate: physical processes, surface energy exchange and modeling. Cambridge University Press, Cambridge

Barry R (1992) Mountain weather and climate, 3rd edn. Methuen, New York, p 209

Benn DI, Evans DJA (1998) Glaciers and glaciation. Hodder Arnold Publication, London

BIG (Batura Investigations Group) (1979) The Batura Glacier in the Karakoram mountains and its variations. Sci Sin 22(8):958–974

Copeland L (2011) Melting processes. In: Singh VP, Singh P, Haritashaya UK (eds) Encyclopaedia of snow, ice and glaciers. Springer, Dordrecht, pp 733–735

Godwin-Austen HH (1864) On the Glaciers of the Mustagh Range (Trans-Indus). Proc Roy Geogr Soc 34:19–56

Grenfell TC (2011) Albedo. In: Singh VP, Singh P, Haritashaya UK (eds) Encyclopaedia of snow, ice and glaciers. Springer, Dordrecht, pp 23–35

Han J, Nakawo M, Goto-Azuma K, Lu C (2006) Impact of fine-dust air burden on the mass balance of a high mountain glacier: a case study of the Chongce ice cap, west Kunlun Shan, China. Ann Glaciol 43:23–30

Hewitt K (1968) The freeze-thaw environment of the Karakoram Himalaya. Canad Geogr 12(2):85–98

Hewitt K (1999) Quaternary moraines vs. catastrophic rock avalanches in the Karakoram Himalaya, northern Pakistan. Quat Res 51(3):220–237

Hewitt K (2005) The Karakoram anomaly: Glacier expansion and the 'elevation effect', Karakoram Himalaya, Inner Asia. Mt Res Dev 25(4):332–340

Hewitt K, Wake CP, Young GJ, David C (1989) Hydrological investigations at Biafo Glacier, Karakoram Himalaya, an important source of water for the Indus River. Ann Glaciol 13:103–108

Inoue J, Yoshida M (1980) Ablation and heat exchange over the Khumbu Glacier. J Japan Soc Snow Ice (Seppyo) 41:26–33

Jenkins M (2010) True colours: There's a meltdown on the ice sheet – and optimism among Greenlanders. National Geogr. June, http://ngm.nationalgeographic.com/2010/06/melt-zone/jenkins-text

Kaser G, Osmaston H (2002) Tropical glaciers. Cambridge University Press, Cambridge, UK

Khan MI (1989) Ablation on Barpu Glacier, Karakoram Himalaya, Pakistan: a study of melt processes on a faceted, debris-covered ice surface. Master's thesis, Department of Geography, Wilfrid Laurier University, Waterloo

Kirkbride MP (2011) Debris-covered glaciers. In: Singh VP, Singh P, Haritashaya UK (eds) Encyclopaedia of snow, ice and glaciers. Springer, Dordrecht, pp 190–192

Margesin R, Zacke G, Schinner F (2002) Characterization of heterotrophic microorganisms in alpine glacier cryoconite. Arct Antarct Alp Res 34(1):88–93

Mattson LE, Gardner JS (1989) Energy exchanges and the ablation rates on the debris-covered Rakhiot Glacier, Pakistan. Z Gletschkd 25(1):17–32

Menzies J (1995) Modern glacial environments: processes, dynamics and sediments, vol 1. Butterworth-Heinemann, Oxford
Mihalcea C, Mayer C, Diolaiuti G, Lambrecht A, Smiraglia C, Tartari G (2006) Ice ablation and meteorological conditions on the debris-covered area of Baltoro Glacier, Karakoram, Pakistan. Ann Glaciol 43(1):292–300
Mihalcea C, Mayer C, D'Agata C, Diolaiuti G, Lambrecht A, Smiraglia C, Vuillermoz E, Tartari G (2008) Spatial distribution of debris thickness and melting from remote-sensing and meteorological data, at debris-covered Baltoro Glacier, Karakoram, Pakistan. Ann Glaciol 48(1):49–57
Ming J, Xiao C, Cachier H, Qin D, Qin X, Li Z, Pu J (2009) Black Carbon (BC) in the snow of glaciers in west China and its potential effects on albedos. Atmos Res 92:114–123
Müller F (1962) Zonation in the accumulation areas of the glaciers of Axel Heiberg Island, NWT, Canada. J Glaciol 4:302–313
Nakawo M, Raymond CF, Fountain A (2000) Debris-covered glaciers. International Association of Hydrological Sciences (IAHS) Publication 264, Wallingford, Oxfordshire
Paterson WSB (1994) The physics of glaciers, 3rd edn. Pergamon, New York, pp 57
Pecci M (2008) Snow cover on the mountains: still white and pure. Mt Res Dev 28(3/4):222–225. doi:10.1659/mrd.1041
Shroder JF Jr, Owen L, Derbyshire E (1993) Quaternary glaciation of the Karakoram and Nanga Parbat Himalaya. In: Shroder JF Jr (ed) Himalaya to the sea: geology, geomorphology and the Quaternary. Routledge, London, pp 132–158
SIHP (Snow and Ice Hydrology Project) (1990) Annual report, 1989. Cold Regions Research Centre, Wilfrid Laurier University, Waterloo
Suzuki R (2011) Debris thermal properties and impact of ice ablation. In: Singh VP, Singh P, Haritashaya UK (eds) Encyclopaedia of snow, ice and glaciers. Springer, Dordrecht, pp 188–190
Untersteiner N (1957) Glazial-meteorologische Untersuchungen im Karakoram. Arch Met Geoph Biokl Ser B 8(1):1–30, and 2:137–171
Van den Broeke M, Fettweis X, Mölg T (2011) Surface energy balance. In: Singh VP, Singh P, Haritashaya UK (eds) Encyclopaedia of snow, ice and glaciers. Springer, Dordrecht, pp 1112–1123
Workman WH (1913/14) Nieve penitente and allied formations in Himalaya or surface forms of neve and ice created or modelled by melting. Z f Gletsch 8:289–330
Yafeng S, Wenying W (1980) Research on snow cover in China and the avalanche phenomena at Batura Glacier, Pakistan. Ann Glaciol 13:289–293
Zichu X, Jiankang H, Chaohai L, Sciyin L (1999) Measurement and estimation of glacier mass balance in China. Geogr Ann 81A(4):791–796

# Chapter 6
# Glacier Mass Balance Regimes

**Abstract** The two parts of mass balance, inputs and outputs, are now considered together. Budget estimates are provided for Biafo Glacier and from partial data for Batura, Baltoro and some others. Overall quantities reflect the greater moisture availability and mass inputs at high elevations in the Karakoram. From glacier basins of the Central Karakoram, the total annual water yields appear to be of the order of 700–900 mm; less than maximum snowfalls measured in glacier source zones but three or more times greater than precipitation reported from valley weather stations. The mass balance regime is characterised as an all-year accumulation and summer ablation type, a distinctive form of Sub- or Outer Tropics regime. The vast areas of rock wall in source zones and predominance of avalanche nourishment and the significant 'inputs' in conventional ablation zones create problems largely ignored in mass balance assessments anywhere. Wind-fed and wind-stripped areas high up and disconnected tributaries above the snowlines add to the unique problems of input estimation. Wind redistribution and avalanches generate complicated patterns not clearly separated into accumulation and ablation zones or according to linear change with elevation. Mass balance gradients depart widely from those described elsewhere and point to distinctive constraints. An S-shaped vertical profile arises even in Alpine-type Karakoram glaciers reflecting reduced losses under debris-covered ice low down and input concentration where maximum snowfall occurs and especially through avalanche deposition. Further complications arise in profiles of Turkestan-type glaciers, where all positive as well as negative components for main ice masses are located in the conventional ablation zones. For such reasons, the notions widely employed in mass balance work are compromised. Equilibrium line altitudes (ELAs), thresholds between positive and negative mass balance, mostly occur below where firn limits or climatic snowlines are expected. These are, however, among the distinguishing features of the region that arise from the regional climatic regime, the exceptional extent of high elevations and verticality relations.

K. Hewitt, *Glaciers of the Karakoram Himalaya: Glacial Environments, Processes, Hazards and Resources*, Advances in Asian Human-Environmental Research,
DOI 10.1007/978-94-007-6311-1_6, 

**Keywords** Mass balance • Accumulation regime • Wind regime • Avalanche regime • Mass balance gradients • ELA • AAR

## 6.1 Glacier Systems

The conditions which nourish and sustain the glaciers have been examined, and losses through ablation and meltwater yield. The larger questions are how these are combined in the mass balance of whole glaciers and what drives their movement and redistribution of mass down slope, the field of glacier dynamics (Chap. 7)?

## 6.2 Mass Balance Estimates

Direct measurement of glaciers remains the preferred strategy for determining mass balance and tracking changes (Zemp et al. 2009; Haeberli 2011). The most well-developed mass balance programs in mountain regions are rooted in a history of field monitoring. The work generally relied upon, and that shapes assessments of glacier change worldwide, involves some decades of repeated surveys in accumulation and ablation zones (Oerlemanns 2002; WGMS 2008). It is, however, time-, cost- and labour-intensive, unusually demanding in logistical and safety matters and patience in the face of adverse weather and other impediments. Where possible, it is being replaced by remote sensing and automatic recording stations.

In South Asia, terrain and high-altitude environments create exceptional problems, but in addition, work on glaciers has been frustrated or prevented by security concerns and armed conflicts, colonial history and political turmoil. All but a few estimates of mass balance or glacier change are based on terminus fluctuations, more rarely on short-term observations for the lowest parts of glacier tongues, usually in summer (Raina 2009). Most existing estimates of what is happening to mass balance on Karakoram glaciers are not based on measurements from the glaciers. Some quantified estimates deduce 'inputs' and 'outputs' from precipitation and temperatures at more or less distant weather stations and stream discharge stations that tend to be at lower elevations, in most cases far downstream (Archer and Fowler 2004). To arrive at results for higher elevations and in glacier basins, models or indicators must be used (Kayastha et al. 2000; Aizen 2011). Those employed in the Karakoram are generally derived from work in other regions with distinctly different climates, ice masses and terrain (Bakke and Nesje 2011; Cogley 2011). Efforts to use satellite imagery offer expanding insights and into hitherto inaccessible aspects of snow and ice conditions in glacier basins (Bolch and Kamp 2006; Naz Bibi et al. 2009; Immerzeel et al. 2009; Scherler et al. 2011; Gardelle et al. 2012). Ground control and experience with conditions on the ice masses remain critical but are not very extensive in time or space. Such observations as are available from the glaciers themselves are given priority here.

**Table 6.1** Estimates of meltwater yields from ablation of ice at Biafo Glacier. Mean ablation on exposed ice was taken as 60 mm per day and for heavily debris-covered ice, 30 mm per day. These averages are based on measured rates for selected cross sections in 1985 and 1986 (see Chap. 4. and Hewitt et al. 1989)

| Period | Exposed ablation area ($km^2$) | Heavy debris mantle ($km^2$) | Meltwater production ($\times 10^6$ $m^3$ $a^{-1}$) | Lower limit of on-ice snow cover (m) |
|---|---|---|---|---|
| 15–30 June | 70 | 40 | 81 | 4,200 |
| 1–31 July | 100 | 40 | 223 | 4,500 |
| 1–31 Aug. | 120 | 40 | 260 | 4,650 |
| 1–15 Sept. | 110 | 40 | 117 | 4,500 |
| Total | | | **681** | |

### 6.2.1 *Estimates for Biafo and Other Glaciers*

Observations at Biafo Glacier in the 1980s, described in previous chapters, support two types of mass balance calculation. Accumulation and ablation measurements at particular sites and glacier cross sections can be extended to the whole glacier, based on basin hypsometry and some assumptions about input and output gradients. Alternatively, the total turnover can be estimated from the depth cross section and movement data near the equilibrium line (Hewitt et al. 1989).

Over the main accumulation zone of about 300 $km^2$ from 4,600 to 5,800 m, snowfall inputs were 0.9–1.9 m water equivalent (Wake 1987). Some 30 $km^2$ of accumulation zone exists at higher elevations, including a few connected tributaries. There are 130 $km^2$ of perennially frozen rock walls. Assuming an average annual snowfall of 1.4 m (w.e.), the total input should be at least 0.6 $km^3$ $a^{-1}$ (details in Chap. 4).

Outputs were extrapolated from ablation measurements between June 15 and September 15 (Table 6.1). The results indicate slightly more than $680\times10^6$ $m^{-3}$ removed from the ablation zone or 0.68 $km^3$ $a^{-1}$ of water. Estimates take into account the changing area of exposed ablation zone ice due to on-ice snow cover. For exposed areas, daily ablation rates were typically 50–90 mm and a mean rate of 60 mm $d^{-1}$ was adopted. The heavily debris-covered area, concentrated over the lower tongue, changed little in these months and is treated as 40 $km^2$ throughout[1] and with a mean ablation rate of 30 mm $d^{-1}$.

Ice flux between the accumulation and ablation zones, was calculated from surface velocity and a radar depth profile (Hewitt et al. 1989, p. 107). The cross section, 2.3 km wide and maximum depth of 1,400 m, was near the estimated equilibrium line at 4,650 m, inferred from firn limits in the years investigated. Measured velocities ranged between 0.7 and 0.85 m/dy. A mean velocity was estimated using Nye's (1965) solution. Results indicate an ice flux of about 0.6 $km^3$ $a^{-1}$, which is of

[1] Some values in Hewitt et al. (1989) are different from those reported in this text, including reducing the heavy debris-covered portion, but the estimates were conservative and the differences well within likely errors.

a similar order to the first estimate. However, it does not include basin areas of the tributaries that join below the cross section, at least 25 % of the whole. Meanwhile, neither estimate takes account of seasonal snowfall and snowmelt over the ablation zone and an ice-free area of about 230 km$^2$. Seasonal snow cover mostly occurs below the zone of maximum precipitation and declines towards the lower elevations. It is likely to be higher everywhere than most valley weather stations and varies from about 0.4 m (w.e.) at 3,000 m, near the terminus, to about 0.75 m (w.e.) at 4,800 m. As such, snowmelt would add 0.15 km$^3$ a$^{-1}$ or 15–20 % of the glacier contribution.

These estimates would give water yields of around 0.7–0.8 km$^3$ a$^{-1}$ for the 1980s. At that time, the glacier was retreating slowly but close to the furthest extent of the 1970s advance of up to 1.2 km (Chap. 10). The regime and verticality relations for Biafo are looked at below.

At Batura Glacier the 1970s mass balance estimate, on the basis of an annual discharge of meltwater, was $930 \times 10^6$ m$^{-3}$ (BIG 1979, p. 971), equivalent to yield of about 1.3 million m$^3$ per square kilometre for the whole basin (710 km$^2$), 3.6 million m$^3$ for the entire glacier cover (about 260 km$^2$) and 8.5 million m$^3$ from all ablation zone areas (about 110 km$^2$). The estimates are higher than for Biafo, in part perhaps because Batura's ablation zone extends 400 m lower, giving measured ablation 20 % higher or more at equivalent distances from the terminus. Where debris was thin or absent, 10 km above the terminus, annual losses of as much as 10–15 m were measured (ibid, p. 970), elevations below Biafo's terminus or where its heaviest debris covers occur. The total elevation range for Batura is 1,000 m greater than for Biafo. Nevertheless, the whole basin results are of a similar order and, again, much higher than precipitation at long-term weather stations in Hunza Basin (Haserodt 1984).

For Baltoro Glacier ice flux estimates at 4,600 m, just below the Concordia section, were equivalent to 0.35 km$^3$ a$^{-1}$ over the upper basin (Mayer et al. 2006). Uncertainties were, possibly, as high as 35 %, but short-term ablation measurements supported a similar estimate (ibid, p. 129; Mihalcea et al. 2006). Good agreement was found between measured ablation and cumulative positive degree days, using Kayastha et al.'s (2000) method. Snowfall in the upper basin was assumed similar to Biafo (ibid, p. 127). However, its estimated yield per unit area is half or less those for Biafo or Batura. This may be because other, considerable inputs and outputs were not considered. The measurements did not include some 60 km$^2$ of relatively debris-free ablation zone areas above the Concordia cross section and many tributary ice stream areas where debris is thin or absent. Seasonal snowfall contributions were not estimated, yet the combined ablation zone and seasonally snow-free area is a third larger than Biafo's. Assuming similar average annual snowfalls between 3,000 and 4,800 m, the snowmelt contribution should be around 0.4 km$^3$ a$^{-1}$ (w.e.).

The termini of the two glaciers are at similar elevations and the total glacier areas are almost the same. However, as a Mustagh-type glacier, Baltoro's accumulation zone is barely half of Biafo's, and it has much higher watershed heights and total elevation range. Areas contributing to Baltoro through avalanching are much larger, indeed a much bigger total source than Biafo's accumulation zone. It is difficult to believe these sources contribute significantly less mass to the glacier; rather, that avalanching concentrates mass more efficiently. Baltoro's ablation zone is nearly twice that of Biafo. Its heavy debris mantle is almost three times larger and continues to almost 1,000 m higher. Yet, it still has an even greater area with light or debris-free ablation zone, mainly on the substantial tributaries or at comparatively high elevations. In all, these features suggest Baltoro's annual mass balance should not be less and could be close to or larger than those of Biafo and Batura.

The water yields from these, among the largest basins in the region, are not trivial. Biafo comprises about 0.09 % of the upper Indus Basin, the all-important ablation zone a third less again. Yet water yields are equivalent to nearly 1 % or ten times more. Similar values apply to Batura and, in all likelihood, to Baltoro. They represent a significant fraction of the glacier cover in support of the view that it is a much greater contributor to upper Indus flows than its 20 % perennial snow and ice share or about 10 % glacier fraction of area might suggest. The balance, mostly snowmelt coming from the remaining 80 % of the basin, is much less concentrated. Indeed, on the arid valley floors below the glaciers and the great, high elevation arid plateau and mountains around and east of the Karakoram, much of the moisture from snowmelt does not reach the rivers. In sum, glacier roles in moisture storage and transfer, and yields of water from them, far exceed what regional climate statistics or average water yields indicate for the upper Indus and Yarkand Rivers.

## 6.3 The Mass Balance Regime: 'All-Year Accumulation and Summer Ablation'

Mass balance regime, the dominant pattern of inputs and outputs over the budget year, is useful for predictive and comparative purposes. Globally, several distinct regimes have been identified for valley glaciers, usually defined in terms of latitude or zonal climates (Kaser and Osmaston 2002, p. 25). They include the following:

(i) An 'Inner Tropics' type, with two-season or year-round accumulation and year-round ablation
(ii) An 'Outer Tropics' type, with most accumulation *and* strong ablation in summer, but little or none of either in winter, essentially a 'cool' monsoonal regime
(iii) A 'Mid-latitude' type, with winter accumulation and summer ablation

The Greater Himalayan range, from Kashmir to Sikkim, fits the Outer Tropics type, with accumulation mainly from the summer monsoon. The Karakoram is commonly portrayed as a winter accumulation regime or 'Mid-latitude' type along with the Hindu Kush and Pamirs. This fits valley weather station precipitation patterns at Gilgit and Skardu reasonably well. However, some other weather stations, and the critical Biafo data, show summer and winter precipitation are nearly equal (Chap. 4). Inputs to the most heavily glacierised Central Karakoram are intermediate between the summer accumulation of the Greater Himalaya (Ageta and Fujita 1996) and the winter accumulation of the Pamirs or European Alps. Thus, the Mustagh Karakoram, at least, has a *year-round accumulation regime.* It is distinctly different from the 'Outer Tropics' Himalayan and the 'Mid-latitude' types, while ablation differs from the 'year-round, Inner Tropics' type (Ageta and Fujita 1996; Benn and Evans 1998,

p. 86). Some melting can occur on the lowest Karakoram ice tongues in every month, but is negligible in winter (Hewitt 1968). Ablation is concentrated between mid-June and early September, *a summer ablation regime* similar to mid- to high latitudes. As a result, the Karakoram identifies the fourth distinctive regime:

(iv) An Outer Tropics type 'B', with year-round accumulation and summer ablation

Moreover, this does not simply reflect zonal climate conditions, but the high mountain context, mainly an orographic effect on precipitation and ablation.

Small and minor glaciers high up in the Karakoram perennial frigid zone and disconnected from main glaciers have no ablation zone in the usual sense. Losses are by avalanching down steep walls from terminal ice cliffs. They resemble the 'Polar' type where ice travels until it calves into the sea. No inventory of such masses or data on their mass balance exist, but there are countless examples comprising substantial amounts of ice in total (see 8.3).

## 6.4 Vertical Gradients

Mass balance gradients, profiles of gains and losses at different elevations, are of special interest in mountains. Most of those reported appear to vary systematically with elevation. Ablation losses are shown as greatest near the terminus and decline upwards to a well-defined ELA, where ablation exactly balances accumulation (Benn and Evans 1998, pp. 78–79). Above that, accumulation is also generally shown to increase linearly with elevation. Gradients emerge as nearly straight lines, seeming to mimic vertical temperature and precipitation lapse rates and the upward shortening of the ablation season. Differences between glaciers and regions appear mainly in gradient slope or spread. Smaller differences with elevation give a narrower spread, said to be typical of drier, colder and more continental settings. Larger differences and a greater turnover of ice are identified with more humid, especially maritime, glaciers.

Mass balance gradients for Karakoram glaciers depart more or less widely from this picture and have considerable diversity. Biafo, an alpine-type glacier, might be expected to have a conventional profile but does not (Fig. 6.1). Around the terminus ablation zone losses are among the lowest, not the greatest, and increase fairly slowly through the lower 15 km where affected by a nearly continuous heavy debris mantle. Inoue (1977) showed the same elsewhere in the Himalaya (Nakawo et al. 2000). At Biafo, ablation losses then expand quickly through the mid-ablation zone, reaching a maximum about a thousand metres or more above the terminus where there are broad expanses of ice with thin or scattered debris. Annual ablation declines again sharply in the upper ablation zone, with its short ablation season and greater snow covers. Inputs increase sharply through the lower accumulation zone to the snowfall maximum which probably occurs between 4,800 and 5,800 m (Chap. 4). It is likely that the curve then reverses in the highest areas because snowfall and glacier surface area decline and the prevailing avalanche component at these elevations concentrates inputs down slope.

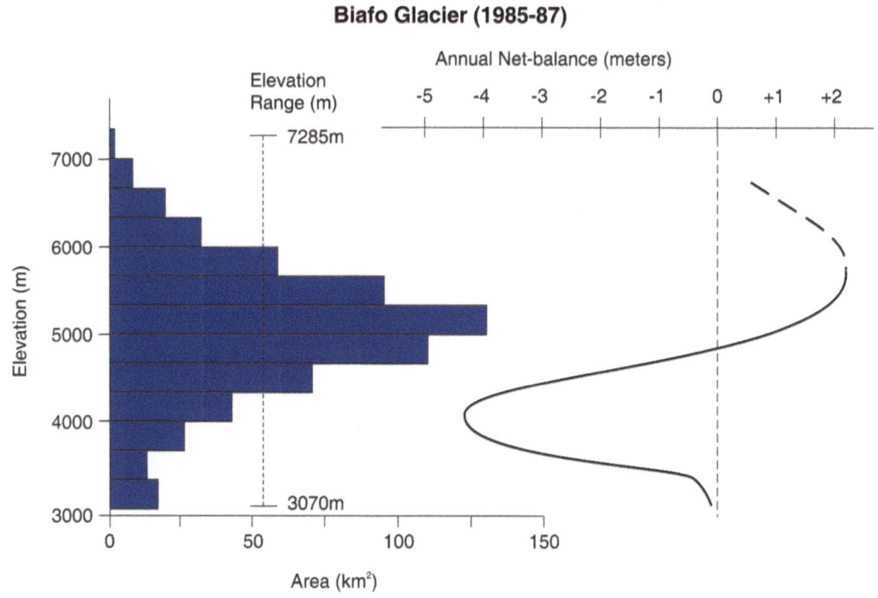


**Fig. 6.1** Vertical mass balance gradient for Biafo Glacier reconstructed from the 1980s measurements (see text, Hewitt et al. 1989; Hewitt 2011)

In sum, a distinctly S-shaped profile applies, much different from typical Alpine-type glaciers elsewhere. Biafo Glacier is the only Karakoram example whose gradient can be bracketed by actual measurements of inputs and outputs. It highlights some features that apply more generally, although major differences must arise in glacier types with different nourishment regimes.

## *6.4.1 Mustagh- and Turkestan-Type Gradients*

> …the redistribution of snowfall by avalanching from steep slopes, and wind scouring from exposed areas, can result in accumulation patterns that differ markedly from original climatically controlled distributions. (Benn and Evans 1998, p. 79)

In the conventional approach mass balance inputs refer to snowfall onto the glacier that becomes glacier ice. Snow falling beyond the glacier margins or seasonal snowfall on the ablation zone is not considered part of *glacier* mass balance, but may be figured into water yields. Specific mass balance is generally reported for the single, main, connected glacier.

Direct snowfall has been shown as a lesser and even negligible input for a majority of Karakoram glaciers. Conventional accumulation zone areas are frequently small or absent. Avalanches comprise virtually all the sources for Turkestan types. Novel consequences follow and some distinctive mass balance gradients (Fig. 6.2).

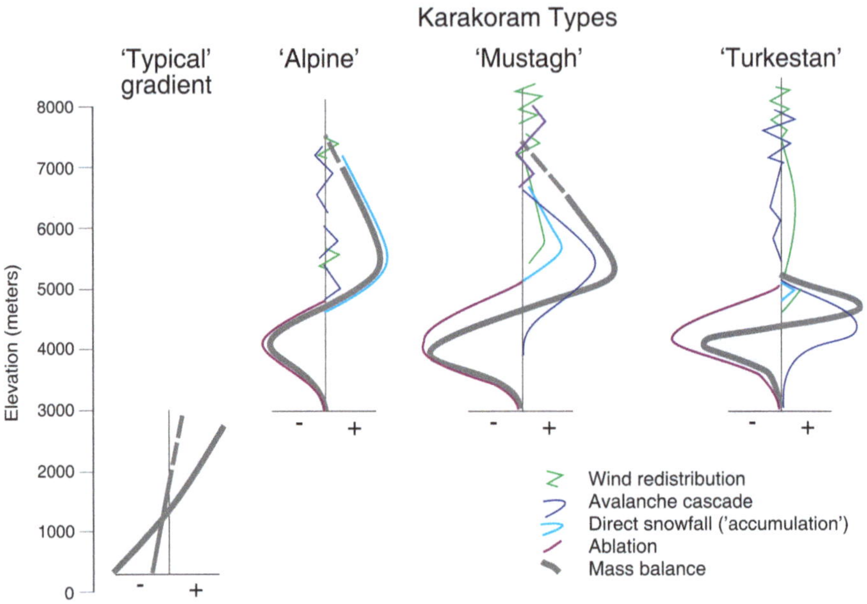


**Fig. 6.2** Schematic vertical mass balance gradients reconstructed from inputs, outputs and transfers described in the text for Alpine-, Mustagh- and Turkestan-type glaciers. These are compared with typical gradients reported in the literature, which are at much lower elevations (Benn and Evans 1998, p. 79)

Since measurements for these glacier types are, at best, only from the ablation zones, their gradients have to be estimated using terrain classes making up the basins. In the 42 largest basins, conventional accumulation zone areas are, on average, only 17 % of the perennial snow zone, while 80 % consist of steep rock walls and disconnected ice masses, the main sources of inputs. Vertical distributions will be affected by orographic and topoclimatic effects on precipitation but also elevations where avalanche paths and icefalls connect with the main glaciers and concentrate inputs. Debris covers tend to be even larger on these glaciers and distort ablation zone gradients even more.

## *6.4.2 Mustagh-Type Gradients*

Compared to Alpine types like Biafo, Mustagh-type glaciers usually have a broader, heavy debris zone above the terminus. It leads to relatively greater losses concentrated in mid- to upper ablation zones which exaggerates the S-shaped ablation gradient. Their upper elevation zone, supplying a critical share of avalanche inputs, is missing from the conventional mass balance gradients based on snowfall inputs only. Likewise, complicated patterns of stripping and deposition of mass by wind and avalanche introduce novel relations. In the Karakoram these are combined with significant snowfall accumulation but truncated towards the head by increasing steepness. Avalanches displace snow mass to much lower elevations compared to precipitation, including into the conventional ablation zone. Wind and avalanche redistribution create positive and negative segments side by side or interwoven. The gradient profile must have a more exaggerated S-shape than Biafo, displaced downwards and more compressed vertically.

### *6.4.3 Turkestan-Type Gradients*

By definition, Turkestan-type glaciers receive nearly all net inputs *below* where perennial snow or firn limits are usually placed. Snowfall inputs come from source zones beyond and above the ice, where wind and avalanche transport prevail. The mass balance gradient still has an S-shape, but both the negative *and positive* curves are in the 'ablation zone'. Inputs overlap outputs through much of the profile and are displaced to even lower elevations than for the Mustagh type. There is no conventional equivalent or measurement basis to define what happens in the frigid source zones, but they must be considered critical for inputs to mass balance and are shown schematically in Fig. 6.2.

The typical 'fall' glaciers, small and minor masses in the lesser ranges also have mass balance gradients displaced well below where firn or snow limits would be expected and compressed into a relatively small elevation range. The avalanches that feed the ice mostly originate in areas near and above the glaciation limit (see Chap. 2). Winter snow falling below the perennial snow zone is also funnelled down steep slopes and may contribute to mass in small receiving basins.

In principle, avalanche-fed glaciers have an elevation where inputs balance outputs. Under the conditions described above, however, the notion of an ELA or how it might be determined is challenged. Given its importance in discussions of mass balance globally, this needs more attention.

## 6.5 Equilibrium Line Altitudes (ELAs)

The notion of an *equilibrium line altitude or* ELA refers to a line, preferably a contour on the glacier, where annual gains equal annual losses. More exactly it is the projection of a surface within the glacier where losses down glacier are exactly balanced by inflows from up-glacier. In mountain glaciers this is expected to intersect the glacier surface at a particular elevation, typically treated as the boundary between the accumulation and ablation zones.

Identifying ELAs is a convenient and powerful tool, widely used and well regarded in many regions (Bakke and Nesje 2011). The ELA is generally considered a more precise and meaningful concept than either firn limits or climatic snowlines (Hawkins 1985, p. 206), but the latter have been found, or are expected to be, close to the ELAs. Given the cost of direct measurement, the possibility of identifying ELAs indirectly is attractive, especially in less accessible regions. Various indirect ways have been proposed to estimate ELAs (Cogley 2010). However, well-established monitoring, on which methods to identify ELAs are based, tends to be for small glaciers of relatively simple morphology, much smaller elevation ranges and steepness compared to Himalayan glaciers, a majority being in Europe (Haeberli and Hoelzle 1995). Conditions and requirements in the Karakoram are very different.

According to Braithwaite and Draper (2010, p. 127): ‘The concept of ELA is only meaningful if the specific balance is a more or less definite function of altitude so that the isoline of zero balance is roughly parallel to ... contour lines on the glacier....’ In the Karakoram such requirements are rarely, if ever, met. Estimating snowlines or firn limits faces great or insurmountable problems and, even where they exist as on Biafo Glacier, are spread over hundreds of metres elevation (Chap. 4). At Batura Glacier, based on observed late summer firn limits, the ELA was placed between 4,700 and 5,300 m (BIG 1979). This does assume the firn limit approximates the ELA, which is open to doubt on a Mustagh-type glacier like Batura.

Mass redistribution by avalanches and wind action raise huge problems, being largest at elevations where Karakoram climatic snowlines and firn limits are said to occur, roughly between 4,600 m and 6,000 m (von Wissmann 1959; Mercer 1975). The notion that ELAs reflect an isotherm at a certain contour line hardly applies and has been questioned elsewhere (Cogley and McIntyre 2003). In the Karakoram there are highly irregular thermal controls such as avalanche source elevations, composition and run-out patterns. There are complex wind fields across slopes and snow that has been wind redistributed, packed and thermally altered (Bakke and Nesje 2011, p. 270). As noted, avalanches generate substantial inputs in conventional ablation zones and almost all inputs to Turkestan types. Icefalls were shown to have a major role in the same elevations. For such reasons, the interface between loss and gain becomes, essentially, a statistical abstraction, impossible to determine with established methods.

Indirect methods, used to estimate ELAs, also seem to have little or no reliability in this region (Table 6.2). Some work uses median elevation of glacier basins as an indicator, but these are generally much higher than snowline estimates in the Karakoram, suggesting they are not helpful (cf. Meierding 1982). The interface between concave and convex or ‘Hess’ contours, where they could be determined on relatively good topographical maps, is commonly 500–1,000 m lower than snowline estimates which seems more promising. However, it is unlikely actual ELAs ever follow a particular contour. Distinctions between area-weighted mean altitudes and the ‘median’ dividing the glacier into equal halves are very difficult to establish for Karakoram glaciers (Braithwaite and Draper 2010). Their value if determined is likely to be compromised by the way so much of the highest areas either are not glacierised or consist of ice masses disconnected from the main glacier. The practice of identifying former ELAs from lateral moraines suffers from problems of dating and, in many cases, accurately identifying the deposits (Fort and Peulvast 1995; Hewitt 1999; Benn and Lehmkuhl 2000). A common practice of inferring or

**Table 6.2** Some Karakoram results of methods used elsewhere to estimate to ELAs. 'TSA' refers to the total elevation ranges from 'T' the 'toe' or terminus to ('S') the highest summit. AAR refers to the ratio of accumulation area to total glacier area. Snowlines are based on von Wissmann (1959) and other data from the 43 main set inventory where possible (see text and Richmond 1965; Meierding 1982; Hawkins 1985; Cogley and McIntyre 2003; Braithwaite and Draper 2010; Bakke and Nesje 2011)

| Glacier | Snowlines | Firn line | TSA mean ele | Hess contour | AAR |
|---|---|---|---|---|---|
| Bualtar | 4,800–5,000 | | 4,783 | 3,200–3,600 | 0.22 |
| Sokha L | 4,800 | | 4,834 | – | – |
| Chiantar | 4,800–5,000 | 4,700 | 4,922.50 | – | 1.0 |
| Sosbun | 4,800 | | 5,003.50 | – | – |
| Braldu | 5,200 | | 5,008.50 | – | 0.45 |
| Virjerab | 5,000–5,200 | 4,800 | 5,008.50 | – | 0.28 |
| Karambar | 4,800 | 4,500 | 5,019 | | 0.19 |
| Kukuar | 4,800 | 4,700 | 5,024 | 3,400–3,600 | 0.08 |
| Kutiah | 4,800 | | 5,098.50 | – | 0.29 |
| Chogo L | 4,800 | 4,600 | 5,109 | | 0.39 |
| Barpu | 4,800–5,000 | | 5,144 | 4,100–4,500 | 0.14 |
| Hasanabad | 4,800–5,000 | 4,700 | 5,146 | 3,500–3,900 | 0.11 |
| Yashuk Y. | 4,800–5,000 | | 5,154 | | 0.15 |
| Biafo | 4,800–5,000 | 4,650 | 5177.5 | 4,300–4,600 | 0.72 |
| Batura | 4,800–5,000 | 4,800 | 5182.5 | 4,1,80–4,220 | 0.34 |
| Ghondo'ro | 5,200 | 4,900 | 5,299 | | 0.24 |
| Toltar–Bal. | 4,800 | 4,600 | 5367.5 | | 0.04 |
| Gyong | 5,200 | | 5,380 | | – |
| Charakusa | 5,200 | 4,700 | 5,381 | | 0.21 |
| Momhil | 5,000 | | 5,390 | | – |
| Panmah | 5,000–5,200 | 4,700 | 5392.5 | 4,200 | 0.37 |
| Malangutti | 5,000 | | 5397.5 | | – |
| Hispar | 4,800–5,000 | 4,700 | 5417.5 | | 0.29 |
| Khurdopin | 5,000 | 4,900 | 5,480 | | 0.32 |
| Kondus-K. | 5,200–5,400 | | 5492.5 | | 0.29 |
| Yazghil | 5,000 | | 5,541 | | – |
| Skamri | 5,400–5,600 | | 5587.5 | | 0.29 |
| Bilafond | 5,200 | | 5,604 | | – |
| Gharesa | 4,800–5,000 | | 5,614 | | 0.4 |
| Sherpigang | 5,200 | | 5,626 | | – |
| Sarpo Laggo | 5.400–5,600 | 5,100 | 5,630 | | 0.53 |
| Siachen | 5,400 | 5,000 | 5,671 | | 0.56 |
| Terong NS | 5,400–5,800 | | 5,680 | | 0.6 |
| Singhi | 5,800–6,000 | | 5882.5 | | – |
| Kyagar | 5,600–5,800 | | 5992.5 | | – |
| S. Shukpa | 5,400–5,600 | | 6006.5 | | – |
| Baltoro | 5,200–5,600 | 4,900 | 6,105 | 4,600–4,900 | 0.47 |
| Chong Kh | 5,600–5,800 | | 6,115 | | – |
| N. Gasherbr | 5,600–5,800 | 5,100 | 6138.5 | | 0.48 |
| Rimo | 5,600–5,800 | | 6152.5 | | 0.8 |
| Urdok | 5,600–5,800 | | 6,164 | | – |
| N. Shukpa | 5,400–5,600 | | 6,171 | | 0.49 |

confirming ELA changes from terminus fluctuations raises unresolved problems of irregular ice movement, with the many surge-type glaciers and surging tributaries (Chap. 9).

## 6.6 Critical Elevations: Area–Altitude Relations of Mass Balance

Conditions that seem most critical to mass balance tend to be concentrated in particular elevation bands, albeit with some large overlaps and critical transitions (Fig. 6.3). This produces marked differences from the unidirectional, linear gradients seen in conventional mass balance models for mountain glaciers. There are zones that assume particular importance in some or all classes of ice mass and that impose distinctive spatial regimes on the inputs and outputs, and the dynamics of the glaciers. The dominant processes and approximate elevation ranges are as follows:

1. Wind regimes, 5,000–8,000 m
2. Avalanche regimes, 4,000–8,000 m
3. Major, avalanche-fed ice streams, 4,000–6,000 m
4. Heaviest snowfall, roughly 4,800–6,000 mm
5. Direct accumulation or 'firn' areas, if and where they occur, 4,500–6,000 m
6. Ablation sub-Zone I: low losses and water yields due to short ablation season, larger and more frequent, all-year on-glacier snowfalls, 4,500–5,000 m
7. Ablation sub-Zone II: high losses and water yields helped by thin debris, 3,500–4,500 m
8. Ablation sub-Zone III: low loss and yield due to heavy debris covers, below 3,800 m

There has been little or no monitoring of these processes or zones, let alone the longer-term observations necessary to sort out their full and exact roles or, even more important perhaps, how they overlap and interact. Nevertheless, the survey thus far indicates that each zone can involve more than 20 % of the terrain in Karakoram glacier basins. Each drives or reflects key glaciological and hydrological conditions. Each needs to be investigated as potentially critical in responses to climate change. Some aspects that can guide yet-to-be-done analyses are summarised below (Fig. 6.3).

### *6.6.1 Wind Regimes, 5,000–8,000 m*

Wind action can redistribute and alter snow at all elevations, but is a dominant factor along and close to interfluves (Plate 6.1). For mass balance its main role is in the perennial snow zone where it transports snow across interfluves, strips or compacts snow in windward areas and creates cornices and other build-ups on lee slopes The main regional impact on mass balance relates to priming of avalanches, and the two work together to

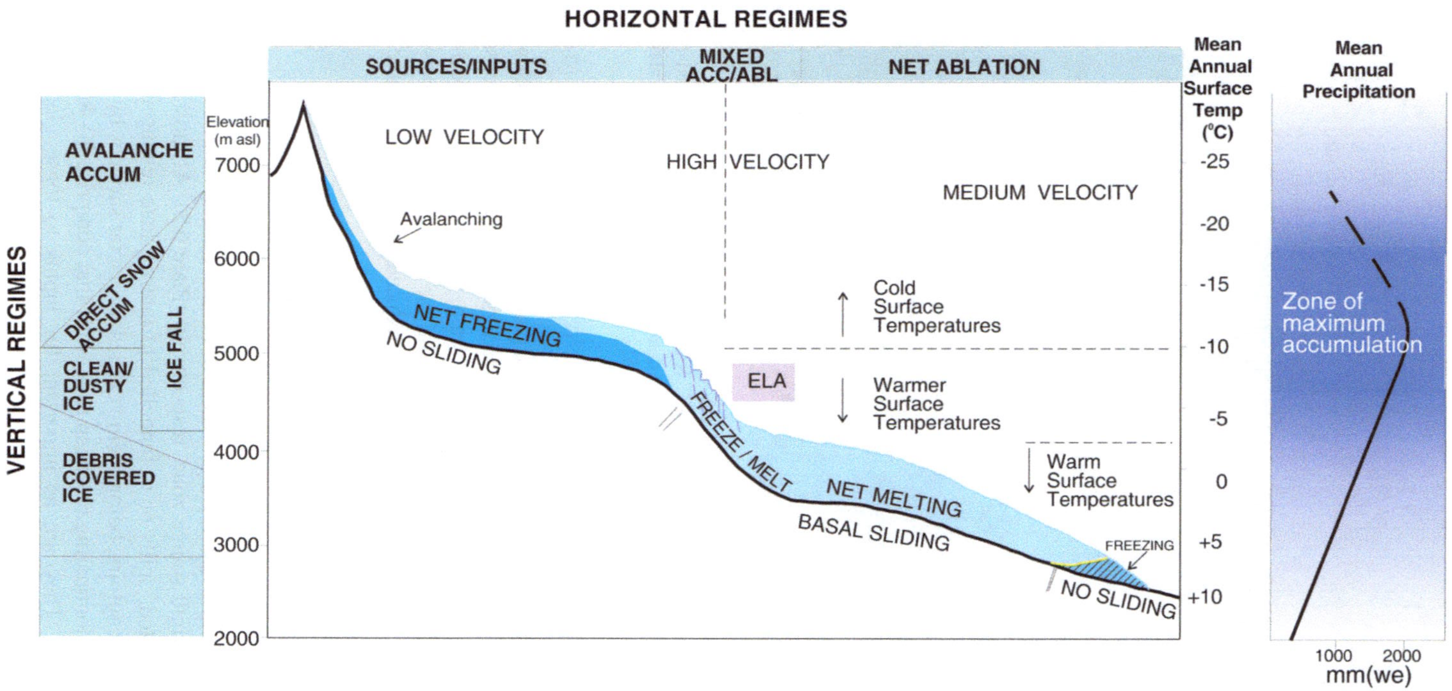


**Fig. 6.3** Main elevation-related regimes and horizontal relations of conditions for the larger Karakoram valley glaciers (see text; Selby 1993, p. 16; Menzies 1995, p. 116). The connected 'source and input' regime applies to Alpine- and Mustagh types, but such processes are detached from the main glacier in Turkestan types. In the latter especially, avalanches may extend far into the 'mixed accumulation/ablation regime' and even the 'net ablation' regime. Vertical gradients of mean annual surface temperatures and precipitation for the Central Karakoram are at the right side (Hewitt 1993, 2005)

**Plate 6.1** Wind redistributed snow around and below interfluves, Drangra Peak (6,325 m), Godoghoro Glacier, Masherbrum Mountains, EC Karakoram. The relations of wind action to avalanche-swept and fluted slopes below should be noted and its roles in priming avalanches, in uppermost glacier initiation, and icefalls (Hewitt 2005)

control inputs to ice masses (see Chap. 7). In lesser ranges, redistribution of snow across interfluves leads to concentrations of lee-slope glaciers, for example, in the Depsang and Aksai Chin Mountains of the far eastern Karakoram. On the north slope basins from Chapursan and Shimshal to the upper Shaksgam, the concentration of snow on lee slopes appears to be as important or more so than reduced exposure to sunshine.

### *6.6.2 Avalanche Regimes, 4,000–8,000 m*

For Mustagh- and Turkestan-type glaciers, there is no 'accumulation' on more than half or all of the frigid source zone, respectively (Plate 6.2). At most, snowfall is temporarily deposited, spills or is funnelled to intermediate resting places down slope, in chutes or couloirs, before being carried to glacier surfaces by avalanches. This is the regime dominating the steep rock walls that comprise 50–80% of frigid upper basin areas. They are as important for inputs to these glaciers as the

**Plate 6.2** Frigid source zone and avalanche-swept rock walls that nourish Charakusa Glacier, a Turkestan-type glacier, between K6 (7,281 m) and Link Sar (7,041 m), Masherbrum Mountains, Central Karakoram (Hewitt 1999)

accumulation zones of the most widely studied valley glaciers, but evidently the processes, spatial redistribution and qualities of snow inputs are radically different.

### *6.6.3 Avalanche Nourishment of Ice, 4,000–6,000 m*

In avalanche-fed glaciers nourishment occurs where avalanches build up on glacier surfaces. The net contributions are shifted downslope compared to precipitation. Inputs are much more concentrated than with snowfall itself or even wind redistribution and occur well below where snowfall, including the zone of maxima snowfall, occurs (Yafeng and Wenying 1980). Avalanched snow is also much denser than snowfall, leading to mass concentration through density. Avalanche deposits have distinctive morphologies associated with terrain geometry and thermal properties. Apart from thin, widely dispersed powder avalanches, there is nothing like the blanket cover of snowfall. Avalanche-fed ice streams commence as large cones and aprons of avalanche snow, often at the base of major chutes and couloirs or, in Mustagh types, of icefalls (Plate 6.3).

**Plate 6.3** Concentrated avalanche deposition in cones that feed a Turkestan-type glacier at around 4,000 m, that is, well below the perennial snow zone and likely location of conventional snowlines or firn limits. Bazhin Glacier receives avalanches descending from under the summit of Nanga Parbat (8,125) in Rupal Gah (Hewitt 2010)

### 6.6.4 *Zone of Highest Snowfall, 4,800–5,800 m*

The zone of maximum precipitation is entirely within the perennial snow zone and sources of glacier nourishment. Precipitation of 1,000–2,000 mm (w.e.) declines downward through ablation zone areas and to as little as 150–300 mm at valley weather stations (Chap. 4). There is no reason to assume precipitation gradients and amounts at given elevations found at Biafo, an Alpine type, do not apply in Mustagh- and Turkestan-type glaciers. This is one of the most important ways in which orographic effects help to concentrate glacierisation or glacier mass by elevation.

### 6.6.5 *Direct Accumulation Areas, 4,500–6,000 m*

Although relatively few in number, large glaciers with substantial snowfall basins at high elevations, notably Siachen, Rimo, Biafo, Braldu and Chiantar Glaciers,

**Plate 6.4** Upper ablation zone of Panmah Glacier, Choktoi Branch between 4,500 and 5,500 m. The transition to the perennial snow zone occurs where avalanches build out from the base of rock walls around Baintha Brakk (7,285 m) and occasional icefalls from high up (Hewitt 2005)

comprise an important fraction of the total Karakoram glacier cover. With reservations described earlier, these could be described as 'firn' areas and occur mainly in the zone of maximum snowfall between 4,800 and 6,000 m. When the incised accumulation basins of Mustagh-type glaciers are included, and smaller ones on disconnected tributaries of some Turkestan types, direct snowfall accumulation probably nourishes as much as one third of total Karakoram ice mass. These are the only places where area–altitude inputs reflect precipitation gradients, but the smaller share except for Alpine types.

### *6.6.6 Ablation Sub-Zone I, 4,500–5,000 m*

This is to identify areas with relatively low losses and water yields due to the short ablation season (Plate 6.4). The highest on-glacier seasonal snowfalls, avalanche deposition and superimposed ice have their greatest ablation zone contributions here. Year-to-year, these inputs and ablation losses will vary widely according to season length and summer weather. Ablation is complicated by the development of the freeze–thaw carapace and extent of firn subzones discussed in Chap. 4 and by the runoff through and out of the snowpack (see 8.5.1). In Mustagh- and Turkestan-type glaciers, major complications arise from the concentrations of avalanche and icefall activity in the same elevation band.

### 6.6.7 Ablation Sub-Zone II, 3,500–4,500 m

Areas with the largest summer expanses of glacier ice and thin or scattered, ablation-enhancing debris. Ablation seasons last from 16 weeks to about 3 weeks at the transition to sub-Zone I and comprise the time when glacier ice has been fully exposed from under winter snowfall and the freeze–thaw carapace and when diurnal temperatures are mostly above zero. The relative extent of the zone diminishes from Alpine- through Mustagh- to Turkestan-type glaciers as the proportion with heavy debris cover increases. For glaciers of similar size, the zone occurs some hundreds of metres lower and is broader in the western, compared to the eastern, Karakoram and is lower and broader in south slope basins compared to north-facing.

### 6.6.8 Ablation Sub-Zone III, Below 3,800 m

This identifies the main areas with heavy debris mantles (5.6). Net ablation rates are a third or less that of the zone above, despite a longer ablation season. The extent of the zone increases in proportion and in elevation range from Alpine- through Mustagh- to Turkestan-type glaciers.

## 6.7 Concentration of Mass Balance Elements

The most important implication of these area–altitude zones lies in spatial and/or temporal concentrations of critical processes. Compared to the total area and elevation range in glacier basins, redistribution of mass by wind, avalanche and icefalls serves to intensify glacier processes, in particular elevation bands and consequences of the vertical cascade of moisture and debris. They raise intriguing, hitherto unexplored, and potentially dangerous issues in relation to climate change (UNESCO 1998). Some conditions, such as avalanche nourishment in ablation zone areas, or heavy debris covers, can buffer or conserve glacier mass against climate warming. Conditions in ablation sub-Zone II seem very sensitive to climate change and could accelerate response (Chap. 12). The regimes identified suggest that changes in prevailing wind strength and direction, or in the zone of maximum precipitation, could be as critical as temperature changes. However, unlike such regional climatic responses to global climate change, important feedbacks are identified that depend upon glacier morphology and processes rather than climate directly. They require attention to glacier dynamics, the focus of the next chapter.

## References

Ageta Y, Fujita K (1996) Characteristics of mass balance of summer-accumulation type glaciers in the Himalayas and Tibetan Plateau. Z Gletscher Glazialgeol 32:61–65

Aizen V (2011) High elevation glacio-climatology. In: Singh VP, Singh P, Haritashaya UK (eds) Encyclopaedia of snow ice and glaciers. Springer, Dordrecht, pp 507–510

Archer DR, Fowler HJ (2004) Spatial and temporal variations in precipitation in the Upper Indus Basin, global teleconnections and hydrological implications. Hydrol Earth Syst Sci 8(1):47–61

Bakke J, Nesje A (2011) Equilibrium-line altitude (ELA). In: Singh VP, Singh P, Haritashaya UK (eds) Encyclopedia of snow ice and glaciers. Springer, Dordrecht, pp 268–277

Benn DI, Evans DJA (1998) Glaciers and glaciation. Arnold, London

Benn DI, Lehmkuhl F (2000) Mass balance and equilibrium-line altitudes of glaciers in high mountain environments. Quat Int 65–66:15–29

BIG (Batura Investigations Group) (1979) The Batura Glacier in the Karakoram Mountains and its variations. Sci Sin 22(8):958–974

Bolch T, Kamp U (2006) Glacier mapping in high mountains using DEMs, Landsat and ASTER data. Grazer Schr Geographie Raumforsch 41:13–23

Braithwaite RJ, Draper SCB (2010) Estimating equilibrium-line altitude (ELA) from glacier inventory data. Ann Glaciol 50(53):127–132

Cogley JG (2010) A more complete version of the World Glacier Inventory. Ann Glaciol 50(53):32–38

Cogley JG (2011) Present and future states of Himalayan and Karakoram glaciers. Ann Glaciol 52(59):68–73

Cogley JG, McIntyre MS (2003) Hess altitude and other morphological estimates of glacier equilibrium lines. Arct Antarct Alp Res 35(4):482–488

Fort M, Peulvast J-P (1995) Catastrophic mass movements and morphogenesis in the Peri-Tibetan ranges: examples from West Kunlun, East Pamir and Ladakh. In: Slaymaker O (ed) Steepland geomorphology. Wiley, New York, pp 171–198

Gardelle J, Berthier E, Arnaud Y (2012) Slight mass gain of Karakoram glaciers in the early 21st century. Nat Geosci 10, 1938/ngeo1450

Haeberli W (2011) Glacier mass balance. In: Singh VP, Singh P, Haritashaya UK (eds) Encyclopedia of snow ice and glaciers. Springer, Dordrecht, pp 399–408

Haeberli W, Hoelzle M (1995) Application of inventory data for estimating characteristics of and regional climate-change effects on mountain glaciers: a pilot study with the European Alps. Ann Glaciol 21:206–212

Haserodt K (1984) Abflussverhalten der Flüsse mit Bezügen zur Sonnenscheindauer und zum Niederschlag zwischen Hindukusch (Chitral) und Hunza-Karakorum (Gilgit, Nordpakistan). Mitteilungen der Geographischen Gesellschaft, München 96:29–36

Hawkins F (1985) Equilibrium-line altitudes and palaeoenvironments in the Merchants Bay area, Baffin Island, NWT, Canada. J Glaciol 31:205–213

Hewitt K (1968) Geomorphology of the mountain regions of the upper Indus Basin, 2 vols. Dissertation, University of London

Hewitt K (1993) Altitudinal organization of Karakoram geomorphic processes and depositional environments. In: Shroder JF Jr (ed) Himalaya to the sea: geology, geomorphology and the Quaternary. Routledge, New York, pp 159–183

Hewitt K (1999) Quaternary moraines vs. catastrophic rock avalanches in the Karakoram Himalaya, northern Pakistan. Quat Res 51(3):220–237

Hewitt K (2011) Glacier change, concentration and elevation effects in the Karakoram Himalaya, upper Indus Basin, Mt Res and Dev 31(3):1–13, http://dx.doi.org/10.1659/MRD-JOURNAL-D-11-00020.1

Hewitt K, Wake CP, Young GJ, David C (1989) Hydrological investigations at Biafo Glacier, Karakoram Himalaya: an important source of water for the Indus River. Ann Glaciol 13:103–108

Immerzeel WW, Droogers P, de Jong SM, Bierkens MFP (2009) Large-scale monitoring of snow cover and runoff simulation in Himalayan river basins using remote sensing. Remote Sens Environ 113:40–49
Inoue J (1977) Mass budget of Khumbu Glacier. Seppyo 39(Special Issue):15–19
Kaser G, Osmaston H (2002) Tropical glaciers. Cambridge University Press, Cambridge, UK
Kayastha RB, Ageta Y, Nakawo M (2000) Positive degree-day factors for ablation on glaciers in the Nepal Himalaya: case study on Glacier AX010 in Shorong Himal, Nepal. Bull Glaciol Res 17:1–10
Mayer C, Lambrecht A, Belo M, Smiraglia C, Diolaiuti G (2006) Glaciological characteristics of the ablation zone of Baltoro Glacier, Karakoram, Pakistan. Ann Glaciol 43:123–131
Menzies J (1995) Glaciers and ice sheets. In: Menzies J (ed) Modern glacial environments: processes, dynamics and sediments, vol 1. Butterworth-Heinemann, Oxford, pp 101–38, Chap 4
Mercer JH (1975) Glaciers of the Karakoram. In: Field WO (ed) Mountain glaciers of the Northern Hemisphere, vol 1. Cold Regions Research and Engineering Laboratory (CREEL), Hanover, pp 371–409
Meierding TC (1982) Late pleistocene equilibrium-line altitudes in the Colorado Front Range: a comparison of methods. Quat Res 18(3):289–310
Mihalcea C, Mayer C, Diolaiuti G, Lambrecht A, Smiraglia C, Tartari G (2006) Ice ablation and meteorological conditions on the debris-covered area of Baltoro Glacier, Karakoram, Pakistan. Ann Glaciol 43:292–300
Nakawo M, Raymond CF, Fountain A (eds) (2000) Debris-covered glaciers. International association of Hydrological sciences (AHS) Publication no. 264, Wallingford, Oxfordshire
Naz Bibi S, Bowling LC, Diffenbaugh L, Owens P, Ashfaq M, Ur-Rehman S (2009) Hydrological sensitivity of the Upper Indus River to glacier changes in the Karakoram Himalaya Region. Unpubl Poster # C31C-0455, AGU Meeting, November
Nye JF (1965) The flow of a glacier in a channel of rectangular, elliptic or parabolic cross-section. J Glaciol 5(41):661–690
Oerlemanns J (2002) Glaciers and climate change. Swets and Zeitlinger BV, Lisse
Raina VK (2009) Himalayan glaciers: a state-of-art review of glacial studies, glacial retreat and climate change. Ministry of Environment and Forests, New Delhi
Richmond GM (1965) Glaciation of the rocky mountains. In: Wright HE, Frey DG (eds) The quaternary of the United States. Princeton University Press, Princeton, pp 217–30
Selby MJ (1993) Hillslope materials and processes, 2nd edn. Oxford University Press, Oxford, UK
Scherler D, Bookhagen B, Streker MR (2011) Spatially variable response of Himalayan glaciers to climate change affected by debris cover. Nat Geosci 4:156–159
UNESCO (1998) Into the second century of worldwide glacier monitoring: prospects and strategies. UN Educational, Scientific and Cultural Organization, Paris
von Wissmann H (1959) Die heutige Vergletscherung und Schneegrenze in Hochasien mit Hinweisen die Vergletscherung der letzten Eiszeit.: Akademie der Wissenschaften und der Literatur Mainz, Germany. pp 1103–1431
Wake CP (1987) Spatial and temporal variation of snow accumulation in the Central Karakoram, northern Pakistan. Thesis, Wilfrid Laurier University, Waterloo
WGMS (2008) Fluctuations of glaciers 2000–2005. World Glacier Monitoring Service (ICSU, IUGG, UNEP, WMO), vol IX, Zurich
Yafeng S, Wenying W (1980) Research on snow cover in China and the avalanche phenomenon of Batura Glacier in Pakistan. J Glaciol 26(94):25–30
Zemp M, Hoelzle M, Haeberli W (2009) Six decades of glacier mass balance observations: a review of the worldwide monitoring network. Ann Glaciol 50:101–111

# Chapter 7
# Dynamics of Snow and Ice in Glacier Basins

**Abstract** This chapter examines the processes that regulate glacier behaviour and how ice masses respond to environmental conditions. Glacier movement is considered and some distinctive features in larger Karakoram glaciers. Rates of movement along main ice streams are typically between 150 and 300 m annually but may be over 1,000 m on large icefalls. Among different styles of ice movement, available observations suggest sliding motion is a large factor in most Karakoram glaciers. The so-called *Blockschollen* or 'block motion' is widely observed, a slab-like movement of large sections of the glacier with sharp lines of shear at the margins. It is typical of relatively gentle as well as steep sections in ablation zones. Strong fluctuations in glacier movement show up in detailed measurements and on all timescales: day to day and minute to minute as well as with season and from year to year. Summer ablation zone velocities are commonly 20–50 % greater than in winter, suggesting sensitivity of the sliding component to meltwater availability. The variety of flow instabilities may reflect the roles of basin steepness and icefalls, debris content and ice thermal conditions. Such ice temperature measurements as are available show that thermal regimes include cold, warm and polythermal types. Verticality is again an important factor and different thermal regimes occur in the same glacier at different elevations. Icefalls are a major but neglected concern, responsible for much of the fastest movement and the larger part of transfers of ice mass vertically. Ablation zone icefalls create environments where rapid modifications may penetrate far into the thinner more broken ice mass and faster responses to climate variability. The great extent of ogives on main ice streams, seasonal forms generated in icefalls, is another indication of their importance. Drivers of ice dynamics, as of mass balance, include processes that commence off-ice, notably vertical mass transfers of snow and ice by wind and avalanche. They intervene in the density, structure and thermal properties of inputs to the glaciers, their timing and spatial organisation. This can be expressed as an 'input concentration factor' (ICF), derived from the ratio between conventional accumulation zone areas and steep walls in glacier source zones. The highest concentration values are for Turkestan-type glaciers, lowest for Alpine and Mustagh types are intermediate. Glacier dynamics

K. Hewitt, *Glaciers of the Karakoram Himalaya: Glacial Environments, Processes, Hazards and Resources*, Advances in Asian Human-Environmental Research,
DOI 10.1007/978-94-007-6311-1_7, © Springer Science+Business Media Dordrecht 2014

and responses to climate are shown to involve the whole contributing basin and a range of processes that affect snow and ice, not solely glacier mass balance.

**Keywords** Verticality • Glacier movement • Block motion • Icefalls • Ogives • Avalanche and wind transport • Vertical cascade • Concentrated glacierisation

## 7.1 Process Regimes and Dynamics

Mass balance directs attention to glacier maintenance and change, quantifying and linking them to climatic controls and the distribution of ice. However, the mass balance or budget approach only deals indirectly, if at all, with snow and ice dynamics. As already indicated, the processes that link glacier systems to climate and terrain in the Karakoram, and in different elevation zones, assume special importance. They are likely factors helping to explain the apparent anomalies reported in recent glacier change (Chap. 12).

Elsewhere, glacier movement has been tied to mass balance. Again, singular departures are found in the Karakoram. In part it may be because the extreme elevation ranges, the steepness and the rugged terrain recast the dynamic relations of mass transfer. Glacier movement is only one of the systems that transfer snow and ice in glacier basins, not necessarily the largest in area affected or net vertical transport. Glaciers are the largest ice masses, in part because they are *relatively slow parts* of the overall systems to which they belong. This is most obvious for the predominant Mustagh- and Turkestan-type glaciers. Snow delivered to their basins by weather systems is transferred laterally and downslope by wind action and avalanches in volumes of a similar order to those transferred down valley by glacier movement. Net vertical displacements are likely to be much larger and are certainly more rapid. Even so, glacier movement is of unique interest and will be dealt with first.

## 7.2 Glacier Movement

Glacier dynamics concern the mechanics and movement of glacier ice (Jiskoot 2011). In mountain glaciers, movement is largely directed down valley. Three main forms of motion have been established: internal deformation, sliding at the base and deformation of subglacial sediments.

Glacier ice responds to applied stress above a certain threshold by permanent deformation. Microscopic melting and recrystallisation in the ice structure are involved, micro-fracture and micro-shearing. The resulting internal deformation or *creep* is the characteristic form of ice motion in response to gravity and superimposed load – the reason why ice streams appear to flow (Jiskoot 2011, p. 249). The thickness of overlying ice is important since creep occurs at depth under relatively strong confining pressures. Other things equal, rates of flow are higher if ice is thicker, or warmer, or on steeper slopes. The upper few tens of metres of a glacier are generally brittle, however, with low bending strength and prone to fracture. This is shown by the crevasses at breaks of slope, in icefalls, and at the glacier margins.

Although dealt with later in this chapter, ice temperature warrants introduction here because it affects rates of creep, and whether sliding is likely to occur (Menzies 1995, p. 153). Two types are of interest: *warm* ice at the melting point and not frozen to the bed and *cold* ice, with temperatures below the melting point and frozen to the bed. Since many glaciers have both warm and cold ice, their configuration is important. Paterson (1994) recognises four types according to whether warm or cold ice is at the bed or in various horizons above it. These regimes are commonly identified with different climates: polar, temperate, maritime and continental. The Karakoram hardly fits this picture. All four regimes are present in this one region, usually at least two in the same glacier, sometimes all (BIG 1979; Hewitt 2005). Ice thermal conditions are diversified by the great elevation range, the differing rates at which snow and ice are transferred and their paths, and by large seasonal variations in air and snow temperatures. Although there are no active volcanoes, hot springs are widespread across the Karakoram. This could be a factor causing warming and instability of some glaciers, but there are no data that might test the idea.

A second form of movement is by sliding of ice over its bed. Whereas creep occurs in both cold and warm ice, tending to be faster as ice temperature rises in the former, sliding is mainly expected in warm ice and glaciers not frozen to the bed. Some experiments have shown limited sliding in cold ice but it seems to be of minor importance. However, patches of frozen and thawed bed may coexist and influence sliding rates. Local melting and refreezing are often present in otherwise warm ice or in cold ice approaching the melting point. This tends to be concentrated around obstacles over which the ice must pass.

The presence of meltwater, its quantity and drainage patterns are major influences on sliding and its velocity. Water from surface ablation may be important if it penetrates to the bed, generally relevant in warm glaciers or sections. Rates of sliding movement are usually greater in the ablation season, a response to higher meltwater availability. The exact relationships are complicated by the form and development of drainage conditions or conduits in and beneath the ice, and the presence of debris under or incorporated into basal ice. The latter may either retard or enhance sliding, depending on particle sizes, temperature and degree of saturation (Drewry 1986, pp. 10–14).

The importance of basal sliding emerges in profiles of movement across Karakoram glaciers (see Sects. 7.2.1, 7.2.2, 7.2.3, and 7.2.4 below). Most show sharp increases at or near the margins, and uniform or nearly uniform rates across the central parts. There have been observations of moulded and fluted basal till left behind by glacier retreat, suggestive of basal sliding. In those few cases where all-year movement rates have been measured, summer rates are generally much higher than winter, reflecting the importance of meltwater (see below).

Another form of glacier motion can result from deformation in soft, subglacial sediments (Paterson 1994, pp. 158–172). It is thought to apply mainly if not only to unfrozen beds and sediments, to be most likely in sediments of sand to clay grades,

**Plate 7.1** Pronounced thrust planes with dirt bands exposed in the ice cliff of the Chhatiboi (Darkot) Glacier, where it reaches and has dammed the upper Yarkhun River (Hewitt 1998)

or where a fine-grained matrix supports coarser particles. Such material can have a lower strength than glacier ice, leading sediments to deform more readily under ice loading. The quantities of debris carried by Karakoram glaciers and evidence of unfrozen beds in ablation zones suggest that deformation of basal sediments could be important. Many glacier tongues sit on thick ramps of moraine with fines-rich horizons. There is some direct evidence of subglacial sediment deformation in proglacial areas overridden during the winter season advance (Hewitt 1967). Whether this happens beneath the major ice streams is not yet confirmed by observation. In the surface movement profiles available, it is difficult or impossible to distinguish this from sliding between ice and a rigid bed.

Finally, differential motion can give rise to inclined thrust planes. They are observed as upper layers shearing over ice below that may be slower moving, stalled or dead. It is particularly evident at the termini, at bends in ice streams, and along the margins where glaciers are thickening (Plate 7.1). Near the terminus, especially, the shear planes can reach to the bed and become zones along which subglacial debris is transported to the glacier surface (Hewitt 1967).

From a mass balance viewpoint, glacier movement appears as the way inputs and losses are maintained in equilibrium. Much of the literature reports or assumes rates of movement that increase systematically through the accumulation zone to a maximum around the equilibrium line altitude (ELA), decreasing again systematically through the ablation zone to the terminus (Benn and Evans 1998, pp. 77–85). Valley glaciers where this applies are usually shown with concave long profiles in the accumulation zone and convex towards the terminus (Drewry 1986, p. 16). Movement rates, or advance and retreat of termini, are seen as due to mass balance change and useful for tracking it. This has been shown to apply in many glaciers with well-established records and as a helpful simplification of glacier budget concerns (Haeberli and Hoelzle 1995). However, while the same fundamental principles of ice behaviour must apply in the Karakoram, its glaciers depart more or less strongly from this picture.

In those larger glaciers for which there are measurements, rates of movement do not increase or decrease in a systematic way along the glacier. They are not neatly symmetrical around an ELA, to the extent that one can be identified. Rather, glacier dynamics are complicated, presumably due to the ruggedness, verticality relations, glacier thermal regimes and variety of mass balance types already outlined.

### *7.2.1 Movement Records for Karakoram Glaciers*

Velocity profiles based on stakes drilled into the ice reveal rates and patterns of surface movement and the relative importance of internal deformation, sliding or some combination of both. Most attempts to measure glacier movement in the region have been short term, some over several months, rarely over a year or more (Finsterwalder and Pillewizer 1939; Pillewizer 1957; Goudie et al. 1984). More recently, important developments involve tracking of surface velocities comparing high-resolution satellite images over several years and for different seasons (Luckman et al. 2007). A few of the more intensive full-year or longer records are outlined below.

### *7.2.2 Surface Velocity Profiles at Batura Glacier*

The most intensive, direct velocity measurements were carried out in 1974 and 1975 at Batura Glacier (BIG 1979). Twelve profiles and 129 points were observed in its lower 20 km in the ablation zone, comprising about one-third of the total glacier. A few scattered measurements were made at higher locations. The lowest reported on the main glacier was 0.3 m $a^{-1}$ at 2,550 m and a short distance above the terminal cliff. The highest rate reported for the main ice stream was 517.5 m $a^{-1}$ at 3,435 m some 20 km above the terminus. While velocities tended to increase up-glacier, they did not change systematically. Two sections had higher velocities than profiles immediately above or below them. Velocity increased in a down-glacier direction by 6 m $a^{-1}$ between profile X and profile IX, then declined again. A similar region of accelerated flow occurred around profile III, 2.5 km above the terminus. The same probably applied to the very high velocities at Station XII, explained by the influx of ice from a huge icefall (BIG 1979, p. 966). The tributary concerned descends from the south wall and from the highest peak, Batura I (7,795 m), the steepest fall within the basin. On the tributary itself, identified as the 'Batura First Ice Flow', velocities exceeded 1,000 m $a^{-1}$. It is likely that velocities on the main glacier decline upwards of this profile but then increase again towards the ELA.

Relations between tributaries and main ice streams on all the larger glaciers show a variety of such complex interactions and there are reports of sudden changes in those relationships. These, and movement profiles generally in this steep and rugged terrain, are likely to be complicated, even if also responding sensitively to mass balance changes.

### *7.2.3 Biafo Glacier*

In 1986, lines of stakes were drilled into the ice on Biafo's main ice stream at four cross sections and at right angles to movement between 3,730 m and 4,650 m. A profile was also measured on Sim Gang tributary in the accumulation zone at 4,950 m (Hewitt et al. 1989, p. 103). The three upper profiles suggest fairly typical relations to mass balance, with highest rates at 4,650 m, the approximate ELA profile, declining at higher and lower profiles (Table 7.1). The change from a summer average of almost 0.79 m $d^{-1}$ at the ELA profile to 0.56 m $d^{-1}$ at the Baintha, 16 km further down glacier, is partly explained by widening of the glacier by over a kilometre. According to radar depth soundings, the ice is less than half as thick. However, velocities for the same period at Mango, although almost 20 km further down glacier, were slightly faster than Baintha, or 0.61 m $d^{-1}$. There is a 1.3 km narrowing between the two sites and the glacier steepens around the Mango profile. Velocities change little across the glacier in any of the profiles suggesting mainly sliding motion summer movement and a critical role for meltwater.

Due to heavy crevassing across the west/right flank of the ice stream, the ELA profile is incomplete, but a more parabolic profile suggests ice creep may be more important there. Temperatures in the firn zone imply the ice could still be cold and frozen to the bed here (Chap. 4). However, though the narrowest of the profiles, the ice is as much as 1,400 m thick, or two to three times greater than at the other profiles.

Velocities were also determined for the Biafo by Copland et al. (2009), using surface feature tracking on ASTER satellite images. The study confirmed a similar general increase in velocities for 32 km above the terminus. Highest values of more than 270 m $a^{-1}$ were found near the 1986 ELA profile, compared to 288 m $a^{-1}$ in the earlier study. A comparison was made at the Baintha profile, using images for July 2006 and June 2007 and six feature-tracking points near the 1985–1986 stakes of Hewitt et al. (1989). The 2006–2007 velocities were in the range from 111 to 145 m $a^{-1}$, meaning that differences from 1985 to 1986 were within a stated 15 m $a^{-1}$ error range. Thus, no significant change in movement could be detected after 20 years in which the terminus did retreat about 0.7 km (Chap. 12). The feature-tracking results also revealed alternating regions of somewhat higher and lower flows over distances of one or two kilometres, evidently resembling the Batura data. The stake profiles at Biafo would not have been sufficiently dense to pick this up if it was present in the 1980s. Biafo is an

**Table 7.1** Summer velocity profiles from stake surveys at selected cross sections on Biafo Glacier in 1986, from unpublished data and compilation by Cameron Wake (1987, p. 86; Hewitt et al. 1989)

| Cross-profile | Survey days | Glacier width (profile) | Stake | Movement total (m) | Movement mean daily (m) |
|---|---|---|---|---|---|
| Sim Gang (4,800 m) | 41 | 3 km (2.8 km) | 1 | 10.7 | 0.26 |
| | | | 2 | 12.9 | 0.31 |
| | | | 3 | 11.8 | 0.29 |
| | | | 4 | 9.6 | 0.23 |
| *Mean velocity* | | | | | *27.3* |
| Equilibrium line (4,680 m) | 41 | 2.1 km | 1 | 27.0 | 0.66 |
| | | (left 0.8 km) | 2 | 30.4 | 0.74 |
| | | | 3 | 32.6 | 0.8 |
| | | | 4 | 35.0 | 0.85 |
| | | | 5 | 36.5 | 0.89 |
| *Mean velocity* | | | | | *78.8* |
| Baintha Base (4,080 m) | 63 | 3.3 km (3 km) | 1 | 30.8 | 0.49 |
| | | | 2 | 39.5 | 0.63 |
| | | | 3 | 39.0 | 0.62 |
| | | | 4 | 38.0 | 0.6 |
| | | | 5 | 37.8 | 0.6 |
| | | | 6 | 37.7 | 0.6 |
| | | | 7 | 37.3 | 0.59 |
| | | | 8 | 34.2 | 0.54 |
| | | | 9 | 25.3 | 0.4 |
| *Mean velocity* | | | | | *56.3* |
| Mango (3,885) | 39 | 1.9 km (1.1 km) | 1 | 22.4 | 0.58 |
| | | | 2 | 24.1 | 0.62 |
| | | | 3 | 23.7 | 0.61 |
| | | | 4 | 23.7 | 0.61 |
| *Mean velocity* | | | | | *0.61* |

Alpine-type glacier so movement might be expected to be more consistent with mass balance-based profiles yet, in detail, it seems quite irregular.

### *7.2.4 Baltoro Glacier*

Baltoro Glacier was investigated for 1993–2008 in another use of cross-correlation feature tracking (Quincey et al. 2009). Profiles were observed over a distance of about 40 km from the terminus at 3,483 m to almost 4,900 m on the South Baltoro or Abruzzi ice stream above Concordia (4,650 m). Longitudinal profiles show an irregular rise in velocity for about 33 km, with highest values just below the exit from Concordia, and a marked decrease in movement around this great junction of major tributaries, increasing again through the highest profile. Velocities ranged from 10 m $yr^{-1}$ or less near the terminus to maxima between 145 and 200 m $yr^{-1}$.

These results can be compared with cross-valley profiles for the lower Baltoro comparing field and satellite-based tracking (Copland et al. 2009). Their velocity profile 'A-A', 5 km above the terminus, had a parabolic cross-glacier shape much like 'f1' of Quincey et al. (2009) indicating that ice creep prevails in the lowest, slowest, most heavily debris-covered section. Above this, at 'B-B' and 'f2' to 'f5', respectively, the box-like profiles of sliding motion were found, or what has been called 'block' motion (see below). This prevails from within about 8 km of the terminus through the rest of the ablation zone.

A dozen other glaciers examined in the same general area of the Karakoram, all within the ablation zones, reinforce an impression of chronic irregularities (Copland et al. 2009). Higher velocities tend to be at some distance from the termini but regions of acceleration and slowing recall the Batura and Biafo data. The highest rates appear mainly around the base of icefalls and at the entrance of steeper tributaries, rather than a well-defined relation to possible ELAs. Some anomalous regions of higher velocities occur along main ice streams, seemingly unrelated to valley form or tributaries. In some cases they occur on known surge-type glaciers such as Skamri and tributaries of Panmah (Chap. 9). The highest velocities measured were, however, one or two orders of magnitude *smaller* than peak velocities reported for Karakoram surges and barely half of those reported for the Batura icefalls. In all, substantial fluctuations in movement rates, spatially and on all timescales, suggest complex and unstable dynamics.

## 7.3 'Block' Motion

A common feature of Karakoram ablation zone margins is a sharp line of shear between moving ice and stationary moraine (Plate 7.2). The cross-glacier profiles seem very like what is called 'plug flow' (Jiskoot 2011, p. 249). However, plug flow, as defined in fluid dynamics and employed in nuclear reactors and filters, is hardly what is observed and cannot occur with glacier ice. Rather than a quasi-plastic or viscous fluid in which return flow vectors are blocked, the ice behaves as a slab with characteristics more like a landslide block (Figs. 7.1a-c).

The phenomenon has been addressed mainly in German-language studies and called *Blockschollenbewegung* (Finsterwalder et al. 1935). *Blocksbewebung* ('block movement') identifies the motion, or *Blockschollen* ('block clod'), which conveys a sense of the slab-like geometry (von Klebelsberg 1938). Block flow and examples of it in the region were identified and well established by a series of expeditions beginning almost eighty years ago (Visser and Visser-Hooft 1935-38: Finsterwalder 1937; Finsterwalder and Pillewizer 1939; Pillewizer 1969). Studies in English have paid it much less attention.

With block motion, a sharp line of shear, commonly observed at the margins, is particularly evident in a thickening ice stream but may be buried by moraine in a wasting one (Plate 7.3). On the Biafo and Baltoro Glaciers, from about 10 km above their termini, velocity varies little across the glacier (Hewitt 1998, 2005, 2007; Quincey et al. 2011). Complications arise at bends, narrows and locally steep zones, where inclined shear planes can emerge at the surface. Such areas are associated with recurring episodes of ice thickening and overriding of lateral moraines, possibly but not necessarily associated with terminus advance (Chap. 10).

**Plate 7.2** Block motion as it affects ice margins at Malangutti Glacier, Shimshal, where the ice rides up on lateral moraines along a single, steep line of shear (Hewitt 2002)

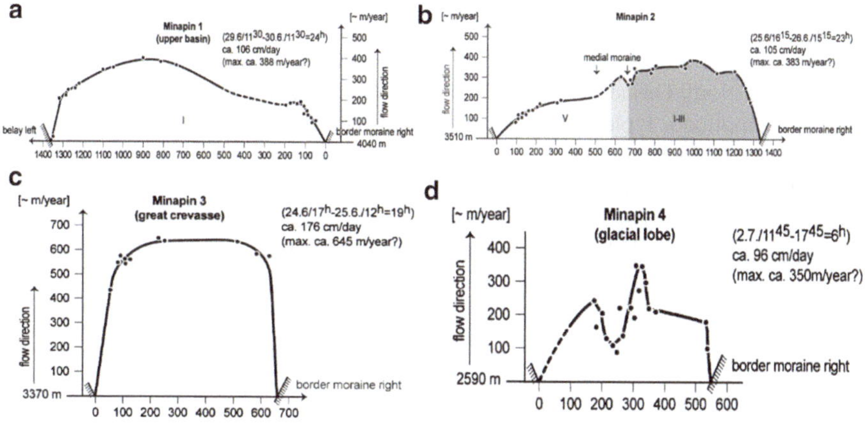


**Fig. 7.1** (**a, b, c** and **d**) Velocity profiles exhibiting block motion at three cross sections in the ablation zone tongue of Minapin Glacier, Hunza Karakoram (After Schneider 1969; cf. Pillewizer 1957)

In glaciers with block motion, thinning by some metres or tens of metres leads to stepwise and inwards change of the marginal line of shear. Slices of dead ice can be left within the lateral moraines. Marginal shear planes are partly if not wholly ice sliding against ice, rather than the moraine-ice contacts seen at the surface. In areas of heavy debris towards the termini, as at Batura, Biafo and Baltoro, block motion is less evident

**Plate 7.3** Biafo Glacier looking up valley towards Baintha from near Mango station, showing block motion lines of sheer at the ice margin related to successive episodes of thickening, mainly related to advancing right-bank tributaries up valley (Hewitt 1998)

or ceases, and movement reflects only internal deformation. Block motion may continue to the terminus as shown, for example, in feature-tracking experiments for Kyagar and North Gasherbrum Glaciers, Karakoram north slope (Mayer et al. 2006). Fast block motion is typical of surges and can also extend to the terminus (Chap. 9).

In general, block motion seems to apply over a majority of ablation zones and, with Mustagh-type glaciers, into the perennial snow zone. It can involve the highest velocities although these sections are susceptible to large seasonal and short-term fluctuations. Most icefalls exhibit block motion and it involves a large part of glacier contribution to the vertical cascade of moisture.

## 7.4 Fluctuations in Movement

Detailed measurements invariably show fluctuating rates of movement on timescales from minutes to decades. As indicated, velocities rarely follow the simple increase downwards in source zones towards an ELA or systematic decrease below that. Marked seasonal variations are widely reported, usually with higher summer rates. At Batura Glacier, movement was generally fastest from June to September (BIG 1979, p. 967). Excepting the terminal lobe, an average decrease of 20 % was found in winter velocities and as much as 42 % lower at 5 km above the terminus (2,985 m). There were other complications. Across part of profile II (2,750 m), between January and March 1975, the velocities suddenly accelerated by five times and then subsided to previous low rates. More generally, local regions of more rapid movement suggest multiple waves of accelerated movement, presumably triggered by events in the upper glacier.

For the Baintha (4,050 m) profile on Biafo Glacier, at two stakes left in place during winter 1985–1986, velocities averaged one-half to one-third less than the summer rates or between 0.22 and 0.3 m $d^{-1}$ (Hewitt et al. 1989, p. 106). For 63 days of summer in 1986, velocities at the same stakes were 0.63 and 0.60 m $d^{-1}$, respectively (Wake 1987, p. 87). It suggests that ablation season movement accounted for almost two-thirds of an annual movement of about 132 m. The seasonal difference is greater than any at Batura Glacier, perhaps because the Baintha profile is 1,000 m higher. Such seasonal fluctuations are attributed to meltwater availability affecting basal sliding (Benn and Evans 1998, p. 166).

At Baltoro seasonal, annual and inter-annual differences in velocities were also identified (Quincey et al. 2009). The greatest seasonal differences were observed around Concordia (4,100 m), with summer velocities as much as 65 % higher than winter. Above this summer velocities were generally 35 % greater (Quincey et al. 2009, their Fig. 6). A larger seasonal spread occurred in centre-line velocities than cross sections. Seasonal variations differed in the middle ablation zone compared to the lower and upper sections, and there were substantial changes from year to year. Summer velocities in 2005 were as much as 20 % greater near Concordia than in 2004. In the mid-ablation zone a 25 % increase in winter velocities was found between winter 2003–2004 and 2007–2008, while values were greatest for Concordia. The movement is described as 'distinctly 'blocky' [in] nature…with little variation in flow speed across almost the entire width of the glacier' (Quincey et al. 2009, p. 7).

The Karakoram observations recall Müller's (1958) pioneering work around Mount Everest (8,848 m). He reported an increase in summer velocities of 20–50 % for Khumbu Glacier and some others, although not in all of those measured. The largest increase occurred just beyond the base of the great Khumbu icefall and was less above it and lower down. He attributed this to meltwater entering the glacier through the extensive and heavily crevassed icefall to collect and accelerate movement below it.

Glaciers in basins of such extreme elevation and steepness are likely responding erratically because of a variety of influences. They include seasonal and short-term fluctuations in water drainage into and through the ice, avalanching and icefall activity, debris production and thermal instabilities, soft sediments at the base and the consequences of movement itself in rugged terrain. Responses are not necessarily transferred smoothly as systematic adjustments in thickness, sliding and flow velocities down the glacier. For the Karakoram, the number of surge-type glaciers is a compelling indication of flow instabilities (Chap. 9).

## 7.5 Glacier Thermal Regimes

At present the overall distribution and roles of cold and warm ice are poorly known, likewise ice temperatures at depth and in the critical basal areas. There are a few near-surface temperature measurements in shallow bore holes and snow pits. For Batura Glacier at elevation of 2,560 m in summer, the ice was at 0 °C at a depth of 8 m; warm ice commonly identified with 'temperate' glaciers (BIG 1979, p. 968). However, at 3,300 m and a depth of 13 m, the temperature was –0.8 to –0.9 °C, and at 4,500 m and 4.7 m depth, it was between −1.5 and −1.7 °C. This is probably warmer than ice in most of the source zone above 5,000 m.

Ice temperatures have usually been found to mainly reflect snow temperature in the accumulation zone, which could apply to Alpine-type glaciers in the Karakoram. At 5,450 m on Biafo Glacier, within the zone of maximum measured snowfall, the temperature at the bottom of a 20 m borehole was –4.5 °C (Wake 1987, p. 49). The annual isotherm at this elevation is probably at or below −10 °C. Warmer snow temperatures reflect higher-than-average temperatures during snow storms, the contribution of warmer summer snowfalls and percolating moisture from surface ablation refrozen in the snowpack. At higher elevations, colder snow and ice would be expected, and above 6,000 m, temperatures may approach those of polar conditions.

However, it must be emphasised that other processes intervene and affect thermal regimes in the majority of cases. Avalanches must be a decisive factor. Unlike normal accumulation, snow is derived from huge 'snow shed' areas and concentrated where it greatly exceeds precipitation amounts, by processes that affect density and probably thermal properties. Wind-redistributed snow is made much denser and may have, as yet unknown but possibly large, thermal consequences according to site, weather and season (Seppällä 2004). For glacier ice, the main influence would be on thermal and density properties of wind-moved snow that enters the avalanche stream. However, no data exist to show whether avalanched snow, descending quickly from higher elevation zones, is (always or sometimes?) colder than ambient temperatures or how far it is warmed by the friction and compression in the descent. On the glaciers are observed sandwiches of alternating winter and summer deposits, introducing further complications.

## 7.6 Icefalls and Ogives

Icefalls are among the most dynamic parts of glaciers. Rates of movement tend to be much higher and the ice heavily crevassed. Icefalls are typically swept by, sometimes smothered in, avalanched snow and ice from surrounding slopes and icefall collapses. Seasonal or secular changes in temperatures, snowfall, avalanching and rockfall debris more readily affect the thinner, heavily crevassed ice (Plate 7.4). The steepest parts have close-spaced, intersecting crevasses that generate *seracs,* unstable pillars of ice and firn liable to sudden collapse (Plate 7.5).

**Plate 7.4** Heavily crevassed icefall with collapsed seracs and broken medial moraines. The ice stream descends southwest from the 'Hunchback' (6,553 m) and 'Mitre' peaks, south spur of Masherbrum (7,821 m), a tributary of Aling Glacier (Hewitt 1999)

**Plate 7.5** Severely broken seracs and ice margins of the Sherpigang icefall below Saltoro Kangri 7,742 m. This is the source of the ogives in Plate 7.8 suggesting all is not chaotic (Hewitt 1998)

**Plate 7.6** Heavily crevassed conditions on icefalls of small glacier descending from south wall of Nangma Valley, Hushe (Hewitt 2012)

Icefalls may occur at all elevations, but tend to increase, proportionately, as altitude increases. The greater part of the descent of larger Karakoram valley glaciers occurs in upper, steep sections and icefalls. Small, high elevation ice masses often consist solely of icefall sections (Plate 7.6), the 'slope' and 'hanging' glaciers of Kotlyakov et al. (2010, F4). Along with glacier ice, icefalls become rare in the very highest and steepest terrain (Chap. 3). Icefall slopes only rarely exceed 25° and can be as little as 7° where connecting relatively gentle sections.

In the main set of 42 glaciers, only two have no icefalls on the main ice stream (Table 7.2). Only one of the others has no icefall in the ablation zone, most originating in the perennial snow zone or forming the initial parts of the main ice streams of Turkestan-type glaciers. A quarter of the icefalls descend more than 1,000 m, two-thirds more than 500 m. Alpine-type glaciers like Biafo, Siachen and Rimo do have large, low angle ice streams passing directly from accumulation zone basins into the ablation zone, but many tributaries with major icefalls descend into their ablation zones (Plate 7.7).

Relatively good topographical maps make it possible to identify most icefalls in upper sources of Biafo Glacier. There are at least 76 connected to the main glacier that are more than 0.5 km wide and descend for distances of more than 2 km. Many icefalls comprise steeper, crevassed areas around the Lukpe Lawo and Sim Gang Basins in the perennial snow zone.

Icefalls are second only to avalanching in carrying and accelerating the downslope movement of ice and snow. It is likely that more than half, possibly three-quarters, of the ice of Karakoram glaciers passes through one or more icefalls, crucial contributions to the vertical cascades of moisture and debris. They help in the concentration of ice by elevation. Where icefalls descend below the perennial snow zone, their effect on glacier areas is most evident in the development of *ogives*, transverse,

**Table 7.2** Icefall (IF) incidence on thirty Karakoram glaciers from the main set. The figures identify the top and base of the lowest major icefall on the main stream, or lowest of main connected ice streams if more than one

| Glacier | Location | Lowest IF top (m) | Lowest IF base (m) | Glacier type |
|---|---|---|---|---|
| Rimo[a] | | | | 'Alpine' |
| Biafo | 36°04′; 75°36′ | 5,300 | 4,900 | '.' |
| Chiantar | 36°44′; 73°49′ | 5,000 | 4,750 | |
| Siachen | 35°88′; 76°51′ | 5,900 | 5,300 | |
| Sarpo Laggo | 35°51′; 76°12′ | 5,800 | 5,000 | |
| N. Shukpa | 34°51′; 77°48′ | 6,300 | 5,600 | 'Mustagh' |
| N. Gasherb'm | 35°48′; 76°36′ | 6,200 | 5,300 | '.' |
| Baltoro[b] | 35°41′; 76°39′ | 5,900 | 5,100 | |
| Panmah | 35°56′; 75°47′ | 5,100 | 4,700 | |
| Batura | 36°32′; 74°38′ | 4,500 | 3,400 | |
| Khurdopin | 36°09′; 75°31′ | 5,100 | 4,300 | |
| Hispar | 36°01′; 75°30′ | 4,900 | 4,600 | |
| Chogo Lungma | 35°59′; 75°00′ | 4,900 | 4,200 | |
| Kutiah | 35°48′; 74°58′ | 5,200 | 4,000 | |
| Skamri | 36°03′; 76°01′ | 5,400 | 4,900 | |
| Kondus | 35°44′; 76°42′ | 5,100 | 4,700 | |
| Virjerab | 36°11′; 75°38′ | 5,150 | 4,850 | |
| Ghondoghoro | 35°36′; 76°22′ | 5,050 | 4,250 | |
| Bualtar | 36°07′; 74°44′ | 4,600 | 3,500 | |
| Charakusa | 35°26′; 76°29′ | 5,400 | 4,600 | |
| Braldu (Sh.) | 36°04′; 75°48′ | 5,100 | 4,900 | |
| Barpu[c] | 36°07′; 74°50′ | 4,800 | 3,500 | |
| Karambar | 36°36′; 74°14′ | 4,800 | 4,050 | 'Turkestan' |
| Hinarche | 36°07′; 74°34′ | 4,700 | 3,400 | '.' |
| Surgin[d] | | | | |
| Masherbrum | 35°34′; 76°17′ | 5,200 | 4,200 | |
| Hasanabad | 36°26′; 74°37′ | 5,000 | 3,900 | |
| Minapin | 36°09′; 74°35′ | 4,400 | 3,600 | |
| Kukuar | 36°33′; 74°15′ | 4,800 | 4,300 | |
| Toltar–Baltar | 36°29′; 74°26′ | 4,600 | 4,150 | |

[a]There are no substantial icefalls on the Rimo and none below the snowline, in this sense the most complete Alpine type

[b]Gasherbrum branch and tributary

[c]Lowest last icefall on Miar branch (see text)

[d]Surgin Glacier commences in large cones of avalanched snow and ice well below the perennial snow zone. There is no icefall on the main tongue, but many small ones in detached ice masses of the upper basin

arcuate bands, convex down valley. They commence at the base of icefalls that occur in or descend to ablation zone ice streams (Plate 7.8).

Ogives are observed in two forms: topographic undulations called *wave-ogives* with alternating ridges and troughs or *Forbes bands*, alternating lighter and darker ice (Fisher 1962; Benn and Evans 1998, p. 215). Both often occur together. They are annual features attributed to conditions uniquely associated with icefalls, although

**Plate 7.7** Steep tributaries of Biafo Glacier above Hoh Bluk (4,100 m) with icefall sections descending from west wall to the ablation zone (Hewitt 2010)

**Plate 7.8** View down on ogives forming the lower tongue of Sherpigang Glacier, EC Karakoram (Hewitt 1998)

not present in every case. Each wave or couplet of a light and dark band represents annual flow through an icefall. Longitudinal variations in ice velocity are involved and consequences of summer versus winter mass balance (Nye 1958; Waddington 1986; Paterson 1994, pp. 185–186). In winter there is little or no ablation and snow collects, especially in crevasses. This ice forms the wave crests or light bands. Summer ice is affected much more by ablation, by wind-borne dust and by mass movements from the cliffs that surround icefalls, creating darker bands. Below the icefalls ogives have an arcuate form due to differential rates of movement between the margins and centre of the glacier, often exaggerated by the way the glacier widens below icefalls.

Ogives are well marked in satellite images and found on more than half the main set of 42 Karakoram glaciers. They involve at least as large an area of ablation zones as heavy debris mantles and persist far down the glacier as an influence on albedo and dirtiness of surfaces, their relief and drainage patterns. They are obscured by heavy debris mantles.

## 7.7 Avalanche Regimes

> Nowhere else have I heard such uninterrupted avalanches' thunder.... This incessant roar provides the solution of the problem of how this long [Batura] glacier...gets its fodder.... (Visser and Visser-Hooft 1935–1938; quoted, Yafeng and Wenying 1980, p. 25)

In the previous chapter, avalanches and wind redistribution of snow were shown to be hugely important for mass balance and different elevation bands (Chap. 4). They are also critical for processes shaping the glacial environment and morphology. In the larger basins tens of thousands of avalanches must occur annually, millions over the whole region. In glacier source zones it is an all-year phenomenon, and in the Mustagh Karakoram, it must reach global extremes in scale, frequency and concentration. The Himalayan literature, both mountaineering and scientific, contains many testimonies to the extent and scale of avalanches, but this has not led to investigations of them in glacier basins. A preliminary sense of their importance is gained from area–altitude relations of rock walls and snowfall estimates.

On average, rock walls comprise about 62 % of the source zones of the 42 largest glaciers, proportionately more and even over 80 % in Mustagh- and Turkestan-type glaciers. An avalanche potential applies in virtually all of this terrain. The walls are generally steeper than 35° and predominantly over 45°; therefore, they are avalanche prone whenever snow falls. Snow will only remain in place at irregularities, if at all.

In round figures, avalanche-prone perennial snow areas of the main set of glaciers comprise some 5,000 $km^2$, half as much again for the Karakoram as a whole. The extent of these walls suggests that at least three-quarters of all snowfall supporting Karakoram glaciers descends to them in avalanches (Chap. 4). Avalanches make the larger contribution to thousands of smaller valley glaciers and minor ice masses. An added role is how huge volumes of avalanched snow form perennial aprons, cones and fans in ablation zones, especially where avalanche-fed ice streams commence.

The seasonally snow-free, off-ice areas are also mainly cliffs and comprise roughly 3,650 $km^2$ of the main set of basins. While making a small direct contribution to glacier mass, most of the snow falling in these areas is also avalanched. This intervenes in seasonal meltwater contributions from glacier basins, delaying them until later in the melt season, to judge from observations outside the glacier basins (de Scally and Gardner 1988, 1989).

The scope of larger avalanche events is indicated by a report of Yafeng and Wenying (1980). On August 9, 1974, an avalanche from Batura Mustagh cirque above Batura Glacier had a downslope displacement of 2,500 m and lasted more than 12 min. The volume of deposits was estimated to be 30,000 $m^3$. Rather than exceptional, however, they considered the event 'quite representative' (ibid, p. 26). An observation programme started in June for the same slopes reported between two and seven similarly large avalanches *daily* – four on average. Most occurred in the afternoon, usually when weather turned fine after a snowfall. Maximum vertical displacements were as much as 3,500 m, more than half the 5,225 m total relief of Batura Basin. Furthermore, another expedition in June 1959, thought to be a year of unusually heavy snowfall, reported six avalanches per day in the same area (Edwards 1960). The quote from the Vissers above refers to the same glacier.

In August 1986 and 1987, I observed similar scales and frequencies of large avalanches from the north slope of Gannish Chissh/Spantik Peak (7,030 m) onto the upper Barpu Glacier (Plate 4.7). Avalanches from both flanks of the glacier make this part of the basin completely inaccessible from mid-September through late June.

Avalanches concentrate mass by density as well as by area and elevation. The dry snow in snowstorms high up has densities of 50–200 kg $m^{-3}$. Airborne parts of dry avalanches may settle out with densities of around 400 kg $m^{-3}$, close to accumulation zone firn at depths of several metres (Wake 1989, p. 281). Concentrated avalanche deposits commonly have densities of 750 kg $m^{-3}$ or more, compared to the 830–910 kg $m^{-3}$ for glacier ice (Benn and Evans 1998, p. 68). Snowflakes are crushed and reconstituted as ice grains, air has been forced out and, when the material comes to rest, most of the density increase towards glacier ice has already occurred. Frictional and compressive warming occurs and the further an avalanche travels the greater the chances some or all the snow will reach melting point. Wet snow tends to be more compact than dry and is more likely to undergo percolation and refreezing of meltwater.

Avalanche science itself is highly developed, but mainly for hazards to mountain settlements, infrastructure and winter sports. In the Karakoram the threat to infrastructure and recreational visitors and to the widespread military presence has been investigated (Said 1998; Derbyshire and Fort 2006; Noorilhudah 2012). Observations focus on damaging events, mainly in settled areas and along highways beyond the glaciers. The implications of even heavier, all-year avalanching in glacier basins remain to be addressed.

## 7.8 Wind Regimes

Wind affects glacier conditions to some extent at all elevations but mainly around interfluves and in the uppermost areas of glacier basins where it strips, erodes, redistributes, deposits and compacts snow (Plates 4.9 and 6.3). Snow tends to concentrate on lee slopes. The more obvious consequences are cornices, but the total mass of snow falling out of wind fields to slide and fall down steep walls is probably much greater. In the highest parts of the Karakoram the extent and maze-like character of interfluves, mean losses on windward flanks can be made up on leeward slopes of the same basins. As already noted, in the lesser ranges and those Inner Asian ranges along the Karakoram northern flanks, wind action favours and supports greater, or only, lee-slope glaciers.

There are no actual measurements to predict just how much of snowfall is affected by wind. Satellite images for the Central Karakoram offer a way to explore the extent of wind-modified features along crest lines. Wind-scoured and moulded windward slopes, crest line cornices and, on lee slopes, cornice glaciers record wind effects. The width of the belt with such features was measured at 25 sites with interfluves above 5,500 m and including some above 7,500 m in the K2 Massif. It was found to average about 0.3 km around 5,500 m, increasing to over 2 km above 8,000 m. The approximate length of interfluves above 5,500 m averages 0.4 km per km$^2$, probably amounting to some 8,000 km for the whole range. Total interfluve length does diminish sharply upwards. Assuming a minimum average width of strongly wind-affected interfluves of 0.5 km, this would comprise a total of some 4,000 km$^2$ – twice the combined ablation zone areas of the main set of 42 valley glaciers, and five times their combined heavy debris-covered areas – mentioned because they have received so much more attention. These wind effects are occurring in the zone of highest precipitation. They must, in fact, extend much further out, but visible effects are truncated by the avalanche activity in the transects examined. Meanwhile, wind action is a critical influence on avalanche preparation.

At lower elevations, valley wind systems develop along the glaciers and influence conditions in ablation zones. Most of the snow falling here is redistributed to fill hollows and pile up on lee flanks of medial moraines and among heavier debris covers. A highly irregular and compacted cover results.

One way to think of the roles of wind and avalanches is as a series of 'line-ups' or vertical 'queues' that redistribute, store and release snow mass in the downslope cascade. There are line-ups for wind-redistributed snow, mainly in the uppermost areas as temporary storage in cornices and couloirs, and longer term in minor, disconnected ice masses at breaks of slope. Major avalanches gather snow high up and quickly pass through 'check out' to deposit it where glacier ice will form. A series of other avalanche line-ups arise where snow is temporarily stored in chutes and along glacier margins or ends up in off-ice basin areas. Seasonal weather can speed up or retard the through-flow.

## 7.9 Inputs and the Concentration of Glacier Cover

A striking result of these conditions is the relative smallness of the glacier ice area, but large contributing areas off-ice. The ratio of the accumulation zone to perennial snow and ice area reflects how important wind and avalanche action in source zones are to glacier nourishment (Table 7.3). A useful indicator is to divide the entire source area by the conventional accumulation zone, accepting it hardly exists in Turkestan-type glaciers (Table 7.4). This provides an *input concentration ratio* or ICR which clearly distinguishes glaciers according to nourishment type. Values are two to ten times larger in Mustagh types than Alpine types and seven to thirty times larger in Turkestan types. The ICR is highly correlated with the relative extent of rock walls in the perennial snow zone; in effect, they measure the same thing.

## 7.10 Concentration of Critical Glacier Hydrological Factors

Processes that concentrate hydrological and dynamic functions in space and time are characteristic of these mountain glaciers and potentially critical for their survival (Hewitt 2011). Only one significant spatial concentration is usually noted in valley glaciers described in the literature; that ablation zone areas tend to be smaller than accumulation zones – which, strictly speaking, is rarely the case for Karakoram glaciers. Water yields from ablation zones are several times greater than the highest, let alone the average, precipitation and, typically, increase towards the terminus.

**Table 7.3** Ratios of zones in glacier basins to their perennial snow and ice (PSI) area for selected Karakoram glaciers. It highlights how the accumulation zone ratio is the most useful indicator of difference or glacier type and of the actual importance of the entire source zone for glacier maintenance

| Glacier | Basin area $km^2$ | PSI $km^2$ (%) | Source Z $km^2$ (%) | Glacier area $km^2$ (%) | Accum. zone $km^2$ (%) | Ablation zone $km^2$ (%) | Glacier type |
|---|---|---|---|---|---|---|---|
| Biafo | 855 | 645 (100) | 515 (80) | 540 (84) | 335 (52) | 135 (21) | Alpine |
| Rimo | 612 | 480 (100) | 390 (81) | 400 (83) | 315 (66) | 85 (18) | Alpine |
| Baltoro | 1,270 | 910 (100) | 610 (67) | 535 (59) | 180 (20) | 260 (29) | Mustagh |
| Batura | 710 | 535 (100) | 295 (55) | 260 (49) | 90 (17) | 170 (32) | Mustagh |
| Kondus–K. | 490 | 365 (100) | 280 (77) | 130 (36) | 35 (10) | 95 (26) | Mustagh |
| Toltar–B. | 205 | 150 (100) | 85 (57) | 49 (33) | 3 (2) | 46 (31) | Turkestan |
| Barpu | 198 | 65 (100) | 36 (55) | 32 (49) | 4 (6) | 30 (46) | Turkestan |

**Table 7.4** Input concentration ratios (ICRs) for selected glaciers

| Glacier | Source Z $km^2$ (%) | Accum. Z $km^2$ (%) | ICR | Glacier type |
|---|---|---|---|---|
| Rimo | 390 (81) | 315 (66) | 1.2 | Alpine |
| Biafo | 515 (80) | 335 (52) | 1.5 | Alpine |
| Batura | 295 (55) | 90 (17) | 3.3 | Mustagh |
| Baltoro | 610 (67) | 180 (20) | 3.4 | Mustagh |
| Barpu | 36 (55) | 4 (6) | 9.0 | Turkestan |
| Toltar–B. | 85 (57) | 3 (2) | 28.3 | Turkestan |
| Kondus–K. | 280 (77) | 35 (10) | 35.0 | Mustagh |

In the Karakoram it was shown that the largest spatial concentrations occur in inputs, mainly from the intervention of avalanche and wind action. Ablation measurements revealed a further concentration of losses in mid- to upper ablation zones and a reversal lower down where debris protects the ice. The combination of short ablation seasons higher up and debris-protection lower down conserves the ice cover but concentrates and intensifies meltwater yields. Some 90 % of annual ice ablation occurs in 2 months or less, leading to extreme concentrations of runoff.

It will be useful to conclude by again summarising conditions that concentrate glacier ice and related processes:

1. Snowfall concentration: an orographic effect with the zone of maximum precipitation, roughly 4,800–5,800 m in glacier accumulation zones, and with 1,000–2,000 mm water equivalent compared to 150–300 mm at valley weather station.
2. Storm-dominated precipitation: a temporal or event effect of Monsoon and winter depression storms (Barros et al. 2009).
3. Avalanche nourishment: a terrain or ruggedness effect transporting, compacting and concentrating snow at lower elevation, mainly nourishing glaciers between 6,000 and 4,500 m.
4. Icefalls: accelerating down-valley transfer of ice mainly between 6,000 and 4,000 m.
5. Concentrated ablation: mid-ablation zones where the most extensive exposed ice areas occur and thin or scattered debris enhance ablation, mainly 3,800–4,800 m.
6. Ablation suppressed by heavy debris covers, mainly below 3,800 m and down to the lowest termini at about 2,500 m.

## References


Barros AP, Chiao S, Lang TJ, Burbank D, Putkonen J (2009) From weather to climate-Seasonal and interannual variability of storms and implications for erosion processes in the Himalaya. In: Willett SD, Hovius N, Brandon MT, Fisher DM (eds) Tectonics climate, and landscape evolution. Geol Soc Am Special Pap. vol 398, pp 17–38

Benn DI, Evans DJA (1998) Glaciers and glaciation. Arnold, London

BIG (Batura Investigations Group) (1979) The Batura Glacier in the Karakoram Mountains and its variations. Sci Sin 22(8):958–974

Copland S, Bishop MP, Shroder JF Jr, Clendon P, Bush A, Kamp U et al (2009) Glacier velocities across the central Karakoram. Ann Glaciol 50(52):41–49

de Scally FA, Gardner JS (1988) The hydrological importance of avalanche snow, Kaghan Valley, Himalayan range, Pakistan. In: Proceedings of the international science workshop, Whistler, October

de Scally FA, Gardner JS (1989) Evaluation of avalanche mass determination approaches: an example from the Himalaya, Pakistan. J Glaciol 40:248–252

Derbyshire E, Fort M (2006) Geomorphology and mountain hazards of the Hunza Valley. In: Kreutzmann H (ed) Karakoram in transition: culture, development and ecology in the Hunza Valley. Oxford University Press, Oxford, pp 73–95

Drewry D (1986) Glacial geologic processes. Edward Arnold, London

Edwards JI (1960) The Batura Mustagh expedition 1959. Alp J 65(300):48–52
Finsterwalder R (1937) Die Gletscher des Nanga Parbat, Glaziologische Arbeiten der Deutschen Himalaya-Expedition 1934 und ihre Ergebnisse. Z Gletsch 25:57–108
Finsterwalder R, Pillewizer W (1939) Photogrammetric studies of glaciers of High Asia. Himalayan J 11:107–113
Finsterwalder R, Raechl W, Misch P, Bechthold F (1935) Forschung am Nanga Parbat: Deutsche Himalaya-Expedition 1934. Hannover
Fisher JE (1962) Ogives of the Forbes type on Alpine glaciers and a study of their origins. J Glaciol 4(34):5361
Goudie AS, Jones DKC, Brunsden D (1984) Recent fluctuations in some glaciers of the Western Karakoram mountains, Hunza, Pakistan. In: Miller KJ (ed) International Karakoram Project, vol 2. Royal Geographical Society, London, pp 411–455
Haeberli W, Hoelzle M (1995) Application of inventory data for estimating characteristics of and regional climate-change effects on mountain glaciers: a pilot study with the European Alps. Ann Glaciol 21:206–212
Hewitt K (1967) Ice-front sedimentation and the seasonal effect: a Himalayan example. Trans Inst Br Geogr 45:93–106
Hewitt K (1998) Glaciers receive a surge of attention in the Karakoram Himalaya. EOS Trans Am Geophys Union 79(8):104–105
Hewitt K (2005) The Karakoram anomaly? Glacier expansion and the 'elevation effect', Karakoram Himalaya, Inner Asia. Mt Res Dev 25(4):332–340
Hewitt K (2007) Tributary glacier surges: an exceptional concentration at Panmah Glacier, Karakoram Himalaya. J Glaciol 53(181):181–188
Hewitt K (2011) Glacier change, concentration and elevation effects in the Karakoram Himalaya, upper Indus Basin. Mt Res Dev 31(3):1–13, doi:http://dx.doi.org/10.1659/MRD-JOURNAL-D-11-00020.1
Hewitt K, Wake CP, Young GJ, David C (1989) Hydrological investigations at Biafo Glacier, Karakoram Himalaya, an important source of water for the Indus River. Ann Glaciol 13:103–108
Jiskoot H (2011) Dynamics of Glaciers. In: Singh VP, Singh P, Haritashaya UK (eds) Encyclopaedia of snow, ice and glaciers. Springer, Dordrecht, pp 245–256
Kotlyakov et al (2010) Glaciers of the former Soviet Union. In: Williams RS, Ferrigno JG (eds) Satellite image atlas of glaciers of the world: Asia, Professional paper 1386-F. U.S. Geological Survey and U.S. Government Printing Office, Washington, DC
Luckman A, Quincey DJ, Bevan S (2007) The potential of satellite radar interferometry and feature tracking for monitoring flow rates of Himalayan glaciers. Remote Sens Environ 111(2–3):172
Mayer C, Lambrecht A, Belo M, Smiraglia C, Diolaiuti G (2006) Glaciological characteristics of the ablation zone of Baltoro glacier, Karakorum, Pakistan. Ann Glaciol 43:123–131
Menzies J (1995) The dynamics of ice flow. In: Menzies J (ed) Modern glacial environments: processes, dynamics and sediments, vol 1., pp 139–196
Müller F (1958) Eight months of glacier and soil research in the Everest region. In: The mountain world 1958/59. Swiss Foundation for Alpine research. Harper, New York, pp 191–208
Noorilhudah (2012) Cover story: into the valley of death. Newsline, Karachi, May 18–23. http://www.newslinemagazine.com/2012/05/into-the-valley-of-death-the-siachen-conflict
Nye JF (1958) Surges in glaciers. Nature 181:1450–1451
Paterson WSB (1994) The physics of glaciers, 3rd edn. Pergamon, New York
Pillewizer W (1957) Bewegungsstudien an Karakorum-Gletschern. Petermanns Geographische Mitt 262:53–60
Pillewizer W (1969) Untersuchungen an Blockströmen der Ötztaler Alpen. Abh Geogr Inst d FU Berlin 5:37–50
Quincey DJ, Copland L, Mayer C, Bishop M, Luckman A, Belo M (2009) Ice velocity and climate variations for the Baltoro Glacier, Pakistan. J Glaciol 5(194):1–11

Quincey DJ, Braun M, Glasser NF, Bishop MP, Hewitt K, Luckman A (2011) Karakoram glacier surge dynamics quantified by cross-correlation feature tracking. Geophys Res Lett 38(18): L18504. doi:10.1029/2011GL049004

Said M (1998) Natural hazards of Shigar Valley, northern areas of Pakistan. In: Stellrecht I (ed) Karakorum-Hindu Kush-Himalaya: dynamics of change. Rudiger Koppe Verlag, Koln, pp 251–268

Schneider HJ (1969) Minapin – Gletscher und Menschen im NW-Karakorum. Die Erde 100:266–286

Seppällä M (2004) Wind as a geomorphic agent in cold climates. Cambridge University press, Cambridge

Visser C, Visser-Hooft J (1935–1938) Wissenschaftliche Ergebnisse der niederländischen Expeditionen in den Karakorum und die angrenzenden Gebiete in den Jahren 1922, 1925 und 1929–30. E.J. Brill, Leiden

von Klebelsberg R (1938) Die Zusammensetzung des Talgletschers. Z Gletschkd Glazialgeol 26:22–43

Waddington ED (1986) Wave ogives. J Glaciol 32(12):325–334

Wake CP (1987) Spatial and temporal variation of snow accumulation in the central Karakoram, northern Pakistan. Thesis, Wilfrid Laurier University, Waterloo

Wake CP (1989) Glaciochemical investigations as a tool for determining the spatial and seasonal variation of snow accumulation in the central Karakorum, northern Pakistan. Ann Glaciol 13:279–284

Yafeng S, Wenying W (1980) Research on snow cover in China and the avalanche phenomena at Batura Glacier, Pakistan. Ann Glaciol 13:289–293

# Chapter 8
# Landforms of the Karakoram Glacierised Areas

**Abstract** The geomorphology of Karakoram glacier basins involves high mountain terrain and surface processes responding to exceptional gravitational forces. Relations between high elevations, steepness, great relief, strong seasonal variations and their vertical migration help to organise the landscape. There are distinct sets of landforms in different elevation bands and zones of transition between them. An approach is developed in terms of landsystems, highlighting assemblages of processes and forms in the various zones. The analysis moves broadly downwards, beginning in glacier source zones where perennial sub-zero or frigid conditions prevail. Rugged interfluves and rock walls are picked out by wind- or avalanche-generated snow forms. Primary erosion of rock is shown to be fierce and former notions of the survival of ancient erosion surfaces in accordant summits unlikely. Below is a second transitional zone where the most extensive glacier surfaces form. Main ice streams are initiated by the run-out and deposits of avalanches, by icefall activity and, to a more limited extent, in broad accumulation zones. On-ice ablation zone landscapes are treated as a third system responding especially to seasonal rhythms. The neglected shoulder seasons are underlined, when freeze–thaw and wind action modify glacier surfaces. In the better-known summer conditions, the varieties of debris-covered ice assume great importance, including medial and lateral moraines, cryoconite forms, facets, on-ice ponds and streams. An off-ice periglacial zone compliments the ablation zone. Here, seasonal conditions, especially freeze–thaw, are again key to the scope and timing of surface processes. Near and beyond glacier margins, there is permafrost with a seasonal active layer and a wide variety of talus forms, hummocky and patterned ground. Though small in relative area, the lateral and terminal margins of the glaciers, with complex sediment assemblages and valley-side troughs have received the fullest attention of researchers. They are treated as a further landsystem. Effects of the cold season advance of termini are noted. There are distinctive developments related to high debris loads from the glaciers and surrounding slopes. Most larger glaciers have substantial areas of pro-glacial outwash sediments, mainly controlled by highly concentrated meltwater yields in a few weeks or months of summer. These, and other processes

K. Hewitt, *Glaciers of the Karakoram Himalaya: Glacial Environments, Processes, Hazards and Resources*, Advances in Asian Human-Environmental Research,
DOI 10.1007/978-94-007-6311-1_8, 

affecting the glaciers or their margins although mainly operating beyond them, introduce compound or transglacial landforms. They are important links in a web of processes and developments that serve to configure landforms and cryosphere relations to climate and tectonics.

**Keywords** Verticality • Seasonality • Arêtes • Icefalls • Avalanche deposits • Chutes • Medial moraines • Cryoconite forms • Landform assemblages • Transglacial forms

## 8.1 Glacial Landscapes

Landforms record the action of earth surface processes that may expose bedrock, transport or deposit rock materials, shape sets of terrain features and control the pace of landform evolution, the field of geomorphology. The Karakoram is identified, above all, by the scale of its high mountain terrain, exceptional number of high peaks and the glaciers that surround them. Much of the existing work on the glaciers is devoted to their geomorphology (Paffen et al. 1956; Hewitt 1968a, 1993; Miller 1984; Searle 1991; Meiners 1998). However, few studies offer a comprehensive view of present-day glacial landscapes. It is almost 80 years since anyone attempted an extended study of Karakoram glacier landscapes (Dainelli and Marinelli 1928; Visser and Visser-Hooft 1935–1938). To do it justice now would exceed the scope of this study, but a perspective on the glacial geomorphology is needed, in particular for landforms in glacier basins and their neglected upper areas.

The greatest interest to date has been in depositional features around and beyond today's glacier margins and the unravelling of glacier changes in the Quaternary (Owen 1988, 2006; Owen and Derbyshire 1993; Kalvoda 1992; Shroder 1993; Kamp and Haserodt 2004; Iturrizaga 2006). These topics do, perhaps, have an importance out of proportion to their small share of glacier basin areas and in helping unravel wider and longer-term developments. Nevertheless, a more balanced sense of what needs consideration can be deduced from the terrain classes used to describe the 42 largest glaciers (Table 8.1). For instance, one-third of their basin areas consist of glacier ice. It seems important to pay attention to glacier morphology, the varieties and distributions of on-ice features and processes, as well as how they relate to developments around and beyond the ice. Over one-third of the basins consist of steep slopes, mainly rock walls, again in need of fuller recognition and in relation to glacier activity and landscape development.

## 8.2 Verticality Relations and Landsystems

Verticality reveals the organisation of geomorphic activity and landforms in two major ways (Fig. 8.1). The thousands of metres that Karakoram glacier basins span mean that characteristic sets of landforms occur in distinct elevation zones, mainly reflecting the interrelations of thermal and moisture regimes (Hewitt 1993). Secondly,

**Table 8.1** Terrain types and their share of the 42 largest Karakoram glacier basins (see Sect. 2.3)

| Terrain type | Area (km²) | % Area |
|---|---|---|
| *Basin area (draining to main glacier terminus)* | 15,540 | 100 |
| *Perennial frigid snow and ice cover incl. abl. zone* | 11,530 | 74 |
| *Glacier cover* | 5,790 | 37 |
| *Main connected glaciers* | 5,075 | 33 |
| *Source zones of main glaciers (perennially frozen)* | 7,910 | 51 |
| (a) Firn basins and ice streams | 2,475 | 16 |
| (b) Steep rock walls | 4,875 | 31 |
| *Snow free (off-glacier but draining to glacier terminus)* | 3,950 | 25 |
| (a) Rock walls | 3,630 | 23 |
| (b) Ice-margin complex | 280 | 2 |
| *Steep rock walls (total)* | 9,460 | 61 |

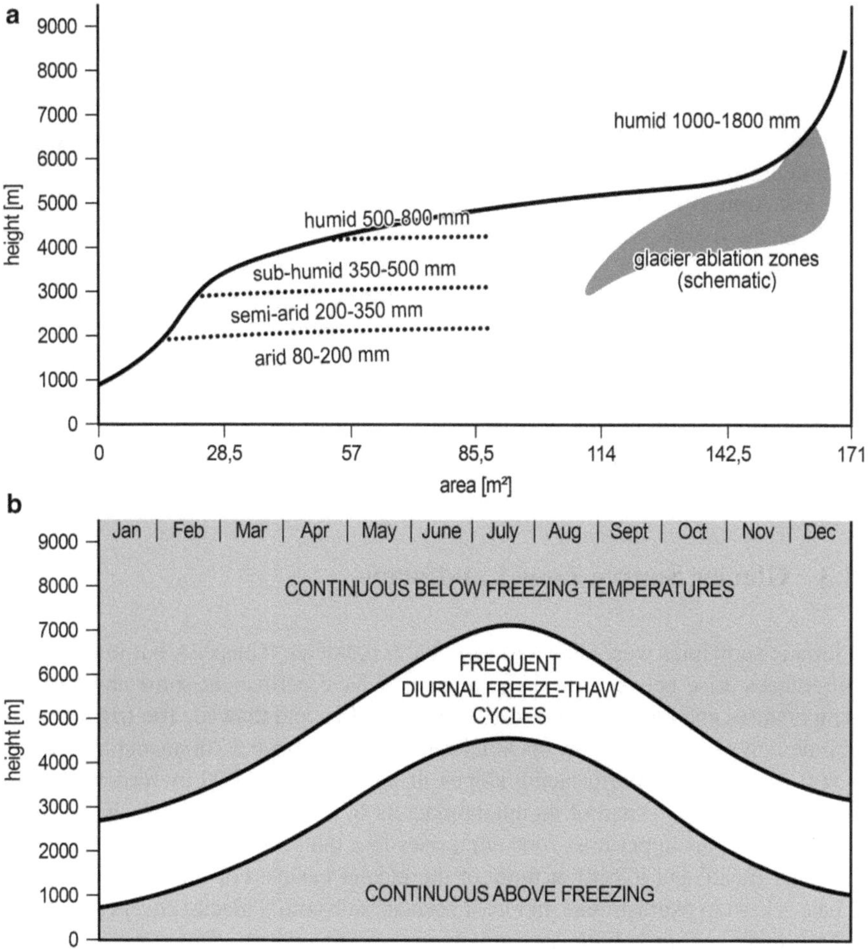


**Fig. 8.1** Verticality relations of glacier basins dependent on key orographic and seasonal climatic conditions (After Hewitt 1968b, 1993): (**a**) Broad vertical climatic zones of the Karakoram overlaid on a hypsometric curve of the upper Indus Basin (**b**) Vertical migration of diurnal freezing and thawing temperatures and frequent freeze–thaw cycles

*transitional* systems follow from strong gravitational forces, especially the downslope cascade of moisture and debris between and across elevation zones and the margins of ice masses. Seasonal conditions and their vertical migration are a major influence on the forms of snow and ice. In some elevation zones, cycles of temporary, transitional developments in each season, notably those involving freeze–thaw, are keys to broader developments and potential climate responses (Troll 1954).

A basis for identifying key conditions, and the combinations of landforms, will be explored mainly in terms of the *landsystems* concept (Evans 2003). Valley glaciers were recognised early on as having distinct types of landsystem, a view pioneered in the Karakoram by Owen and Derbyshire (1993). The approach has been directed mainly at sediment bodies, how they arise from ablation zone conditions and at existing or former ice margins (Eyles 1983). Landsystems types have been based especially on the depositional legacies of past glaciations, notably from the Pleistocene ice sheets of North West Europe and North America. An important concern has been to identify properties relevant to subsurface engineering soils and aggregate resources (Boulton and Eyles 1979). However, the larger parts of the Karakoram glacier basins, their vast middle and upper areas, have few or no visible sediments. Those emphasised in valley glacier landsystems elsewhere; indeed, the few formerly described for the Karakoram involve a tiny fraction of existing basin areas, if somewhat more broadly where active, debris-covered ice has been considered. Otherwise, as noted, these landscapes are dominated by bedrock exposed in steep cliffs and by snow and ice forms. Much of the larger parts of the basins have no depositional features, or only thin, ephemeral, or mobile debris. With this in mind, Karakoram landsystems are explored in terms of broad zonal landsystems, mainly identified by elevation, transitional ones where typical forms represent gradients or interfaces between zones and others where temporal or seasonal rhythms dominate landforms and processes.

## 8.3 Glacier Source Zone Landforms

Climatic snowlines were shown as difficult to delineate (Chap. 1), but fundamental differences arise between the large areas that have permanent snow and freezing temperatures and those that are seasonally snow free and thawed. The frigid zone in Karakoram glacier basins occurs almost everywhere above 5,700 m and from about 4,600 m on steeper north-facing slopes. It spans 2,000–3,000 m vertically, more than even the entire span of mountain glaciers in most other regions. Hypsometric curves show this uppermost zone comprises less than 5 % of the Upper Indus and Yarkand Basins but a third or more of the glacier basins. For the latter, its significance is hard to overestimate, not least because substantial glacier covers *only* occur where there are such high, perennially frozen watersheds. However, glacier ice itself, at these elevations, is generally confined to small and minor forms clinging to steep slopes, trapped in local irregularities and disconnected from main glaciers (Plate 8.1). Snow and ice forms are of special interest, but the prevailing landforms

**Plate 8.1** North face near summit of Gannissh Chissh or Spantik (7,027 m) above upper Barpu Glacier, showing prevalence of rock walls and diverse minor ice masses and wind-scoured snow (Hewitt 1986)

are steep rock walls. Direct snowfall rarely remains on the ground but is stripped away, reworked and redeposited by wind and avalanche. The steep walls may be veneered by snow and ice but are almost entirely swept by avalanches that descend to glacier surfaces and by wind action.

### *8.3.1 Rock Walls in the Perennial Snow Zone*

In the 42 largest basins, rock walls comprise, on average, 37 % of frigid source zones. In half they exceed 60 % and, in some cases, over 80 %. Because they are so steep, the true surface area is two or three times greater, increasing their presence in, and impact on, the landscape.

The precipitous walls culminate above in blade-like interfluves, picked out by snow cornices and aprons on lee flanks. Bedrock is everywhere at or close to the surface and exposed in vast cliffs. Veneers of snow and ice tend to cover even the steepest walls above about 6,000 m, lower on north-facing walls. All-year avalanching is the rule and, in places, extensive areas of avalanche fluting. Little is known about the conditions but rime formation may occur as well as wind packing. Conceivably, frictional heating on avalanche sliding planes and more or less

**Plate 8.2** Upper Panmah, Choktoi Glacier and Latok I (7,145) massif (Hewitt 2005) to illustrate: *Karakoram Valley Glacier Landsystems I: Frigid source zones*: *1* Interfluves: arêtes, peaks, spires and notches. *2* Wind regime forms: cornices, wind scoured, wind slab. *3* Avalanche regime: rock walls, flutings, chutes and couloirs, depositional cones. *4* Glaciers: (disconnected) cornice, niche, hanging, cirque-valley, fall (main). *5* Lower section passes into Transition Zone Landsystem 1 (see below)

immediate refreezing may help create icy veneers. Fierce winds are also a factor. In any case, this could be called the definitive *Himalayan landsystem* of the region, its true 'abode of snow' (Wilson 1875). It is present and necessary wherever the larger valley glaciers occur if most of the glacier ice forms lower down (Plate 8.2).

Avalanches are likely wherever snow falls on slopes greater than at least 30°. The larger, temporary build-ups and mass failures usually occur between about 45 and 50° (Perla 1978). Steeper slopes lose snow sooner and faster. This may limit avalanche size although, in the Karakoram, there are such great vertical spans that snow caught in narrow chimneys, ledges and other local breaks of slope can be swept up in smaller, initiating movements to create much larger events.

High, near-vertical cliffs testify to great rock resistance or 'strength equilibrium slopes' (Selby 1993, p. 103). They are, nonetheless, subject to extreme stresses. In active tectonic zones rock that was compressed deep within the Earth's crust contains varying amounts of conserved elastic strain. Further stresses develop as it is exhumed and exposed in steep slopes. Primary mechanical responses to unloading and stresses that arise in unsupported walls prepare rock for removal, whether by spontaneous collapse in rockfalls and rockslides or by avalanches.

In this zone, permafrost exists at and below the surface. Its influence on rock walls where melting and refreezing do not occur is unclear. It may only be important

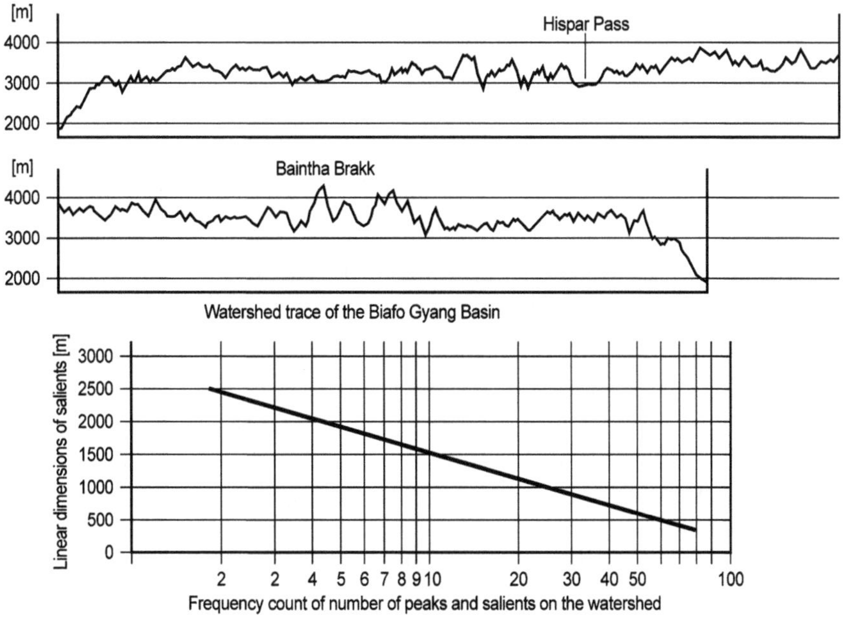


**Fig. 8.2** Watershed trace of the Biafo Glacier basin (After Hewitt 1968a)

in response to seismic events (see below). Where steep walls reach into warmer zones below, freeze–thaw assumes an important role, and permafrost may be critical for geomorphic responses to climate warming (Haeberli and Gruber 2009). Again there are no data for the Karakoram.

## 8.3.2 *Interfluve and Peak Morphology*

An exception to the relative neglect of glacier source zone geomorphology involves long-held notions about 'accordant summits' or 'peak plains', including the highest Himalayan watersheds (De Terra 1932; Paffen et al. 1956; Derbyshire et al. 1984, p. 460). However, establishing their validity is not easy or, if they exist at all, explaining how they could survive. The Biafo Glacier watershed can serve to illustrate the problems.

From high-level viewpoints, watersheds in and around Biafo Basin can create an impression of plateau-like accordance (Hewitt 1968a). The watershed trace has less relief than appears in slopes from the interfluves to glacier surfaces or for the whole basin (Fig. 8.2). Even so, the trace rises and falls 500–1,500 m over short distances; equal to, say, maximum available relief of the British Isles and in a much smaller space. This occurs along jagged, knife-edge interfluves above narrow, blade-like ramparts. As shown earlier, mountains like Baintha Brakk (7,285 m) or the Latok

Peaks – Latok I is 7,145 m – appear massive from some viewpoints but turn out to be very narrow from others (Plate 2.7). In general, these measurements hardly support the notion of accordance in summit elevations.

Meanwhile, the survival of ancient, high-level erosion surfaces seems unlikely unless erosion rates are extremely low (Finlayson et al. 2002; Ollier 2004). The opposite seems to be the case, given the quantities of angular rubble coming from these sources to glacier ablation zones. Then there is geological evidence of great vertical uplift *and* equally high rates of denudation. Rock cooling histories suggest 10–20 km of unroofing in the past 10–15 million years (Searle 1991). In such conditions, how could ancient landscape surfaces be preserved?

An alternative explanation is Penck's (1919) views on such 'Peak plains' or 'Gipfelflur' (Price 1981, p. 229). It explains interfluve forms in terms of depth of valley dissection and channel-to-ridgeline intersection of opposing slopes through systematic slope recession. Even in high mountains, channels and interfluves tend to be organised with roughly uniform spacing, leading to relatively consistent interfluve levels. However, they are a response to ongoing erosion rather than preservation from ancient landscapes. Just how such high, rugged interfluves erode is not well understood. Recently a mechanism has emerged that seems particularly important in the Karakoram.

### 8.3.3 Massive Rock Slope Failures on Mountain Walls

Failure or primary break-up on rock slopes mainly involves relatively small events, giving rise to loosening of material that may be entrained by avalanches or rock fall (Whalley 1984). Given the near absence of freeze–thaw in the frigid source zones and that surface processes mainly erode through the debris they carry with them, primary rock break-up is probably the origin of most debris carried to the glaciers. Large events will be rarer but include massive rock slope collapses. These can carry off large sections of rock wall and generate huge rockslides or rock avalanches (Voight 1978; Evans and DeGraff 2002). They are one of the few, perhaps the only, ways in which significantly large and geometry-changing erosion can occur on steep, frigid cliffs and interfluves (McSaveney 2002).

Until the late twentieth century, few such events had been documented anywhere and none in the Karakoram. That has changed dramatically worldwide. In the Karakoram and Nanga Parbat Ranges since 1985, eleven such failures have been reported from glacier basins (Hewitt et al. 2011a, b). At least six of these originated partly or wholly in frigid source zones (Table 8.2).

This may not seem very much until placed in the time frames for supposed ancient 'peak plains' – millions to tens of millions of years. The numbers of similar prehistoric events support their occurrence throughout the Holocene, and many more must remain to be found (Plate 8.3). If events in the past 25 years are at all representative, they suggest more than 40 massive rock slope failures per century could occur in glacier basins and at least half in the frigid zone. This would translate

**Table 8.2** Massive rock slope failures (MRSFs) identified in glacier basins in the Karakoram and Nanga Parbat Ranges (Updated after Hewitt et al. 2011b). Several recent events were identified on satellite imagery but the prehistoric count applies only to deposits examined in valleys visited, a tiny fraction of all glacierised valleys where such events may occur

| MRSFs | Dates | Number | Frigid zone? |
|---|---|---|---|
| Regional | Prehistoric | 24 | ? |
| Regional | Historic | 12 | 6 |
| Bualtar Glacier | 1890s | 1 | – |
| Bualtar Glacier | 1986 | 3 | – |
| Masherbrum Glacier | 1988 (?) | 1 | Yes |
| Lokpar–Aling Glacier | 1990 (?) | 1 | Yes |
| Chillinji Gl.–Karambar | 1991 | 1 | Yes |
| Buldar Glacier NP | 2002 | 1 | – |
| North Terong Glacier | 2002 | 1 | Yes |
| Chhichi Glacier NP | 2005 | 1 | – |
| KhaiKhai Glacier Baltoro | 2008 | 1 | Yes |
| E. Ching Khang Glacier | 2008 (?) | 1 | Yes |

**Plate 8.3** Gannissh Chissh or Spantik Peak (7,027 m) and a tributary of the upper Barpu Glacier is the site of a prehistoric catastrophic rock slope failure. The breakout zone is on the peak in the upper, centre of the photograph. It generated a rock avalanche that descended down the glacier for almost 11 km to the right of the picture. Remnants of the rock avalanche, composed of the distinctive crystalline limestone from the peak, cover the terrace in the foreground along the left flank of the glacier (Hewitt 2002)

into as many as 4,000 (2,000) in the 10,000 years of the Holocene. Climate warming and permafrost degradation can lead to such collapses (Haeberli and Gruber 2009). This may apply to some of the recent massive failures that occurred partly in the seasonal freeze–thaw zone and could inflate numbers there. However, it hardly seems to apply in the upper perennial snow and ice zone, where collapses identified in the Little Ice Age and widespread prehistoric examples indicate the phenomenon has continued throughout the Holocene at least. In fact, earthquakes must be considered the more likely cause of fracture and collapse of high mountain slopes. They are the most common triggers of massive rock slope failures, with or without the presence of permafrost or its degradation (Keefer 1984; Bull 2007, Chap. 6). Several examples, including the Buldar 2002 event on Nanga Parbat, occurred in relatively small earthquakes. In the past two centuries, however, much of the Karakoram has been relatively quiet, seismically. Other evidence points to rarer Himalayan mega-quakes affecting the region and occurring on 500–1,000 year time frames (Feldl and Bilham 2006). If so, this could greatly increase long-term landslide frequencies. The possibility of much higher frequencies than appear in the recent past is indicated by the dozens of large rock slope failures in the magnitude 9.0 1964 Alaskan earthquake (McSaveney 1978). In general, *under* estimation of such landslides is most likely, given the combination of difficult observing conditions and rapid burial or removal of them in glacier basins.

Of particular significance for landform evolution in these highest areas are the events involving modification of interfluves. The Ghomboro complex in Braldu valley provides an example. It is classed as a *megaslide*, when mass exceeds 1 $km^3$, the original volume in this case estimated as greater than 3 $km^3$ (Hewitt 2013, p. 340). Landslide run-out is entirely in the Braldu River valley, but the detachment zone above cuts into the opposing slope by as much as 0.5 km. This happens to be part of the watershed of Biafo Glacier. It has been displaced by tens to hundreds of metres along several kilometres of the interfluve. It captures a part of the glacier catchment and indicates how watersheds and peaks can be modified by such events (Hewitt 2006b, p. 543; Seong et al. 2007). The uppermost north face of Spantik Peak (7,027 m) was similarly modified and displaced by the prehistoric Gannish Chissh rockslide–rock avalanche (Hewitt 2002a, p. 368) and also the east face of Buni Zom (6,550 m) in the Hindu Raj, by the prehistoric Miragram–Parwak event (Hewitt 2002b, p. 72). Here is a rare but recurring major interfluve-modifying mechanism that adds further doubts about the survival of ancient, peak level landforms in glacier basins.

## 8.4 Transitional Landsystem I: Where Main Ice Streams Develop

The common division of glaciers into accumulation and ablation zone areas at an equilibrium line elevation (ELA) might seem to offer the best division in terms of surface processes and forms. However, it was found that in the Karakoram conventional

**Plate 8.4** Upper Panmah Glacier, Nobande Sobonde Branch (Hewitt 2005). This Mustagh-type main glacier illustrates: *Karakoram Valley Glacier Landsystems II, Transitional system I*: (1) Above 6,000 m; areas of Landsystem I disconnected and connected source areas, tributary icefalls, bergschrund-defined ice streams, subzones in firn and seasonal snow; avalanche inputs above and below firn limits. (2) (*blue arrow*) transitional system I, approximately 4,800 m to 6,000 m. (3) (below 4,800 m) transition to or in ablation zone with emerging medial moraines

accumulation zones are often small or absent. Where they do exist, accumulation zones display extensive and complicated subzones responding to these processes. In most cases, the areas of major inputs to the glaciers are strongly affected by high energy events from uppermost areas just described. The main, continuous ice streams are formed from snow, ice and debris arriving in avalanches, windblown material or descending in icefalls.

In most larger Karakoram valley glaciers, main ice streams typically develop 2,000 m or more below the watersheds and somewhere between 5,500 and 4,500 m. Some involve the convergence of several smaller ice streams, usually deeply incised and with steep icefall sections. All of this leads to a distinctive and critical set forms in a zone of transition between source and ablation zones (Plate 8.4). In the case of Turkestan-type glaciers, the massing of large avalanche cones and aprons is observed where main ice streams commence (Plate 8.5). Icefalls play a large role, and where some weeks of freeze–thaw can occur, wind- and avalanche-delivered snow is interspersed with the subzones described earlier for firn (Chap. 4). These features, although constantly disturbed and modified by processes from above, can extend over many kilometres horizontally and 1,000 m or more vertically. They are hardly typified by ambient conditions at the elevations involved, but can be identified, rather, as a transitional type of landsystem.

**Plate 8.5** *Transition landsystem zone I*, in a Turkestan-type glacier where main ice streams form below source zone walls, fed by snow avalanches, ice avalanches from detached small or minor tributaries and icefalls, Upper Chillinji Glacier, Karambar valley, Gilgit Basin (Hewitt May 1997)

## *8.4.1 Chute Systems*

Off-ice channelised slope processes assume increased importance in this zone. *Chutes,* referring to steep cuts, channels or ravines in bedrock with flat, trough-like floors, are widely present. Matthes (1938), in a pioneering paper, discussed chutes associated with avalanches in particular, important but neglected features of mountain sculpture.

Single channels may occur on the steepest slopes and more massive rock, 'chimneys' and 'couloirs' associated with frequent and deadly avalanche activity. Some chutes reach down from the interfluves without significant changes of width. Others have a distinct fan-wise spreading at the head and may become wide alcoves, including the 'rock funnels' of Rapp (1960) – recesses where two walls meet at a sharp angle. A diversity of branching patterns is observed comparable to streams and suggesting a similar organisation into drainage networks. Activity, however, largely involves intermittent, short-lived events strongly affected by seasonal changes. Markgren (1964a, b) proposed a terminology for chutes, but in conditions very different from the Karakoram. What he referred to as the 'regular, normal chute' is not common, while his 'irregular, non-normal' type prevails (1964b, p. 63). In this transitional landsystem, there is increased development of branching or chute networks descending to the glaciers (Fig. 8.3).

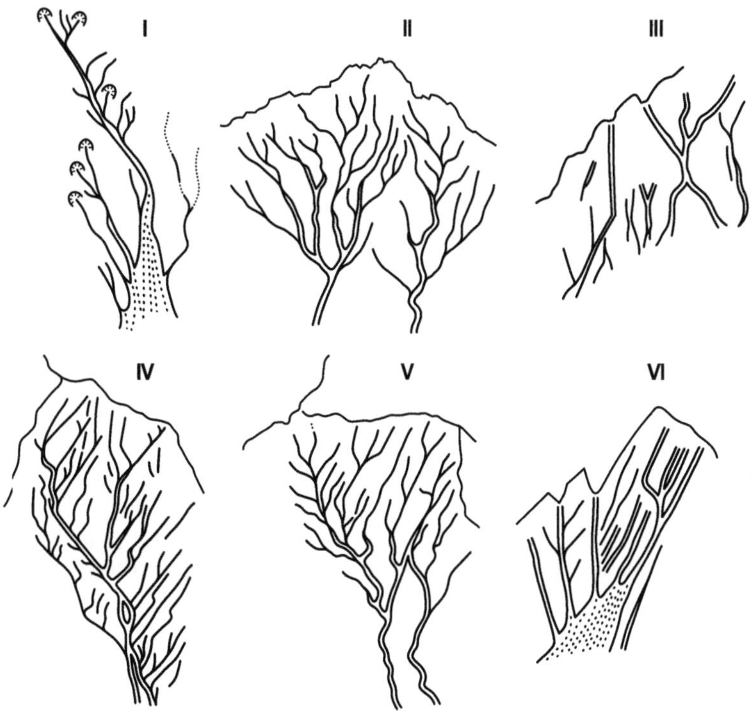


**Fig. 8.3** Branching chute systems sketched in the field in Biafo Glacier basin (After Hewitt 1968a; Markgren 1964a, b). Most are in metamorphic rocks variously exploiting joint, intrusive and foliation planes: (I) Dendritic–pinnate pattern, with fresh rockfall scars, 3,000–4,000 m elevation range, southerly orientation. (II) Dendritic–pinnate pattern, on scarp slope 4,000–5,500 m range, south facing. (III) Trellised–rhomboidal pattern, 4,000–5,000 m range, east facing. (IV) Trellised–angulate pattern with anastomosing, 3,500–5,000 m range, north facing. (V) Trellised–angulate pattern with divergence 3,500–5,000 m north facing. (VI) Trellised–angulate, 4,000–5,500 m, east facing. Sets (II), (IV), (V) and (VI) are common in the transition zone

Chutes prevail in areas with slopes between about 35° and 55° and drain a large fraction of the off-ice areas below the highest interfluves. They concentrate the delivery of snow to glacier surfaces. Lithology and rock structures clearly influence their geometry and development. Chute networks are more typical of the metamorphic and sedimentary terrains where partings in the rock such as bedding planes, joints, faults and foliation become important in chute geometry.

Chute drainage areas and elevation spans are reflected in the scale of avalanche cones and aprons. Chutes in most of this zone are debris free and typically filled with windblown snow. The larger chutes span considerable elevation ranges so that thermal properties and debris content of avalanches vary from the head to the foot. At the lower limits of this landsystem, wet avalanches may occur in summer and debris accumulations start to develop at narrows, bends and breaks of slope (Butler et al. 1992).

**Plate 8.6** Icefalls and the development of major ice streams. Avalanche-fed icefalls from the southeast flank of Masherbrum (7,821 m) join and expand the Gondoghoro Glacier (Hewitt 2005)

### *8.4.2 Icefalls*

The on-ice transitional landforms are also associated with icefalls (Plate 8.6). Most major ice streams were shown to commence around the last or lowest icefalls (see Table 7.1). Many of them commence in the perennial snow zone and terminate in upper ablation zone areas. Some Alpine-type glaciers such as Biafo, Siachen and Rimo have main ice streams that pass directly from accumulation to ablation zones as large, low angle tongues. Most others do not, and in Turkestan-type glaciers, icefalls are common as their disconnected tributaries and high in ablation zones. As already described, icefalls are important in the rapid downslope concentration and transformation of ice (Chap. 7). Transitions from icefalls to main ice streams are typical features of this landsystem.

## 8.5 Seasonal Landsystems I: Glacier Ablation Zones

The most important developments in relation to ablation itself and described in Chap. 5 largely concern summer season events, when ablation generates ice forms sensitive to weather conditions, debris types and volumes (Plate 8.7). Debris-covered glaciers have been said to characterise the Karakoram but, as already shown, two-thirds of ablation zones have relatively clean ice or thin and scattered debris. The overall evolution of on-ice conditions and down-glacier changes reflects how ablation and relief inversion shed and disperse debris laterally from medial moraines.

**Plate 8.7** Ablation zone, ablation season surface conditions in July on Panmah Glacier, Nobande Sobonde branch, showing on-ice relief, stratification, medial moraines (Hambrey and Glasser 2011, p. 992) and incipient penitents from intense sunlight (Hewitt 1998)

Ablation zone verticality is reflected in two landscape phenomena. One is essentially permanent: the down-glacier increase in medial and lateral moraines and their eventual merging into complete debris mantles. These record the total numbers of years given transects have been exposed to ablation and the increasing ablation season length at lower elevations. Such features underline the role of glaciers in ‘debris transport paths’ (Benn et al. 2005, p. 376).

However, a second aspect concerns cyclical, seasonal developments which, due to constant migration vertically, are spatial as well as temporal phenomena. Seasonality affects all parts but assumes singular importance in Karakoram ablation zones. While ablation and water yields are a prime concern, it must be remembered that over most of the ablation zone, summer is the shortest season, while winter and the neglected fall and spring shoulder seasons have longer durations over the middle and upper ablation zone areas. The duration of each season, its timing and intensity, varies with elevation. The processes or dynamic aspects of the on-ice landforms and their margins are largely responses to seasonally varying conditions. Related surface features and changes may seem quite small and local, from millimetre-sized cryoconite holes and winter icy layers a few centimetres thick to glacier margin advances and retreats of some metres. Meltwater channels, ponds and slush flows can be a few tens of metres across. On-ice relief features rarely exceed 100 m vertically. However, the glacier areas over which these recur add up to or exceed those of, for instance, the better-known debris-covered areas. And they are shaped and modified primarily, and strongly, by seasonally varying conditions. For this reason, the idea of a seasonal landsystem will be highlighted.

### 8.5.1 Freeze–Thaw and Seasonal Landforms

Temperature and its gradients influence earth surface processes mainly through the effect on moisture. Continuous below-zero temperatures mean liquid water is unavailable. Where they are above zero, they are subject to water availability. In high mountains of middle and low latitudes, intermediate zones have frequent frost cycles (Troll 1975; Price 1981). Available moisture undergoes repeated thawing and freezing, often on a daily basis, a core concern of periglacial geomorphology if mainly studied in higher latitudes (Rempel 2011). How such conditions affect glacier ablation zones has received little notice. Yet, in the Karakoram they coincide with elevations having the greatest numbers of frost cycles. Like winter freeze-up and summer thaw, freeze–thaw is concentrated in an elevation belt that migrates vertically (see Fig. 8.1b). Diurnal cycles exceeding 20 °C are not unusual if, unlike the tropical mountains, they are not as great as annual variations (Hewitt 1968b).

In mid-ablation zones, at roughly 4,500 m, diurnal freeze–thaw is recorded on as many as 180 days, the longest, continuous period. Below this, the potential number of cycles is unchanged but they occur in two distinct periods, in effect, fall and spring. Above 4,500 m the freeze–thaw period and total numbers of cycles decline, becoming rare above about 5,500 m, if temperatures above zero are occasionally reported even up to 7,000 m. In general, significant freeze–thaw seems to stop below 6,000 m on south-facing slopes, at least in the past 100 years. It continues down to the lowest Karakoram valleys where its significance is limited by aridity.

Conditions at the terminus of Biafo Glacier (3,100 m) help bracket the low elevation and winter elements. Thermograph records showed diurnal freeze–thaw cycles were most frequent in March (29 days) and November (27 days). In the months of January and February, there were just 2 days when temperatures went above zero and none below zero from late April through mid-September. There were 164 freeze–thaw cycles in all. Diurnal temperature ranges averaged about 10 °C, with extremes just over 20 °C. They were roughly symmetrical around zero in months with the most freeze–thaw cycles. Diurnal ranges were greatest on sunny days, much less on cloudy days.

In these high, subtropical mountains, standard weather station records are, at best, indicators. The geomorphic or glacial significance of freeze–thaw applies at and below ground level. Here, frost cycles can be more, or less, frequent than air temperatures show (Barry 1992, p. 84). They are complicated by melting of snow and ice or refreezing within the soil or snow profile.

Around the Biafo terminus in winter during sunny weather, even when air temperatures remained below freezing, snow was observed to melt. On a day in early February, after 1.5 h of direct sunshine, meltwater below a snow patch reached +10 °C after running 1.3 m over rocks. The air temperature was −6 °C (Hewitt 1968b, p. 88).

Winter soil temperatures revealed even more complicated relationships, including frost cycles when none appear on the station record, and vice versa. Results from soil temperature probes in a dry lakebed beside Biafo Glacier illustrate the conditions. With snow cover, little or no diurnal change in the soil temperatures was observed regardless of air temperatures. January 31, for example, began with a snow cover and soil temperatures showed no change until it had melted. After that, at 4 cm depth, the soil warmed from a night-time low of −6°C to a maximum of +4 °C at midday. It fell below zero again at about 4 pm. The thermograph at the nearby weather station recorded *no* above-zero temperatures that day. On the following day, with no snow cover, the soil temperature at 4 cm rose to 11 °C and, again, returned to zero after 4 pm. The weather station recorded a low of −13.5 °C overnight and a high of +0.5 °C during the day.

These data, among the few available for winter conditions around Karakoram glaciers, are indicative of the complexities involved. Summer data from the Baintha profile (4,050 m) stations on Biafo illustrate how freeze–thaw cycles at off-ice weather stations can differ from those on-ice (see Fig. 5.3).

Ablation zones are typically described as having exposed glacier ice, medial and lateral moraines and, in the Karakoram, large debris-covered areas. Between summer ablation and winter freeze-up, however, are lengthy but neglected periods when ablation loses are small or absent and freeze–thaw cycles prevail. Their longest duration and the greatest areas affected are in mid-ablation zones, roughly 3,800–4,600 m, where the four seasons approach roughly equal duration.

### 8.5.2 Shoulder Seasons

Fall and spring are transitional periods which also transit between lower and upper ablation zone areas. They share some common developments, but with an obvious asymmetry. Fall builds towards complete freeze-up; spring, ultimately, removes its effects. At summer's end there are ice surfaces, local topography and debris distributions reflecting the ablation season. Fall snowfall helps shut down ablation, but most of it does not survive into the winter. An irregular patchwork of wind-stripped and wind-deposited snow follows storms. Snow accumulates in depressions where temporary melting and percolation convert much of it to ice. Black ice occurs where wind strips the surface snow, often with debris frozen into it.

Together, wind action and freeze–thaw generate a carapace of black ice, wind-packed and refrozen snow, resembling the 'icy layers' of Müller (1962). They smother and protect glacier ice beneath (Plate 8.8a). The frozen carapace extends progressively down the glacier in fall. In spring, freeze–thaw conditions return and move up-glacier. Eventually thawing will overcome night frosts but for some weeks further icy layers and refrozen snow will be added to the carapace. Spring snowfall is usually high and there are significant additions to wind-redistributed snow.

In the later stages, the thickness of the icy carapace can range from some tens of centimetres to a metre or more. Eventually, however, it is softened and depleted, leading to exposure of ever-larger areas of glacier ice. Early in the ablation season, patches and ridges of carapace ice can be mistaken for emerging glacier ice (Plate 8.8b). Initial glacial meltwater comes from the icy carapace and seasonal

**Plate 8.8** Aspects of the cold season freeze–thaw carapace on Biafo glacier: (a) Midwinter conditions in February, with superficial ice and packed snow at 4,300 m (Hewitt 1962). (b) Early summer conditions, late June at 4,500 m with irregular remnants of the icy carapace as ablation works through it to the glacier ice beneath (Hewitt 1999)

snow; yields are sufficient to reopen englacial drainage systems even before glacier ice ablation is well under way.

An even more complicated and varied picture affects debris-covered areas. Early in fall melt out and redistribution of surface debris ceases. Clasts become frozen into the glacier surface. Thin, finer-grained material may be frozen in with black ice and packed snow layers. Dirt in freeze-dried areas is readily picked up by glacier winds and redeposited. Large boulders quickly lose such snow as falls, but under and in the lee of them blown snow becomes trapped in irregular flags or drifts. Black ice will build up where meltwater runs down from rocks exposed to sunshine and freezes over night. Thicker debris covers insulate the underlying ice from these seasonal changes but facets are affected as exposed glacier ice and create further complications.

Through the shoulder seasons a patchwork of differing subzones is observed, partly reflecting seasonal weather conditions, partly the glacier surface topography and debris covers. Major snowstorms, or the timing and length of runs of cold or warm days, can alter the scope of these developments in any given season. Limits or lines at well-defined elevations or contours should not be expected.

Two significant interactions should be noted between ice-surface and off-ice conditions. As already shown, avalanches can spread over the ice and carry debris onto it far into the ablation zones. Secondly, the great extent of dusty snow and ice reflects aeolian deposition from far and wide. Karakoram valleys are very dusty environments. Rapid slope processes like debris flows and rockfalls introduce dust into the atmosphere. A few days of sunny weather are enough to spread dust everywhere, even well above the snowlines. It is of particular importance for ablation-generated forms where relatively clean ice exists.

### *8.5.3 Longitudinal On-Ice Forms*

In broad sections of the mid-ablation zones, longitudinal patterns predominate, as noted at Batura Glacier in relation to on-ice morphology and ablation rates (Chap. 7). They may continue over 1,000 m of elevation or more and for tens of kilometres in the larger glaciers (see Plate 8.9). Medial moraines and ice streams from tributaries converge and flow parallel, carrying varying amounts and qualities of clean and debris-covered ice. In these sections, features in ice and debris occur in bands reaching far up and down the glacier and parallel to medial moraines. Glacier tables, for example, are concentrated beside medial moraines in marginal troughs where the larger boulders are spread through fall sorting. On-ice meltwater streams tend to occur in the same troughs, developing incised meanders until plunging into glacier mills.

Cryoconite pits, holes and basins cover flatter areas of relatively clean ice streams. Seasonal features from icefalls, the ogives described earlier, affect ice conditions far down valley and include pennant-shaped debris patterns from more intense summer rockfall activity. Longitudinal patterns mean on-ice topography has pronounced transverse relief, which continues until debris spreads across

**Plate 8.9** Biafo Glacier from Baintha section (4,050 m) upwards, fall (October) shoulder season view (Hewitt 2010), illustrating: *Karakoram Glacier landsystems III and IV: seasonal, mid-ablation zone and ice-margin complex*: (1) Main ice streams with seemingly clean ice or scattered debris. (2) Supraglacial moraines: including medial; 'ice-stream interaction' (largest), 'ablation dominant', rockfall or 'avalanche-dominant' and supraglacial lateral moraines (Benn and Evans 1998, pp. 224–228). (3) Transitional Landsystem II: Lateral Margin Sediment–Landform Assemblages (see text)

the whole surface. At Baltoro, less obvious longitudinal patterns were found to continue to affect topography and ablation rates despite a complete debris cover (Mihalcea et al. 2006, p. 125).

## 8.6 Landsystem IV: Debris-Mantled Ice

In lower ablation zone areas supraglacial debris usually expands to cover, protect and interact with ice and helps generate a range of characteristic forms (Plate 8.10). Some describe this as passive debris transport, but it is so only in the sense that, unlike beneath the ice, erosion and comminution are not directly due to glacier dynamics (Benn and Evans 1998, p. 221). However, in the supraglacial environment, other processes modify and redistribute material in debris mantles. The most important is differential ablation already described in detail earlier (Chap. 5).

There is a constant dispersal of debris and related relief inversions through the ablation season, which lasts much longer than in the previous Seasonal Landsystem I. The facets discussed earlier are associated with typical sets of forms and rapid, ablation season developments of on-ice relief (Chap. 5). Surface relief also derives in part from crevasses, supraglacial ponds and drainage lines that develop on the ice.

**Plate 8.10** Lower Barpu Glacier (Hewitt 2002) illustrating *Karakoram Valley Glacier Landsystems IV and V: ablation zone transition to full debris cover:* (1) Merging of medial and (on-ice) lateral moraines. (2) Debris-covered ice: facets, on-ice ponds, ablation-inversion hummocky relief. (3) Lateral margin assemblage with typical 'Great Lateral Moraine' (Meiners 1998) from former high ice levels, well-developed valley-margin trough below ice level but separated from glacier. Note increased aridity low down and contrast with humid, snow-covered source zone

Different types of debris influence the pace of developments and surface patterns. Coarse and angular material derived from mass movements in the upper basin is predominant, but not the only materials. Scattered but considerable quantities of water-rounded stones become increasingly mixed in and more common down glacier (Hewitt 1999). Deposits in englacial tunnels melt out at the surface, the main source of water-rounded material. Towards glacier margins and termini, finer debris is brought to the surface along englacial shear planes. As debris cover increases so do on-ice pools and ponds. These become sites with water-sorted, mainly fine-grained sediment. When drained they give rise to most of the distinctive areas where dirt cones occur.

## 8.7 Transitional Landsystem II: Lateral Margin Sediment–Landform Assemblages

Concentrated deposition occurs at most ablation zone margins. Along the larger Karakoram glaciers, diverse deposits and associated landforms continue for tens of kilometres. These are sites of complex sedimentation between the ice and valley-side slopes. Although less than 3 % of the 42 largest Karakoram basins, they comprise the valley glacier landsystems elements that have received the greatest attention here and worldwide (Eyles 1983; Benn et al. 2005). With the typical 'block' motion, a sharp line of shear is typically found separating the active ice from off-ice margins. However, the landforms and related sediment bodies are dominated and defined by interactions between on- and off-ice processes, another transitional landsystem. Although the scale and diversity of deposits tend to increase down valley, no distinct divisions appear between upper and lower ablation zone margins even down to the terminus (see below). This transition zone is localised at the margins in a relatively narrow band, more or less continuous and spread through as much as 3,000 m elevation range.

Deposition is controlled primarily by the types and rates of debris delivery from the glacier and from the adjacent valley walls, influenced by ice levels and their fluctuations. Where the ice rises above and over rides the margins and assemblage surface, the glacier acts as a source of debris aprons and may push and squeeze adjacent areas. The build-up of lateral moraines dumped from the ice is accompanied by subglacial ice processes that contribute by lodgement of finer-grained material helping create resistant cores beneath the moraine ridges (Benn and Evans 1998, p. 197). Off-ice areas can be underlain by generations of dead ice left by glacier fluctuations. When ice levels drop, steep and near-vertical cliffs in moraine can be tens of metres high.

Seasonal differences affect ablation zone margins too. Winter observations at Biafo Glacier terminus showed active ice advancing as much as 10–15 m and reactivating buried dead ice up to 30 m ahead of the summer ice front (Hewitt 1967). No meltwater flowed from the glacier for about 4 months, but glacitectonic disturbances affected lake and stream deposits beyond the ice margin, and coarse gravels in dry outwash channels. In winter, wind- and avalanche-derived snow fills the valley-side troughs in middle and upper ablation zones.

### *8.7.1 Valley-Side Troughs or 'Ablation Valleys'*

Valley-side troughs between ice or lateral moraines and mountain slopes may continue for kilometres (Plate 8.11). They are rich in vegetation and wildlife and offer a host of resources exploited by local people and provide some of the most convenient routes to higher areas (Plate 8.12). The troughs are underlain by a wedge of glacial, pro-glacial and slope deposits and are shaped according to interactions of glacier and valley slope processes. They have long been considered distinctive of the Karakoram (Hewitt 1994; Iturrizaga 2003).

In the older literature the troughs were called 'ablation valleys'. Oestreich (1911–1912) employed the German term *Ablationsschlucht* at Chogo Lungma Glacier. A chapter under this heading is in Visser and Visser-Hooft (1935–1938). A common explanation of their origin is as products of ablation beginning in a slot between an ice cliff and valley (ibid, pp. 38–41). Such conditions are observed but

**Plate 8.11** Ice-margin complex and extensive valley-side trough or 'ablation valley' at 4,000 m beside Biafo Glacier, showing the combination of lateral moraines, extensive fluvio-glacial and slope deposits (Hewitt 1999)

**Plate 8.12** Ice-margin trough near Baintha, Biafo glacier, showing varieties of vegetation, wetlands and multiple lateral moraines (Hewitt 2010)

lateral margin assemblages do not derive from them. Indeed, ablation is generally suppressed along glacier margins by heavy supraglacial moraine. Most valley-side troughs arise from the combined effects of the following:

(i) Glacier thickening and thinning
(ii) Heavier supraglacial debris at the margins from valley wall avalanche and rockfall sources up-glacier
(iii) The net migration of mobile surface debris towards the margins
(iv) Englacial material brought by the lateral spread of ice responding to ablation losses
(v) Impoundment and channelling of water from the valley side by moraines and ice

## 8.8 Transitional Landsystem III: Terminal Sediment–Landform Assemblages

Processes at and near glacier termini have been studied more than any other process (Benn and Evans 1998, Chap. 7). As such, there is a rich literature on this zone if it actually comprises a tiny part of present-day glacier landforms. Pioneering studies of Karakoram glaciers focus on terminal areas and have inspired the landsystems models developed for the Karakoram to date (Goudie et al. 1984; Owen 1994). Conditions are most evident from the study of terminal fluctuations and past as well as contemporary moraines described in Chap. 12.

In the approach to valley glacier landsystems of Benn et al. (2005), the most basic controls are seen as the ratios of sediment and ice carried to the glacier margins and the efficiency or otherwise of sediment transport through pro-glacial areas. These invoke broader conceptual relationships and sediment transport pathways identified above, but mainly in order to explain conditions at glacier margins, especially in terminal areas.

Major valley glacier landsystems types are described as having either 'coupled' or 'uncoupled ice margins…', depending upon the efficiency or otherwise of transfer of sediment between the glacier and pro-glacial fluvial system (ibid, their Fig. 15.23). Debris-covered glaciers and rock glaciers are portrayed as 'decoupled', with high sediment supply and build-up or inefficient removal at the margins. They are identified mainly with high mountains and/or drier climates. Climate and terrain are said to determine which category is found and changes in them.

Karakoram conditions suggest a need for some modifications of this scheme. On the one hand, there are many examples of glaciers with their 'giant bounding moraines' and 'moraine-dammed' margins, including 'repeated superposition of moraines around the margins' and diversity of sediments around such glaciers (ibid). They are exemplified by the 'Ghulkin type', derived from a Hunza glacier in which the whole ice stream travels over a raised bed of subglacial sediment (Owen 1994). Associated with such moraine-dammed ice are striking examples of 'breach lobes' that burst through the moraine dam to emplace secondary lobes (Plate 8.13).

**Plate 8.13** Terminal lobes of the Yazghil Glacier, Shimshal valley, showing older moraine-dammed and uncoupled, debris-mantled ice lobes, and contemporary breach lobes (Hewitt 1998)

**Plate 8.14** Small tributary glacier in Nangmah valley, Hushe Karakoram. The ice remains relatively debris free although moraine dammed, a result of engorging of debris by frequent crevasses or rapid transfer of boulders to the margins in these steep ice masses (Hewitt 2012)

Moraine-dammed types are common among smaller glaciers that descend steeply to main valleys. However, many of these steeper glaciers have clean ice exposed right to the terminus (Plate 8.14). Where heavy debris mantles occur, the ice tongues are of lesser slope.

**Plate 8.15** Terminal area of Baltoro Glacier, one of the most heavily debris-covered glaciers but clearly of the 'outwash head' type (Benn et al. 2005), with powerful, migratory pro-glacial streams that remove frontal moraines and maintain strong coupling between the glacier and fluvial system (Hewitt 2005)

On the other hand, larger glaciers, despite heavy debris mantles, are rarely uncoupled types. Ghulkin, Yazghil and, until recently, Chilliji are the only moraine-dammed examples in the 42 main set. Most belong, instead, to the outwash head type. They are among those termini strongly coupled to powerful migratory outwash rivers with small or absent moraines (Plate 8.15).

Features common to the terminus areas of the Siachen, Kondus, Baltoro, Chogo Lungma, Hispar and Bagrot Glaciers, and relevant to landsystems criteria, are as follows:

1. *Debris covered*: heavy supraglacial debris over the lower 10–15 km or more of each glacier tongue.
2. *Coupled ice margins*: efficient transfer of sediment to fluvial zones and limited moraine development.
3. *Outwash head type*: pro-glacial outwash lies across the ice front and sandur plains below.
4. *Valley fill*: each terminus rests on or has advanced and retreated over valley fill tens of metres thick, possibly hundreds. It means ice levels are partly constrained by valley fill levels.
5. *Linked stream terrace and latero-terminal moraine geometries*: downstream of the termini are stream terraces with upper levels generally lower than the high stand lateral moraines, but continuous with associated glacier margin troughs.
6. *Landslide interrupted drainage*: in every case, a few kilometres or tens of kilometres downstream of the termini is one or more cross-valley landslide barriers. These have controlled local base levels, stream thalwegs, sediment movement

and storage (Hewitt 2006a, b). In these cases the "elevated beds" and "bounding moraines" of Benn et al. (2005) are associated with interrupted drainage and locally controlled base levels.

It should be noted that terminus features at Ghulkin Glacier in Hunza, and Yazghil Glacier in Shimshal, also relate to more recent and large episodes of aggradation in landslide-fragmented rivers (Hewitt 2006c; 2011).

Some cases, such as Batura and Biafo, do have well-developed sets of terminal moraines, continuous with huge lateral margin complexes. However, they enter main river valleys as tributaries. Their terminal moraines sit on river terraces that continue above and below the termini. Otherwise, these glaciers have the same six conditions as the others.

## 8.9 Seasonal Landsystems II: Periglacial Environments in Glacier Basins

Extensive periglacial areas within glacier basins compliment seasonal ablation zones over similar elevation ranges. They include seasonally snow-free parts of glacier basins that are humid or sub-humid and contain permafrost, hummocky and patterned ground (Harris 1988; Dixon and Abrahams 1992). The larger parts consist of rock walls that feed a wide variety of talus forms. The Karakoram also includes considerable areas where periglacial conditions are constrained by arid or semiarid conditions, mostly below the glaciers or in the lower parts of their basins. On cliffs below about 3,500 m, chemical weathering and salt weathering become important and sources of widespread cavernous weathering features (Hewitt 1968a). These are mainly influential for cliff and talus development at and below the limits of today's glacial zone.

### *8.9.1 Talus Forms*

*Talus* consists of rock fragments that usually accumulate in well-developed cones below chutes and gullies or aprons and ramparts at the base of cliffs. A wide variety of these depositional forms characterise the Karakoram periglacial zone. There are countless classic, *scree* slopes consisting mainly of coarse unstable fragments at or near the angle of repose (Brunsden et al. 1984). Most talus cones contain as much or more finer material, usually below the surface ánd sometimes called *talluvium.* The transitional nature of this zone and seasonal conditions is reflected in the extent and behaviour of talus. The importance of frost action usually becomes apparent with the thaw that accompanies the spring shoulder season and return to cooler temperatures in the fall. The highest incidence of rockfalls occurs as and where the spring season is moving up valley (Hewitt 1968a, Chap. 10).

In the Central Karakoram, talus development between 3,500 and 4,500 m depends upon freeze–thaw, avalanching and ice content. Below 4,000 m sediment delivery and a range of distinctive surface patterns involve vegetation adapted to the steep, unstable talus environment and, indeed, the intermittent, severe freeze–thaw and drought conditions. Rock glaciers, lobes or tongues of coarse, angular debris are closely associated with talus and widespread around the most heavily debris-covered, small glaciers. They are an integral part of these periglacial landscapes but have received so little attention that they warrant a chapter to themselves to stress their place in the Karakoram cryosphere (Chap. 11).

## 8.10 Concluding Remarks: Transglacial and Paraglacial Conditions

At all elevations the steepland environment and abundant rock walls ensure that surface processes transport moisture and debris between off-ice and glacier areas. Avalanche activity is the main factor in glacier basins at high elevations. In mid-elevations there is a seasonal sequence of winter avalanches, spring or summer debris flows and rockfall debris that may cross ablation zone margins. These can be treated as modifiers of given zones or as expressions of the vertical cascade (Hewitt 1993). Rather than a clear separation of the glacial system from others based on fluvial, mass movement or aeolian processes, they are found to interact. The overlap between glacier and other processes can be identified in *transglacial* landforms and influences: glacial developments that arise from and depend in part on non-glacial processes (Iturrizaga 2011a, b).

To date, transglacial features have been identified mainly with mass movements that cross the ice-margin troughs in ablation zones (Iturrizaga 2006). It is suggested here that they are much more widespread and diverse. They include the perennially frigid and seasonal avalanche zones, moisture and debris transferred between ice and ice margins, the impact of local base level or episodes of aggradation and trenching on tongues of glaciers, or how rivers fragmented by landslides and obstructed by glaciers, modify up-valley glacier activity (Hewitt 2006a).

Finally, the landforms and spatial relations addressed in this chapter raise issues of the temporal dimensions of their development, and the pace and sources of landscape change. They relate to the magnitude and frequency of geomorphic events. The relations of present-day conditions to formerly glaciated landscapes must be borne in mind too. These involve *paraglacial* conditions (Church and Ryder 1972); ongoing developments that are affected by past glacier activity and relate to the longer-term Quaternary history of glaciations (Hewitt 2009).

The widespread alluvial or sediment fans, first named from the Karakoram by Drew (1873), arise in part at least from paraglacial and transglacial conditions (Derbyshire and Owen 1990). They are intimately bound up with the glacial history and that of the Indus streams, if mostly lying beyond today's glaciers and the scope of this study.

In the present-day Karakoram, the role of extreme processes is constantly encountered. Certain ones have assumed a particular importance both for understanding and distinguishing the Karakoram glacial environment and as hazards for human communities. They also figure in potentially adverse responses to climatic change. Among them, glacier surges, impoundments and outburst floods will be the main topics of the following chapters.

## References

Barry RG (1992) Mountain weather and climate, 2nd edn. Routledge, London

Benn DI, Evans DJA (1998) Glaciers and glaciation. Hodder Arnold, London

Benn DI, Kirkbride MP, Owen LA, Brazier V (2005) Glaciated valley landsystems. In: Evans DJA (ed) Glacial landsystems. Hodder Arnold, London, pp 370–406

Boulton GS, Eyles N (1979) Sedimentation by valley glaciers; a model and genetic classification. In: Schluchter C (ed) Moraines and Varves. Balkema, Rotterdam, pp 11–24

Brunsden D, Jones DKC, Goudie AS (1984) Particle size distribution on the debris slopes of the Hunza Valley. In: Miller KJ (ed), International Karakoram project, vol 2. Cambridge University Press, Cambridge, pp 536–580

Bull WB (2007) Tectonic geomorphology of mountains: a new approach to paleoseismology. Blackwell, Oxford

Butler DR, Malanson GP, Walsh SJ (1992) Snow-avalanche paths: conduits from the periglacial-Alpine to the Subalpine Depositional Zone. In: Dixon JC, Abrahams AD (eds) Periglacial geomorphology: proceedings of the 22nd annual Binghampton symposium in geomorphology. Wiley, New York, pp 185–202

Church M, Ryder JM (1972) Paraglacial sedimentation: a consideration o fluvial processes conditioned by glaciation. Bull Geol Soc Amer 83:3059–3072

Goudie AS et al (1984) The Geomorphology of the Hunza Valley, Karakoram Mountains, Pakistan. In: Miller KJ (ed) International Karakoram project, vol 2. Cambridge University Press, Cambridge, pp 359–410

Dainelli G, Marinelli O (1928) Le condizioni fische attuali. Relazioni scientifiche della spedizione Italiana de Filippi nell' Himalaja, Caracorum e Turchestan Cinese (1913–1914). Series 1, vol 8. Zanicelli, Bologna

Derbyshire E, Jijun L, Perrot FA, Xu S, Waters RS (1984) Quaternary glacial history of the Hunza Valley, Karakoram Mountains, Pakistan. In: Miller KJ (ed) International Karakoram project, vol 2. Cambridge University Press, Cambridge, pp 456–495

Derbyshire E, Owen L (1990) Quaternary alluvial fans in the Karakoram mountains. In: Rachochi AH, Church M (eds) Alluvial fans: a field approach. Wiley, New York, pp 27–54

Derbyshire E, Jijun L, Perrot FA, Xu S, Waters RS (1984) Quaternary glacial history of the Hunza Valley, Karakoram Mountains, Pakistan. In: Miller KJ (ed), pp 456–495

De Terra H (1932) Geologische forschungen im westlichen Kun-lun und Karakorum Himalaya. Reimer & Vohsen, Berlin

Dixon JC, Abrahams AD (eds) (1992) Periglacial geomorphology: proceedings of the 22nd annual Binghamton symposium in geomorphology, Wiley, New York

Drew F (1873) Alluvial and lacustrine deposits and glacial records of the upper India Basin. Q J Geol Soc 29:441–471

Evans DJA (2003) Glacial landsystems. Hodder Arnold, London

Evans SG, Degraff JV (2002) Catastrophic landslides: effects, occurrence, and mechanisms. Geological Society of America, Rev Eng Geol

Eyles N (1983) Glacial geology: an introduction for engineers and earth scientists. Pergamon, Oxford

Feldl N, Bilham R (2006) Great Himalayan earthquakes and the Tibetan plateau. Nature 444:165–170

Finlayson DP, Montgomery DR, Hallet B (2002) Spatial coincidence of rapid inferred erosion with young metamorphic massifs in the Himalayas. Geology 300(3):219–222
Goudie AS et al (1984) The Geomorphology of the Hunza Valley, Karakoram Mountains, Pakistan. In: Miller KJ (ed) International Karakoram project, vol 2. Cambridge University Press, Cambridge, pp 359–410
Haeberli W, Gruber S (2009) Global warming and mountain permafrost. In: Margesin R (ed) Permafrost soils, soil biology. Springer, Berlin, pp 205–218, Chap. 14
Hambrey MJ, Glasser NF (2011) Sediment entrainment, transport and deposition. In: Singh VP et al (eds) Encyclopaedia of snow, ice and glaciers. Springer, Dordrecht, pp 984–1003
Harris SA (1988) The Alpine periglacial zone. In: Clark MJ (ed) Advances in periglacial geomorphology. Wiley, New York, pp 369–413
Hewitt K (1967) Ice-front sedimentation and the seasonal effect: a Himalayan example. Trans Inst Br Geogr 45:93–106
Hewitt K (1968a) Studies in the geomorphology of the mountain regions of the upper Indus Basin. 2 vols. Unpubl. dissertation, University of London
Hewitt K (1968b) The freeze-thaw environment of the Karakoram Himalaya. Can Geogr 12:85–98
Hewitt K (1993) Altitudinal organization of Karakoram geomorphic processes and depositional environments. In: Shroder JF Jr (ed) Himalayas to the sea. Routledge, London, pp 159–183
Hewitt K (1998) Glaciers receive a surge of attention in the Karakoram Himalaya. EOS Trans Am Geophys Union 79(8):104–105
Hewitt K (1999) Quaternary moraines vs. catastrophic rock avalanches in the Karakoram Himalaya, northern Pakistan. Quat Res 51(3):220–237
Hewitt K (2002a) Styles of rock avalanche depositional complex in very rugged terrain, Karakoram Himalaya, Pakistan. In: In: Evans SG, Degraff JV (eds) Catastrophic landslides: effects, occurrence, and mechanisms. Geological Society of America, Reviews in Engineering Geology XV, Boulder, pp 345–378
Hewitt K (2002b) Postglacial land form and sediment associations in a landslide-fragmented river system: the transHimalayan Indus streams, northern Pakistan. In: Hewitt K, Byrne M-L, English M, Young G (eds) Landscapes of transition: landform and sediment associations in cold regions. Kluwer, Amsterdam, pp 63–91
Hewitt K (2006a) Disturbance regime landscapes: mountain drainage systems interrupted by large rockslides. Prog Phys Geogr 30(3):365–393
Hewitt K (2006b) Rock avalanche dams on the transHimalayan Upper Indus streams: a survey and assessment of hazards-related characteristics. Ital J Eng Geol Environ, Spec Issue I, pp 61–66
Hewitt K (2006c) Glaciers of the Hunza basin and related features. In: Kreutzmann H (ed), pp 49–72
Hewitt K (2009) Paraglacial rock slope failures, disturbance regimes and transitional landscapes, Upper Indus Basin, northern Pakistan. In: Knight J, Harrison S (eds) Periglacial and paraglacial processes and environments. The Geological Society, Special Publications 320, London, pp 235–255
Hewitt K (2011) Rock avalanche dams on the Trans Himalayan Upper Indus Streams: a survey of late quaternary events and hazard-related characteristics. In: Evans SG, Hermanns, RL, Strom A, Scarascia-Mugnozza G (eds) Natural and artificial rockslide dams, Chap. 6, pp 177–191
Hewitt K (2013) Large, topographically-constrained rockslide complexes in the Karakoram Himalaya, Northern Pakistan. In: Magnottini C (ed) Landslide science and practice, vol. 4, Proceedings of the Second World Landslide Forum. Springer, Berlin. doi:10.1007/978-3-642-31337-43, pp 335–346
Hewitt K, John J, Clague JJ, and Gosse J (2011a) Rock avalanches and the pace of late Quaternary development of river valleys in the Karakoram Himalaya. Geol Soc Am Bull, doi:10.1130/B30341.1, 16p. 14 June 2011
Hewitt K, Clague JJ, Deline P (2011b) Catastrophic rock slope failures and mountain glaciers. In: Singh VP, Singh P, Haritashaya UK (eds) Encyclopaedia of snow, ice and glaciers. Springer, Dordrecht, pp 113–126
Hewitt K, Clague JJ, Deline P (2011b) Catastrophic rock slope failures and mountain glaciers. In: Singh VP et al (eds), pp 113–126

Iturrizaga L (2003) The distribution and genesis of lateroglacial valleys in the Karakoram Mountains (Pakistan). Z Geomorphologie 130:510–574

Iturrizaga L (2006) Transglacial landforms in the Karakoram (Pakistan): a case study from Shimshal Valley. In: In: Kreutzmann H (ed) Karakoram in transition: culture, development and ecology in the Hunza Valley. Oxford University Press, Karachi, pp 96–108

Iturrizaga I (2011a) Lateroglacial landform systems. In: Singh VP, Singh P, Haritashaya UK (eds) Encyclopaedia of snow, ice and glaciers. Springer, Dordrecht, pp 704–708

Iturrizaga L (2011b) Paraglacial landscape transformations. In: Singh VP, Singh P, Haritashaya UK (eds) Encyclopaedia of snow, ice and glaciers. Springer, Dordrecht, pp 817–823

Kalvoda J (1992) Geomorphological record of the Quaternary orogeny in the Himalaya and the Karakoram. Developments in earth surface processes, vol 3. Elsevier, Amsterdam

Kamp U, Haserodt K (2004) Quaternary glaciations in the high mountains of northern Pakistan. In: Ehlers J, Gibbard PL (eds) Extent and chronology of glaciation. vol. 3: South America, Asia, Africa, Australia, Antarctica. Elsevier Science, Amsterdam, pp 293–311

Keefer DK (1984) Landslides caused by earthquakes. Geol Soc Am Bull 95:406–421

Kreutzmann H (2006) Karakoram in transition: culture, development and ecology in the Hunza Valley. Oxford University Press, Karachi

Markgren M (1964a) Geomorphological studies in Fennoskandia I: chute slopes in northern Fennoskandia, Regional Studies. Lund Stud Geogr, Ser A, #27

Markgren M (1964b) Geomorphological Stuiesin Fennoskandia II: chute slopes in northern Fennoskandia, systematic studies. Lund Stud Geogr, Ser A, #28

Matthes FE (1938) Avalanche sculpture in the Sierra Nevada, California. Int Assoc Sci Hydrol, Union of Geodesy and Geophysics, Riga, Bulletin 23:631–637

McSaveney MJ (1978) Sherman Glacier rock avalanche, Alaska, U.S.A. In: Voight B (ed) Rockslides and avalanches, I; Natural phenomena. Elsevier Scientific Publishing, New York, pp 197–258

McSaveney MJ (2002) Recent rockfalls and rock avalanches in Mount Cook National Park, New Zealand. In: Evans SG, Degraff JV (eds) Catastrophic landslides: effects, occurrence, and mechanisms. Geological Society of America, Reviews in Engineering Geology XV, Boulder, pp 35–70

Meiners S (1998) Preliminary results concerning historic to Post-glacial stages in the NW-Karakorum (Hispar Muztagh, Batura Muztagh, Rakaposhi Range). In: Stellrecht I (ed) Karakoram-Hindukush-Himalaya: dynamics of change, vol 4. Rüdgers Köppe Verlag, Köln, pp 49–70

Mihalcea C, Mayer C, Diolaiuti G, Lambrecht A, Smiraglia C, Tartari G (2006) Ice ablation and meteorological conditions on the debris-covered area of Baltoro Glacier, Karakoram, Pakistan. Ann Glaciol 43(1):292–300

Miller KJ (1984) International Karakoram Project. 2 vols. Cambridge University Press

Müller F (1962) Zonation in the accumulation area of the glaciers of Axel Heiberg Island, NWT, Canada. J Glaciol 4:302–313

Ollier C (2004) The evolution of mountains on passive continental margins. In: Owens P, Slaymaker O (eds) Mountain geomorphology. Hodder Scientific, London, pp 59–88

Oestreich K (1911–1912) Der tschogletscher in Baltistan. Z Gletsch 6:1–30

Owen L (1988) Terraces, uplift and climate in the Karakoram Mountains, northern Pakistan. Unpubl. dissertation, University of Leicester, UK

Owen L (2006) Quaternary glaciation. In: Kreutzmann H (ed) Karakoram in transition: culture, development and ecology in the Hunza Valley. Oxford University Press, Karachi, pp 12–23

Owen LA, Derbyshire E (1993) Quaternary and Holocene intermontane basin sedimentation in the Karakoram Mountains. In: Shroder JF Jr (ed) Himalayas to the sea. Routledge, London, pp 108–131

Owen LA (1994) Glacial and non-glacial diamictons in the Karakoram Mountains and Western Himalayas. In: Warren WP, Croot DG (eds) The formation and deformation of glacial deposits. Balkema, Rotterdam, pp 9–28

Paffen KJ, Pillewizer W, Schneider HJ (1956) Forschungen im Hunza-Karakorum. Erdkd 10:1–33

Penck A (1919) Die gipfelflur der alpen. Sitzungsberichte preuss Akad 17:256–268

Perla R (1978) Failure of snow slopes. In: Voight B (ed) Rockslides and avalanches, I; Natural phenomena. Elsevier Scientific Publishing, New York, pp 731–751

Price LW (1981) Mountains and man: a study of process and environment. University of California Press, Berkeley

Rapp A (1960) Talus slopes and mountain walls at Tempelfjorden, Spitsbergen. Norsk Polarinstitutt, Skr. 119: Oslo

Rempel AW (2011) Frost. In: Singh VP, Singh P, Haritashaya UK (eds) Encyclopaedia of snow, ice and glaciers. Springer, Dordrecht, pp 303–306

Searle MP (1991) Geology and tectonics of the Karakoram mountains. Wiley, New York

Selby MJ (1993) Hillslope materials and processes, 2nd edn. Oxford University Press, Oxford

Seong YB, Owen A, Bishop MP, Bush A, Clendon P, Copland L, Finkel RC, Kamp U, Shroder JF Jr (2007) Quaternary glacial history of the central Karakoram. Quat Sci Rev 26:3384–3405

Shroder JF Jr (1993) Himalayas to the sea. Routledge, London

Singh VP, Singh P, Haritashaya UK (2011) Encyclopaedia of snow, ice and glaciers. Springer, Dordrecht

Troll C (1954) Über das Wesen der Hochgebirgsnatur. Jahrbuch Deutscher Alpenvereins 80:142–147

Troll C (1975) Vergleichende Geographie der Hochgebirge der Erde in Landscahftsökologischer Sicht. Geographische Rund 27:185–98

Visser PhC, Visser-Hooft J (1935–1938) Wissenschaftliche Ergebnisse der niederländischen Expeditionen in den Karakorum und die angrenzenden Gebiete in den Jahren 1922, 1925 und 1929–30. E.J. Brill, Leiden

Voight B (1978) Rockslides and avalanches, I; Natural phenomena. Elsevier Scientific Publishing, New York

Whalley WB (1984) Rockfalls. In: Brunsden D, Prior DB (eds) Slope instability. Wiley Interscience, New York, pp 217–256

Wilson A (1875) Abode of snow. Repr 1979. Ratna Pustak Bhandar, Kathmandu

# Chapter 9
# Surge-Type Glaciers

**Abstract** The Karakoram is one of a relatively few mountain regions with high numbers of surge-type glaciers, subject to sudden, short-lived accelerations that transfer large volumes of ice down glacier. Some 55 surges have been identified since the 1860s involving 46 glaciers. Various studies suggest many more are surge type, perhaps one-third of Karakoram valley glaciers. Existing observations are reviewed and how they compare with those in other, better-known regions. The glaciers involved are predominantly or wholly avalanche-fed; occur in the highest, steepest parts of the Karakoram; and have large elevation spans. In recent decades, many surge-type tributaries have also been identified; 20 out of 33 events since the 1960s and nearly half (22) of all surges are recognised. In the past, most would have been missed, their numbers underestimated. Surges have impacts out of proportion to their short duration, affecting glacier morphology, surface features, hydrology, erosion and deposition. They are explained by instabilities at the bed of the glacier. Timing and recurrence intervals are peculiar to each case and have little or no relation to climate change or fluctuations in adjacent glaciers. They create huge mass balance anomalies and compromise the use of glacier advances or retreats as climatic signals. They pose a range of hazards for nearby communities.

**Keywords** Surge-type glaciers • Fast flow • Tributary surges • Hasanabad Glacier • Kutiah Glacier • Karambar Glacier • Balt Bare Glacier • Hispar Glacier • Panmah Glacier • Bualtar Glacier

## 9.1 Introduction

Glaciers of *surge type* undergo short episodes of fast flow, alternating with relatively quiet phases lasting decades to centuries (Jiskoot 2011). Surges are usually observed as sudden accelerations that massively disturb the glacier and transfer large volumes of ice from upper to lower sections. Most last a few weeks or months and, in some

K. Hewitt, *Glaciers of the Karakoram Himalaya: Glacial Environments, Processes, Hazards and Resources*, Advances in Asian Human-Environmental Research,
DOI 10.1007/978-94-007-6311-1_9, © Springer Science+Business Media Dordrecht 2014

cases and regions, several years. In the quiet phase, the lower glacier usually retreats and may stagnate. In the upper basin, ice mass is rebuilt preparing conditions for the next surge (Sharp 1988).

It is thought that less than 1 % of glaciers, worldwide, are surge type (Jiskoot et al. 2000). A small fraction of all mountain glaciers are known to surge and only in certain regions. The distribution is highly uneven if seemingly not random. It does include a wide range of environments, from polar to subtropical regions and maritime to extreme continental. Some are in relatively low relief basins near sea level, for example, in Svalbard and southern Patagonia. Others are in extreme high relief terrain such as the Alaska–Yukon, Pamir and Karakoram Ranges. Apart from small concentrations in the north Caucasus and Pamir Ranges (Kotlyakov et al. 2010), none are reported from most Inner Eurasian ranges, including the Greater Himalaya.

The Karakoram has an exceptional concentration (Hewitt 1998a, b; Shroder and Bishop 2010). Some estimates suggest at least 12 % of its valley glaciers are surge type, possibly up to one-third (Kotlyakov 1997; Barrard and Murray 2006). In most of the glaciological literature, this is the only topic under which Karakoram glaciers are mentioned.

The incidence and behaviour of surge-type glaciers alters the character of glaciation and creates unique problems for tracking ice fluctuations and interpreting glacial deposits. Surges and surge cycles are largely independent of climate-related fluctuations and show no relation to rhythms in surrounding glaciers, including others that are surge type. Their cycles of mass balance involve periods of gain that stretch out over many decades or centuries. Major losses are compressed into a few months or years. The sequence and patterns of input and ablation differ substantially from those described in Chaps. 4, 5, and 6. It does seem likely that some of the apparent anomalies in Karakoram glacier responses to climate change have to do with the numbers of surge-type glaciers and, possibly, related flow instabilities in others that have no record of actual surging (Chap. 12).

It is not known whether, or how far, surge-type glaciers declined or expanded, remained the same or ceased over the course of Quaternary glaciations. Some studies suggest recent increases in events are due to global warming (Hewitt 2007a; Smiraglia et al. 2008; Cogley 2010). However, improved observation, especially the aid of satellite coverage, is a large factor too, perhaps the main one. There are no compelling reasons to say recent events are not representative of the Holocene. Evidence elsewhere suggests some glaciers can become or cease to be surge type over time, but today's surge-type glaciers can also have persisted as such for centuries or millennia. It suggests Holocene developments in the Karakoram may well have involved *thousands* of surges, and there is no work to show they were insignificant during the major glaciations.

Modern surge phenomena have been a focus of investigations and theory in glacier dynamics (Clarke et al. 1986; Kamb 1987; Sharp 1988; Paterson 1994; Fowler et al. 2001; Murray et al. 2003). This work can help guide description and interpretation of Karakoram events, but data needed to confirm or challenge proposed surge mechanisms and controls are generally lacking. Critical conditions can only be inferred from surface observations and glacier dimensions. Prior to the 1980s, only one report described surge conditions in the upper basin (Visser and Visser-Hooft 1935–1938, see below). Meanwhile, some evidence suggests there are distinct surge styles and a greater diversity than other regions.

## 9.2 Historical Records of Karakoram Surges

There are confirmed reports of 55 surges since the 1860s, involving 46 glaciers (Table 9.1). Older reports are not always clear about details, although some contain compelling observations (Hewitt 1969, 1998b). It is sometimes possible to reconstruct surges from early descriptions that did not recognise them as such (Hewitt 2007a). Commonly, there are local traditions of sudden, extreme behaviour in the more or less distant past by glaciers with modern surges but also others that seem to lack them.

The surge-type glaciers confirmed here have one or more reports of typical surge events: a sudden, short-lived acceleration, large down-glacier transfer of ice, massive disturbance of the glacier and equally sudden end. Velocities in these cases have reached 10–20 m $d^{-1}$, possibly over 100 m $d^{-1}$. Almost two-thirds occurred after the mid-1970s and one-third in the new millennium, but this seems mainly to reflect improved observation opportunities (Hewitt 1969, 1998a, b, 2007a). Before the 1970s, surges were known almost entirely from terminus advances. Formerly identified surge-type glaciers were those that reach inhabited areas or terminate along trade routes. It meant that almost all reports before the 1990s were of main or trunk glaciers (Table 9.2).

A singular development in recent years concerns many surge-type *tributary glaciers*, 23 recognised and now amounting to almost half of all known cases (Table 9.3; Wake and Searle 1993; Hewitt 1998, 2007a). Comparable events are known in the Pamirs but not comparable numbers (Kotlyakov et al. 2010, F17). No surging tributaries have been reported from most glacierised areas. The near absence of Karakoram reports before the 1990s means that numbers are underestimated. They occur in rarely visited or unvisited valleys and most stall in the main glacier far above the terminus, making observation unlikely.

There are other reasons to expect that events went unreported in the past. Recently, in addition to the less accessible tributaries, main glacier surges have been observed that petered out before reaching or affecting the terminus. New evidence from satellite tracking shows that surge phenomena are not confined to the classic surges emphasised here. Events are reported resembling 'mini-surges', either involving localised acceleration of the glacier concerned or less extreme rates of movement (Kamb and Engelhardt 1987; Shroder and Bishop 2010; Quincey et al. 2009). Velocities may increase by two to ten times but are still an order of magnitude less than the classic surges (Copland et al. 2009; Quincey et al. 2011). Such initiatives will likely extend and may transform the scope and understanding of these phenomena.

**Table 9.1** Inventory of glacier surges in the Karakoram–Himalaya[a]

| Event | Date | Glacier (trib. of) | River basin | Comment | Sources |
|---|---|---|---|---|---|
| 1 | 1860–1861 | Karambar | Ishkoman | Ice dam and GLOF | 1, 2, 3 |
| 2 | 1860–1861 | Maedan (Panmah) | Braldu | Reconstr. | 4, 5, 6 |
| 3 | 1868–1869 | Aqtash | Shyok | 1,600 m in 3 months | 7, 2 |
| 4 | 1886–1887 | Chiring (Panmah) | Braldu | 'immense ice-slip' | 8, 6 |
| 5 | 1890–1892 | Pumari Chhish (Hispar) | Hispar | Major disturbance of main Hispar | 9 |
| 6 | 1892–1893 | Minapin | Hunza | 1,200 m advance | 10, 11 |
| 7 | 1893–1895 | Hasanabad | Hunza | '9.5 km in 2.5 months' | 12, 13 14 |
| 8 | Late nineteenth century | Garumbar (Hispar) | Hispar | 2.5 km adv. | 16 |
| 9 | Late nineteenth century | Barpu | Hispar | Joined Bualtar | 9 |
| 10 | 1895–1905 | Karambar | Ishkoman | Largest GLOF on Gilgit R. | 15, 3 |
| 11 | 1901–1902 | Yengutz Har | Hispar | '2 miles in 8 days', 2.6 km 'in spring' | 16 |
| 12 | 1902–1903 | Aqtash | Shyok | | 17 |
| 13 | 1902–1903 | Marpah Gans (Chogo Lungma) | Basha | 'several miles' | 18 |
| 14 | 1906 | Khurdopin | Shimshal | Major GLOFs | 29 |
| 15 | 1926 | Chong Khumdan | Shyok | Major advance and GLOFs | 31 |
| 16 | 1922–1930 | Bualtar | Hispar | Two advances, 1922–1923 and 1929–1930 | 16 |
| 17 | 1930 | Karambar | Ishkoman | 100 paces in 3 weeks in march | 16 |
| 18 | 1930 | Sultan Chussku | Shyok | 'enormous push… 200–300 million cubic meters of ice….' | 19 |
| 19 | ca. 1930 | Drenmang (Panmah) | Braldu | Reconstructed from photos | 20, 6 |
| 20 | 1928–1939? | Skamri or 'Crevasse' (Gl) | Shaksgam | Massive surge within this time | 2 |
| 21 | 1935–1936 | Aqtash | Shyok | 2.5 km in 7 months | 21 |
| 22 | 1953 | Kutiah | Stak | 12 km in 2 months[b] | 22 |
| 23 | 1974–1975 | Balt Bare | Hunza | 2 km rapid adv. | 23 |
| 24 | 1977 | Fareole (Panmah) | Braldu | | 24 |
| 25 | 1977–1978 | Drenmang (Panmah) | Braldu | From satellite image | 6 |
| 26 | 1987–1990 | Bualtar | Hispar | Two related (?) surge episodes | 25, 26 |
| 27 | 1988–1989 | Pumari Chhish (Hispar) | Hispar | 1.5 km advance into main gl. | 27 |
| 28 | 1989–1993 | Lokpar (Aling) | Hushe | 3 km+ advance of Aling | 28 |
| 29 | 1993 | Karambar | Ishkoman | 3.5 km advance, 7–10 m/day in June | 28 |
| 30 | 1993 | Unnamed (Masherbrum) | Hushe | Surge assoc., rock avalanche | 28 |

(continued)

**Table 9.1** (continued)

| Event | Date | Glacier (trib. of) | River basin | Comment | Sources |
|---|---|---|---|---|---|
| 31 | 1994–1995 | Chiring (Panmah) | Braldu | 2.5 km push into main gl. | 28, 6 |
| 32 | 1990s | Liligo (Baltoro) | Braldu | 2 km advance reaching main gl. | 30 |
| 33 | 1990s | Moni (Sarpo Laggo) | Shaksgam | 2 km rapid advance into main gl. | 28 |
| 34 | 1992 | Sumaiyar Bar (Barpu) | Hispar | | 29 |
| 35 | 1999–2000 | Khurdopin | Shimshal | Main ice surge | 29 |
| 36 | 2001–2002 | Sherpi Kangir (Kondus) | Saltoro | Advance into main glacier | 34 |
| 37 | 2002 | Maedan (Panmah) | Braldu | 3 km advance | 6 |
| 38 | 2002–2010 | Tatulu | Shaksgam | 2.5 km advance | 33 |
| 39 | 2005 | Shingchukpi (Panmah) | Braldu | 3.5 km advance | 6 |
| 40 | 2005–2006 | Drenmang (Panmah) | Braldu | 2 km advance into main glacier | 6 |
| 41 | 2005–2009 | S. Skamri (Skamri) | Shaksgam | 2 km advance | 32 |
| 42 | 2007 | Unnamed | Shaksgam | 36° 08" 22′N 76° 19" 24′E | 34 |
| 43 | 2006–2007 | N. Gasherbrum | Shaksgam | | 33 |
| 44 | 2002–2010 | Sughet | Shaksgam | | 33 |
| 45 | 2007–2009 | Khiang (Hispar) | Hispar | Massive thrust into main glacier | 33 |
| 46 | 2007–2008 | East Kukuar | Hunza | Tributary advance | 34 |
| 47 | 2008–2009 | Murkhun | Hunza | 2008–2009 dead ice along sides, rapid advance | 34 |
| 48 | 2008–2009 | Unnamed (Bilafond) | Saltoro | Rapid advance | 34 |
| 49 | 2008–2009 | Chong Khumdan | Shyok | Rapid advance across river | 34 |
| 50 | 2009 | Unnamed | Shaksgam | 36 07/76 20 | 34 |
| 51 | 2009 | Unnamed | Shaksgam | 36 04/7621 | 34 |
| 52 | 2009 | Aqtash | Shyok | | 34 |
| 53 | 2009 | Nera trib (Panmah) | Braldu | 1 km advance into Chiring, heavy crevassing | 34 |
| 54 | 2010 | Fareole (Panmah) | Braldu | 2 km advance | 34 |
| 55 | 2009–2010 | Balt Bare | Hunza | Rapid advance and break-up | 34 |

[a]Sources: 1 – Hayward (1870); 2 – Hewitt (1969); 3 Kreutzmann (1994); 4 – Schlagintweit and Schlagintweit (1866); 5 – Godwin-Austen (1864); 6 – Hewitt (2007); 7 – Shaw (1871); 8 – Younghusband (1892); 9 – Conway (1894); 10 – Mason (1930); 11 – Mason (1935); 12 – Hayden (1907); 13 – Neve (1907); 14 – Workman and Workman (1911); 15 – Todd (1930); 16 – Mason (1931); 17 – Longstaff (1910); 18 – Workman (1908); 19 – Visser and Visser-Hooft (1935–1938); 20 – Shipton (1938); 21 – Lyall Grant and Mason (1940); 22 – Desio et al. (1961); 23 – Zhang (1984); 24 – Copland et al. (2009); 25 – Gardner and Hewitt (1990), 26 – Hewitt (2009); 27 – Wake and Searle (1993); 28 – Hewitt (1998); 29 – Hewitt (2006); 30 – Pecci and Smiraglia (2000); 31 – Hewitt and Liu (2010); 32 Shroder and Bishop (2010); 33 – Quincey et al. (2011); 34 – KH from satellite image and field notes

[b]See text

**Table 9.2** Locations and some dimensions of known surge-type main valley glaciers, arranged from smallest to largest basin area. Two, the Minapin and Pasu, are added on the basis of indirect evidence of surge-type behaviour (see Shroder and Bishop 2010, F231 and Chap. 12)

| Glacier | Lat.; long. | River basin | Basin area | Length | Main conn glacier | Elevation range |
|---|---|---|---|---|---|---|
| unnamed | 35 13; 77 18 | Nubra–Siachen | 12 | 6.5 | 5.5 | 1,670 |
| Murkhan | 36 35; 75 00 | Hunza | 16 | 7 | 3.4 | 2,585 |
| Sultan Ch | 34 53; 77 57 | Shyok–Indus | 18 | 6 | 5 | 2,140 |
| Balt Bare | 36 20; 74 56 | Hunza | 25.5 | 9.2 | 6.5 | 3,400 |
| Yengutz Har | 36 07; 74 58 | Hunza–Hispar | 32 | 7 | 4.1 | 2,500 |
| Aqtash | 35 05; 77 45 | Shyok–Indus | 35 | 13 | 13 | 2,150 |
| Tatalu | 36 20; 76 33 | Shaks–Yark'd | 42 | 10 | 14 | 2,030 |
| Ghulkin | 36 25; 74 51 | Hunza–Indus | 65 | 20 | 16 | 4,888 |
| Sughet | 35 58; 76 30 | Shaks–Yark'd | 85 | 15 | 38 | 2,830 |
| Kichik KH | 35 07; 77 44 | Shyok–Indus | 86 | 20 | 24 | 2,850 |
| Pasu | 36 27; 74 50 | Hunza | 98 | 24.5 | 25 | 5,060 |
| Minapin | 36 11; 74 33 | Hunza | 110 | 17 | 48 | 4,790 |
| N. 'Crown' | 36 09; 76 11 | Shaksgam | 122 | 18 | 24 | 2,320 |
| Kutiah | 35 48; 74 58 | Stak–Indus | 130 | 17 | 28 | 4,597 |
| Bualtar | 36 08; 74 44 | Hunza–Indus | 139 | 22 | 32 | 4,966 |
| Chong KH | 35 11; 77 40 | Shyok–Indus | 152 | 21 | 39 | 2,830 |
| Urdok | 35 45; 76 45 | Shaks–Yark'd | 162 | 26 | 50 | 3,810 |
| Gasherbrum N | 35 50; 76 41 | Shaks–Yark'd | 190 | 26 | 52 | 3,817 |
| Karambar | 36 36; 74 09 | Gilgit–Indus | 205 | 23 | 36 | 4,298 |
| Toltar–Bal. | 36 29; 74 24 | Hunza–Indus | 212 | 17 | 72 | 4,779 |
| Kukuar | 36 31; 74 12 | Hunza–Indus | 225 | 22 | 80 | 4,072 |
| Hasanabad | 36 25; 74 33 | Hunza–Indus | 320 | 23 | 45 | 5,232 |
| Khurdopin | 36 13; 75 27 | Hunza–Indus | 415 | 41 | 72 | 4,560 |
| Braldu | 36 08; 75 51 | Shaks–Yark'd | 430 | 36 | 165 | 2,937 |
| Skamri | 36 04; 76 08 | Shaks–Yark'd | 495 | 40 | 166 | 3,355 |
| | | **Mean** | **153** | **19** | **43** | **3,539** |

Finally, various studies employ morphological criteria to classify Karakoram glaciers as surge type (Kotlyakov 1997; Barrard and Murray 2006; Shroder and Bishop 2010). These include 'bulging moraine loops', 'convoluted medial and lateral moraines', 'looped' or 'teardrop' medial moraines (Meier and Post 1969; Croot 1988; Jiskoot 2011). The forms certainly occur with known surges, sometimes attaining remarkable scope and complexity (Plate 9.1). The 1998–2000 surge of Khurdopin left a striking, longitudinal zigzag pattern of the medial moraines. Such features occur in dozens of glaciers across the Karakoram and are taken to indicate surge type in many that lack records of actual surging (Kotlyakov 1997). From these data estimates suggest at least 12 % of Karakoram valley glaciers as surge type and possibly a third (Barrard and Murray 2006; Shroder and Bishop 2010). Obviously, if only 44 glaciers are recognised from actual surging, or barely one-third of the main set, the phenomenon has more limited importance than these observations suggest. Instead, it seems certain that many more examples do await identification.

**Table 9.3** Locations and some dimensions of known surge-type tributary glaciers, arranged from smallest to largest basin area

| Glacier | Lat.; long. | River basin | Basin area | Length | Elev range | Gradient | Orient. |
|---|---|---|---|---|---|---|---|
| Pumari Ch (H) | 36 08; 75 12 | Hunza–Hispar | 38 | 10 | 3,950 | 395 | S |
| Nera (P) | 35 54; 76 04 | Braldu | 8 | 4 | 1,710 | 428 | N |
| Fareole (P) | 35 50; 76 03 | Braldu | 29 | 9 | 1,950 | 216 | NE |
| Lokpar (A) | 35 29; 76 14 | Hushe | 15 | 6 | 2,600 | 433 | ENE |
| Marpah G. (CG) | 35 53; 75 10 | Basha–Chigar | 18 | 7.5 | 2,840 | 379 | NE |
| Garumbar (H) | 36 07; 75 01 | Hunza–Hispar | 64.5 | 11.5 | 3,277 | 285 | N |
| Baintha L (B) | 35 52: 76 13 | Braldu | 45 | 11 | 3,075 | 280 | SW |
| Liligo (B) | 35 41; 76 13 | Braldu | 68 | 17 | 3,890 | 229 | N |
| Sumaiyar B (B) | 36 04; 74 55 | Hunza–Hispar | 87 | 21 | 4,093 | 195 | NNW |
| S. Skamri (S) | 36 01; 76 11 | Shaksgam | 14 | 18 | 2,550 | 182 | N |
| Khiang (H) | 36 09; 75 07 | Hunza–Hispar | 175 | 23 | 4,300 | 187 | S |
| Chiring (P) | 35 52; 76 09 | Braldu | 98 | 15.5 | 1,940 | 125 | WNW |
| Unnamed (T) | 36 10; 76 15 | Shaks–Yark'd | 14.5 | 6 | 1,750 | 292 | NE |
| Unnamed (T) | 36 11; 76 11 | Shaks–Yark'd | 39.5 | 12 | 1,865 | 155 | NE |
| Maedan (P) | 35 59; 76 05 | Braldu | 28 | 9.5 | 1,800 | 189 | NNW |
| E. Drenmang (P) | 36 00; 76 05 | Braldu | 40 | 9.3 | 2,420 | 249 | SW |
| W. Drenmang (P) | 36 00; 76 05 | Braldu | 20 | 8.3 | 2,420 | 260 | SSW |
| Shingch'i (P) | 35 48; 76 00 | Braldu | 36 | 11 | 2,450 | 223 | NW |
| Moni (SP) | 35 50; 76 18 | Shaksgam | 42 | 11 | 2,634 | 239 | NW |
| S. Chongpar (SP) | 35 62; 76 19 | Shaksgam | 62 | 15.5 | 2,935 | 189 | NW |
| Chag Tok (SL) | 35 54; 76 15 | Shaksgam | 27 | 8.5 | 2,175 | 256 | SE |
| Unnamed (SLl) | 35 58; 76 20 | Shaks–Yark'd | 22 | 9 | 2,605 | 124 | E |
| Unnamed (B) | 35 17; 76 55 | Saltoro | 15 | 6 | 1,665 | 278 | W |

**Plate 9.1** Surface forms generated by tributary surges into the main Panmah Glacier, Nobande Sobande branch (Hewitt 2007a). Distortions of glacier movement involve at least four known surge events, three in the previous ten years and one from the 1930s (Hewitt 2005)

However, it should also be noted that main ice streams can be surge type *without* displaying such features. They were not observed in surges of the Aqtash and Staghar Glaciers, for instance. When a main glacier surge does occur, former bulging and looped moraines of its tributaries can be completely removed, as at Chiring after the 1994 events (Hewitt 1998). Conversely, tributary surges can create these forms in main glaciers that do not surge. Thus, the main Hispar and Panmah Glaciers have been classed as surge type, but there is no evidence they have surged. Several of their tributaries do and are responsible for observed convoluted moraines and ice streams in the main glaciers.

The discussion here will focus on known 'classic' surge events. Nevertheless, the broader picture should be kept in mind. The possibility of many unrecognised surge-type glaciers, and of a diversity of surge-like behaviours, has implications for any overview of the region's glaciers. They could modify the scope of responses to climate change and complicate recognition of past glaciations being traced through former terminal and lateral moraines, perhaps in glacial sediments far beyond today's glaciers.

## 9.3 Characteristics of Surge-Type Glaciers

> It seems that a glacier which has once made such a rapid advance is extremely likely to repeat the performance after an interval of time…. (Mason 1935, p. 33)

All known surge-type glaciers originate along the main Karakoram crest and a few in the Haramosh Massif, the parts of the range shown to include most of its highest, steepest, coldest and snowiest parts. There are 7 (15 %) that are Alpine types, 20 (42 %) Mustagh type and 21 (44 %) Turkestan-type. This indicates that avalanche nourishment predominates (Chap. 3). On average, rock walls make up 72 % of these basins and 66 % in the perennially frigid source zones. Actual glacier areas comprise just 35 % and heavy debris mantles cover less than 20 % of the ice (Table 9.4). More than half of the glaciers have watersheds rising above 7,000 m, almost 90 % above 6,500 m. Hasanabad Glacier in Hunza, with a maximum basin span of 5,230 m, is notorious for the longest surge length, possibly the fastest advance (see below). With regard to termini, almost 45 % of the main glacier examples end above 4,000 m, almost half descend below 3,000 m and three below 2,500 m. Of tributary glaciers that surge, 70 % terminate above 4,000 m, and two descend slightly below 3,500 m, suggesting they are generally found in somewhat higher locations.

Surge-type glaciers face in all possible directions but a preponderance of north-orientated ice streams is evident due north being the single strongest aspect. Slightly more are of westerly than easterly aspect. The differences are not great and may be altered as more main and tributary surges are identified.

Main surge-type glaciers are, on average, nearly twice as long as tributary ones, the longest known to surge being the Khurdopin (41 km). Only two observed to surge lie within the 'large' category (Khurdopin and Skamri). There are 13 longer than 20 km and five shorter than 10 km. No actual records of surges exist for 32 of

**Table 9.4** Features of surge-type main valley and tributary glaciers identified from surge events

| Dimensions | All surge type | Main | Tributary |
|---|---|---|---|
| No. identified | 48 | 25 | 23 |
| Length (km) | | | |
| *Mean* | 14 | 19.5 | 11 |
| *Longest* | 41 | 41 | 23 |
| *Shortest* | 4 | 6.5 | 4 |
| Basin area (km$^2$) | | | |
| *Mean* | 77.5 | 153 | 43.7 |
| *Largest* | 495 | 495 | 175 |
| *Smallest* | 8 | 12 | 4 |
| Elev. span (m) | | | |
| *Mean* | 3,078 | 3,538 | 2,648 |
| *Largest* | 5,232 | 5,232 | 4,300 |
| *Smallest* | 1,665 | 1,670 | 1,665 |
| Gradients | | | |
| *Mean* | 238 | 207 | 252 |
| Glacier area km$^2$ (*% basin*) | | | |
| *Mean* | 22 (35) | 43 (28) | 15 (42) |
| *Largest* | 166 (34) | 166 (34) | 43 (44) |
| *Smallest* | 2 (25) | 3.4 (21) | 2 (25) |
| Rock wall area km$^2$ (*% basin*) | | | |
| *Mean* | 60 (72) | 110 (69) | 36 (??) |
| *Largest* | 335 (81) | 335 (81) | 115 (66) |
| *Smallest* | 10 (62) | 10 (62) | ?? |
| Heavy debris km$^2$ (% glacier area) | | | |
| *Mean* | 4 (16) | 9 (18) | 2 (15) |
| *Largest* | 32 (44) | 32 (44) | 5 (29) |
| *Smallest* | 0 (0) | 0.5 (10) | 0 (0) |

the set of 42 largest glaciers. The 9 longest glaciers[1] are not known to surge, although most have surging tributaries. Those in the 42 main sets also tend to be steeper than non-surge types, and the tributary ones are especially steep.

These relations seem to differ significantly from the other two regions with high concentrations of surge-type valley glaciers. In Alaska–Yukon Ranges and Svalbard, it has been shown that surging is positively correlated with length and lesser gradients (Clarke et al. 1986; Jiskoot et al. 2000). The statistics for the Karakoram are not as exhaustive as the other two regions, but sufficient to suggest relations to length and steepness are different. The overall picture may change as data for nonclassic type surging or 'mini-surges' expands, as improved satellite imagery suggests it will (Copland et al. 2009; Quincey et al. 2011).

[1] Some have claimed that Hispar, Siachen and Baltoro are surge types. Perhaps they are but there is no record of them surging, and morphological criteria reported are due to the action of surging tributaries (see below).

It should be added that the work cited for Alaska and Svalbard had not recognised or dealt with any surge-type tributary glaciers. With regard to glacier length and *surge length* – the increase that occurs in a surge event – tributary glaciers require different treatment from main glaciers. Many surging tributaries are connected to a main glacier or with termini close to it. When a surge occurs, run-out is generally stalled in main glacier ice which it massively disturbs, as described below. In examples known to date, this has not caused the main glacier to surge. Meanwhile, not all main glacier surges reach the terminus, at least in the Karakoram. Those that stall before reaching it further complicate relations to glacier dimensions and the meaning of surge length.

## 9.4 Surge Phenomena

Surges have dramatic impacts on glacier behaviour and on surrounding landscapes. They can displace some cubic kilometres of ice far down valley. They can transport as much debris in a few months as some centuries of normal activity (Hewitt 2009). Typically, signs of extreme hydrological disturbance include expulsion of turbid water, development of ice-margin ponds and their sudden drainage, destabilisation of surrounding slopes or rapid deposition along glacier margins (Plate 9.2).

The classic picture of surge events emphasises a sudden switch to fast flow and back again and a long quiescent period, essentially two phases. However, more

**Plate 9.2** Impacts along ice margins during the 1987 Bualtar Glacier surge (Hewitt 1987)

detailed investigations have revealed complications. In Karakoram examples, as many as five or, perhaps, six distinct phases can be discerned. Based on terms used by Jiskoot (2011, p. 417), they are:

(i) *A build-up phase*: when the upper glacier is growing and the lower can seem to behave 'normally'.
(ii) *A pre-surge phase*: when the glacier gradually speeds up and advances, possibly over several years.
(iii) *The surge or 'active' phase*: the period of fast flow as described above.
(iv) *Post-surge deceleration and thinning*: in which the glacier slows gradually, crevassing and ice level of the lower glacier declines and a slower advance may continue.
(v) *Quiescent phase*: early in which a substantial section of the glacier may detach and stagnate to persist for decades far below the terminus under heavy debris. The active glacier steadily retreats.
(vi) *Two or more active events*: rather than a single acceleration, two may occur separated by a few months or years, a short time compared to a much longer quiescent phase. This has been observed at Bualtar Glacier (see below).

Many Karakoram surges have been described as classic two-phase events. More recent and sustained observations have shown other developments, especially in tributary surges. Distinctive and extended pre- and post-surge phases occurred in each of five tributary surges at Panmah Glacier (Hewitt 1998a, 2007a) and probably in the Liligo 1990s event (Pecci and Smiraglia 2000). Skamri Glacier displays a variety of such irregular developments (Copland et al. 2009, p. 45). Some glaciers continue to advance for months or years after an active phase, if at slower rates.

Some surges are exhausted before reaching the terminus, although a delayed advance may occur later, as after the Bualtar 1987 and 1990 surges and at Khurdopin since 1999–2000 (Gardner and Hewitt 1990; Hewitt 2006). Surging tributaries have special relations to main glacier termini and mass balance. Surge ice reaching and entering the main glacier during the active phase is stalled, and large volumes build up around the junction. They continue to spread into, thicken and disturb the main glacier for years or decades after the surge proper.

### *9.4.1 Surge Dynamics*

Surge-type glaciers are invariably out of phase with neighbouring glaciers, not least when these are surge type. The exact timing of the active surge may be influenced or triggered by weather conditions, notably episodes of heavy ablation, possibly landslides (Eisen et al. 2005; Hewitt 2009). The length of a given cycle may be affected by variations in snowfall or ablation during intervening years (Harrison and Post 2003). Even so, simultaneous surges in two or more independent glaciers, or tributaries of the same glacier, are virtually unknown.

In his classic paper on 'threatening glaciers', Mason (1935) treated events now recognised as surges as 'accidental advances'. He attributed them to earthquakes or avalanches, perhaps 'obstructions' along the glacier. He and others attempted to interpolate advances from presumed climatic 'cycles', but these are problematic and probably mistaken (Longstaff 1910; Hedin 1910; Mason 1930). Today, explanations look primarily at ice mass morphology and dynamics. The decisive conditions are thought to be instabilities occurring at critical sites at the bed of the glacier. Constraints may include morphology that resists sliding and throughflow, such as at valley constrictions, bends, changes in steepness or interference by inflow of tributaries. However, the sudden switches to fast flow require factors that change more or less quickly or thresholds that trigger catastrophic change.

Elsewhere in the world, two classes are proposed 'Alaska type' and 'Svalbard type' (Jiskoot 2011, p. 419). The former involves sudden onset events, high maximum velocities and equally sudden termination, usually associated with large discharges of water, essentially what was referred to as the 'classic' surge for the Karakoram. In Svalbard, on the other hand, surges typically have a slower rate of onset, rise to maximum velocities more gradually, but may sustain them much longer. Return to quiescent mode may also take years, supporting the notion of three-phase surges.

It has been suggested that these represent two distinct surge mechanisms, one governed mainly by glacier plumbing and hydrology and the other by thermal instabilities (Kamb 1987; Clarke and Blake 1991; Fowler et al. 2001; Kavanaugh and Clarke 2006). Another possibility is that the strength and deformation of basal sediments controls the process (Sharp 1988; Murray et al. 2003). In Iceland and the Pamirs, somewhat different and more diverse types of surge are reported, and even in a few glaciers in Alaska and Svalbard.

In principle, all the requirements of the hydrological, thermal and sediment deformation models could be satisfied in the Karakoram. Thermal regimes are diverse and change down glacier. In the mid- to upper ablation areas where surges appear to commence, highly concentrated ablation can occur. Rugged terrain, vertical climatic gradients and diverse morphology involve obstructions and interactions that can interrupt or disturb the throughput of ice. Vertical gradients help ensure thermal complexity. Avalanche nourishment helps the build-up of subglacial sediment. The diversity of lithologies in glacier basins may be a factor affecting subglacial sediment characteristics. However, there are no data on any of these conditions at the bed of any Karakoram glacier.

Until the 1970s, all reported surges seemed of Alaska type, with sudden onset, high peak velocities, large frontal advance and sudden termination (Hewitt 1969). A few direct observations reported ice-margin ponds and floods of turbid water expelled from the glacier, as also in the Bualtar surge of 1986 and the Karambar surge of 1993. Conversely, the Maedan, Shingchukpi and Fareole tributaries of Panmah seem more like Svalbard types, with gradual advance over several years prior to the sudden fast flow episodes and continued disturbance for years afterwards (see below). Indeed, other cases seem a mix of the Alaska and Svalbard types and some not exactly like either.

## 9.5 Karakoram Main Glacier Surges

Some events, reported by early modern explorers of the region, acquired a legendary status. Accounts of the small Yengutz Har Glacier surge made a great impression (Table 9.1). It descends northwards in a steep valley just below the main Hispar Glacier. In 1892, Conway (1894) found the path descended into Yengutz Har valley which had '…in its bowels some half a dozen [water] mills…' The terminus was glimpsed some 2.5 km higher up. By contrast, Hayden (1907) found that in 1906 the path *ascended* '…over a steep mass of black and slippery ice, the mills are gone, and their ruins hidden under the snout of the advancing glacier…'. The glacier had surged in 1903. A compelling description, touching on some typical surge features, was given by a villager to one Capt. Berkeley:

> The glacier [had been far] above where the present snout is. One day…the water in the irrigation channels was very muddy and coming down in greater quantity than usual. We went up the nullah [= valley]… and saw the glacier advancing. It came down like a snake, quite steadily… At the same time water and mud gushed out from the ice… and flooded our polo ground and some fields. When an obstruction got in the way the ice went round it at first then overwhelmed it. The ice … continued to move for eight days and eight nights and came to a stop about forty yards from the Hispar River. As soon as the ice stopped, the mud and water…stopped too. The ice remained down for fifteen years…. Twelve years ago (1918) the ice began to go back. Every day, a length of about fifteen yards would break off the main ice and was washed away… (qu. Mason 1935, p. 31)

The dangers and shock for local inhabitants are evident. Details point to an Alaska-type event. When I passed through Hispar in 1985 and again in 2000, the Yengutz Har was not visible from the path. The half-dozen water mills had been rebuilt. In 2010, the terminus was 2.9 km from the Hispar River, but ice in the source area appears fully restored and the terminus active. A surge seems likely in the not too distant future.

The Hasanabad Glacier in Hunza is a contender for the furthest and, possibly, most rapid advance of any mountain valley glacier, something usually accorded to the Icelandic Brúarjökull Glacier, with an estimated surge velocity of 125 m $d^{-1}$ and advance of 10 km in a year (Jiskoot 2011, p. 417). However, reports have Hasanabad advancing up to 11.5 km between 1892 and 1903 (Mason 1930). One describes an advance of 'over 6 miles' (= 9.7 km) in 2½ months (Hayden 1907). If true, it comes to 130 m $d^{-1}$. The report does contain contradictions (Goudie 1984).

Meanwhile, Hasanabad Glacier is fed by two large tributaries, the Mouchuhar and Shishpar Glaciers. They join about 8 km north of the Hunza River. In 1892, when observed by Conway (1894), the tributaries were detached, their termini over 14 km from the Hunza River. In effect, there was no Hasanabad Glacier, and except for a few years following the great advance, the same has applied for most of the time since the 1940s. The apparent coincidence in the great surge of rapid advances from the two independent tributaries is difficult to reconcile with surge models now accepted. It had seemed plausible in the days when external triggers such as earthquakes were invoked.

A huge retreat occurred between 1932 and 1954. In the latter year, the Shishpar and Mouchuhar were, respectively, 9 km and almost 12 km from the Hunza or close to the 1892 positions (Paffen et al. 1956; Schneider 1967). The two glaciers rejoined by 1979–1980, but no surge was reported (Goudie 1984). By 2005, the tributaries were again separated. Shishpar lay across the mouth of Mouchuhar whose active terminus was as far back as in 1892. Much stagnant, debris-mantled ice filled its valley, a condition often found after a surge. By 2010, the Shishpar branch had advanced independently about 1.5 km beyond the junction, but Mouchuhar was unchanged. In all, the record and explanation of surges involving separate basins are an unresolved challenge.

## 9.5.1 Kutiah Glacier

A surge of Kutiah Glacier in 1953 was described as exceptional in extent and rapidity (Desio 1954). The glacier descends from the east flank of Haramosh (7,397 m). A year after the surge, it was concluded the advance had been 12 km at '…113 m a day….' (Desio et al. 1961). If so, it would be the furthest advance on record, possibly the highest velocity. However, I visited Kutiah in the 1990s and observations then seemed at odds with the earlier report. There were three sets of well-defined terminal moraines over a distance of some 1.8 km below the active terminus. It seems the Italian map and report mistook those left by earlier events and descriptions of them given by villagers, for the 1953 position and story. One was from the nineteenth century and another records the greatest reported calamity from some centuries earlier. The latter involved both the south-flowing Klars'ng Glacier, which has been disconnected from the main glacier for a century or more, and the main east-flowing Kutiah or 'Khotia Lungma'. Only the latter was involved in 1954 event, but the Italian team was led to believe both ice streams surged and attributed some other conditions to it that actually referred to the much older event. The mid-nineteenth-century Survey of India maps do show the two glaciers joined and Kutiah terminus close to the 1953 surge position, but only the Kutiah was involved in the nineteenth-century surge. In the 1946, 1:250,000 Survey of India topographical sheet, the two glaciers are as far apart as today. The area between the two glaciers has extensive terraced fields and mature trees, old enough to date from well before the 1950s. The much older surge event that involved both glaciers caused the terminal lobe to advance a kilometre beyond that of the mid-nineteenth-century. It left the outermost moraines at the margins of the highest permanent settlements in Stak Valley, and those moraines seem to have been mistaken for the 1953 event. The latter certainly remains one of the larger and faster surges on record, but my informants and survey suggest its outer moraine was almost 2 km further back than the Italian report suggests.

It is of note that Hasanabad and Kutiah had their furthest known retreats *and* greatest advances during the Little Ice Age, the last general phase of glacier expansions. Many other Karakoram glaciers reached their greatest or last advance towards the end of the nineteenth century, but some early in the twentieth, and others at some time before the earliest, mid-nineteenth-century records (Kick 1989). No one has worked on earlier Quaternary sediments to try to determine if and when surges occurred in prehistoric times. It is a complicated, hard-to-resolve record, and many of the apparent anomalies of Karakoram glacier fluctuations may arise from the presence of surge-type glaciers (Chap. 12).

### 9.5.2 *Karambar Glacier*

In the far western Karakoram, several glaciers have reputations for sudden advances and blocking the Karambar River (Hewitt 1998b). Some six glaciers are known to have dammed the river (Hewitt 1998c; Iturrizaga 2011, p. 387). Their past advances are evident on the ground and from glacier lake outburst floods reported in the nineteenth century (Longstaff 1920; Todd 1930; Mason 1931; Hewitt 1969; Kreutzmann 1994). However, only surges of Karambar Glacier have been observed, a Turkestan-type responsible for several dams and outburst floods (Chap.10). Its surges were observed in 1955 and 1993 and inferred from a 1905 ice dam, suggesting a 35–40-year surge interval, or two to three surges per century.

Shepherds from Bhurt first reported the 1993 surge in April. They had already crossed to Karambar Ilag, a temporary pasture and settlement along the glacier, and were trapped there by the surge. I reached the glacier in June when the terminal area was still advancing at 7–10 m $d^{-1}$ (Plate 9.3). Irregular streams of muddy water rushed along the glacier margins. Ahead of the main ice front was a secondary snout of formerly dead ice, reactivated from beneath the pro-glacial plain and pushing steep banks and cones of moraine and outwash materials before it. The dead ice goes back to the 1950s surge, buried and protected from melting by outwash and terminal moraines as the glacier receded. The total advance in 1993 was about 2.5 km. The ice reached the Karambar River but no dam was sealed. In 2006, the active terminus had retreated a kilometre from the river, and the glacier below the first icefall had thinned considerably.

### 9.5.3 *Khumdan Glaciers*

> When the Chong Khumdan glacier is advancing, its neighbour the Kichik Khumdan is degenerate and vice versa. The movements therefore must be...independent of local climatic causes...[and] mainly due to properties possessed by each individual glacier. (Mason 1935, p. 17)

As glacier monitoring spread to the region in the early twentieth century, four glaciers in the upper Shyok valley generated great interest (Hedin 1910; Longstaff 1910; Dainelli and Marinelli 1928; Mason 1930). Surge-type advances have been

**Plate 9.3** Heavily debris-covered terminus of Karambar Glacier during the 1993 surge with the outermost cliff involving dead ice and pro-glacial sediments mobilised ahead of the main front (Hewitt June 1993)

observed for Chong Khumdan, Sultan Chhusku and Aqtash (Table 9.1). Conditions during advances of Kichik Khumdan seem likely to involve surges but they have not been confirmed. A peculiarity of Kichik and Chong Khumdan is how, following a sudden advance, they remain active for up to a decade and enough to reseal ice dams after breaches (Chap. 10).

In 2010, the small Sultan Chussku Glacier was 5–6 km from the Shyok, but in 1930, it sent a huge bulb of ice into the river. Following the surge, the basin was explored by members of the Visser expedition (Visser 1932; Visser and Visser-Hooft 1935–1938), who reported that, in the 'reservoir zone', ice had collapsed as dramatically. Tributaries and large cones of avalanched snow were sheared off and left suspended above cliffs 50–60 m above the glacier.

After the 1930s, no more twentieth-century advances were reported. In 2000, however, the Kichik Khumdan advanced and reached the far shore of the Shyok River, staying there until 2005. Satellite imagery between 1998 and 2002 showed the lower 4 km of Chong Khumdan was nearly stagnant, its north branch detached (Hewitt and Liu 2010). Then, in 2006, the north branch started a sudden forward push, rejoining the west branch and massively distorted its flow to create looped moraines indicative of a surge. By 2009, the terminus was across the Shyok. Velocities of 150 and 250 m annually were measured but higher short-term velocities probably occurred (Quincey, pers. comm. 2011). In 2009, it was Aqtash that

**Table 9.5** Observations of surface velocities of Bualtar Glacier in summer 1986 and during passage of 1987 surge event (After Gardner and Hewitt 1990 and SIHP, unpubl. records)

| Location above Hispar River (km/m) | Dates | Velocities (m $d^{-1}$) | Comment |
|---|---|---|---|
| 4/2,600 m | July 1986 | 0.4 | Hopar–Shishkin profile |
| | May 31, 1987 | 7.6 | Mid-surge zone |
| | Aug. 1987 | 0.3 | Surge passed |
| 6/2,750 | June 20–25, 1987 | 11.5 | Theod. obs. from Hopar of single mid-ice stake |
| 8/2,900 | July 86 | 0.8 | Shaltar profile |
| | Jun.1, 1987 | 3.2 | Trailing part of surge |
| | Aug. 1987 | 0.6 | Surge passed |

suddenly thickened and advanced across the river. A steep terminal cliff shedding ice blocks may indicate a surge. Clearly these glaciers are not acting in conformity with, nor always in the same direction as, the continuous and unprecedented glacier retreats reported from the rest of the Himalaya.

### *9.5.4 The 1986 and 1989 Bualtar Surges*

Bualtar, sometimes called 'Hopar' Glacier, in Nagar, has a reputation for periodic, sudden advances. These have dammed the Hispar River at times before the late nineteenth century (Conway 1894; Workman and Workman 1911). Sudden advances were reported in both 1922 and 1923 and again in 1929–1930 with stagnation and retreat in between (Visser 1926; Mason 1931). There were also two distinct surges in 1987 and 1990 (Gardner and Hewitt 1990; Hewitt 2009). Neither recent surge reached or caused immediate change in the terminus but, over the next few years, it advanced some 2.5 km.

Glacier ablation and movement studies being conducted on Bualtar Glacier in 1985 and 1986 provide some baseline information (SIHP 1989). They showed mid-glacier velocities of 0.8 m $d^{-1}$ in the ablation zone, about 9 km from the terminus and declined to about 0.4 m $d^{-1}$ some 3 km above the terminus. Observations were made between these same sections in June and July 1987. They identified a wave of fast-moving, higher ice. A few measurement stakes that had survived and some large boulders on the ice were surveyed to determine velocities. The highest observed were between 7 and 11.5 m/day (Table 9.5).

In the region of rapid movement, there was intense crevassing and seracs. The ice thickened as much as 15–25 m (Plate 9.4). It overtopped lateral moraines that had been ice-free for some decades and formed an ice edge of vertical or overhanging ice cliffs 10–15 m high. There was a sharp increase in erosion at the ice margins, new collapses from older glacial deposits above the glacier and sudden filling and drainage of ice-margin ponds (Plate 9.5).

**Plate 9.4** Surface of Bualtar Glacier during 1987 surge, as fast flow region travelled from right to left causing heavy crevassing. Collapses and landslides of lateral margin deposits are shown along the Shishkin flank on the far side of Bualtar. Barpu terminus is above them (Hewitt 1987)

**Plate 9.5** Lateral margin impoundment below Hopar villages and heavily crevassed ice during the 1987 surge of Bualtar Glacier (Hewitt 1987)

The 1987 surge occurred six months after a series of landslides descended onto the glacier (Hewitt 1988). They occurred between July 29 and 31, 1986, and covered some 3 $km^2$ of the glacier with rock debris to depths of 5–10 m. The surge appeared to have commenced in the area of the rock avalanche debris, and their combined disturbance of the glacier continues to the present day (Hewitt 2009).

## 9.6 Tributary Surges at Panmah

Between 1994 and 2010, five independent tributaries of Panmah glacier surged, three in quick succession between 2001 and 2005 (Fig. 9.1). It is the sixth largest Karakoram glacier and the surges affected the west-, or south-flowing, east-bank tributaries of the Nobande Sobande branch. All these tributaries excepting Drenmang had detached and retreated from the main glacier sometime after the 1930s. All surges reached the main glacier except, to date (2011), Fareole's. Three of the tributaries had been classified as 'suspected' surge type before the surges occurred (Kotlyakov 1997).

Chiring tributary was much wasted but still connected to the main glacier in the 1930s (Savoia-Aosta 1936). It became detached by the mid-twentieth

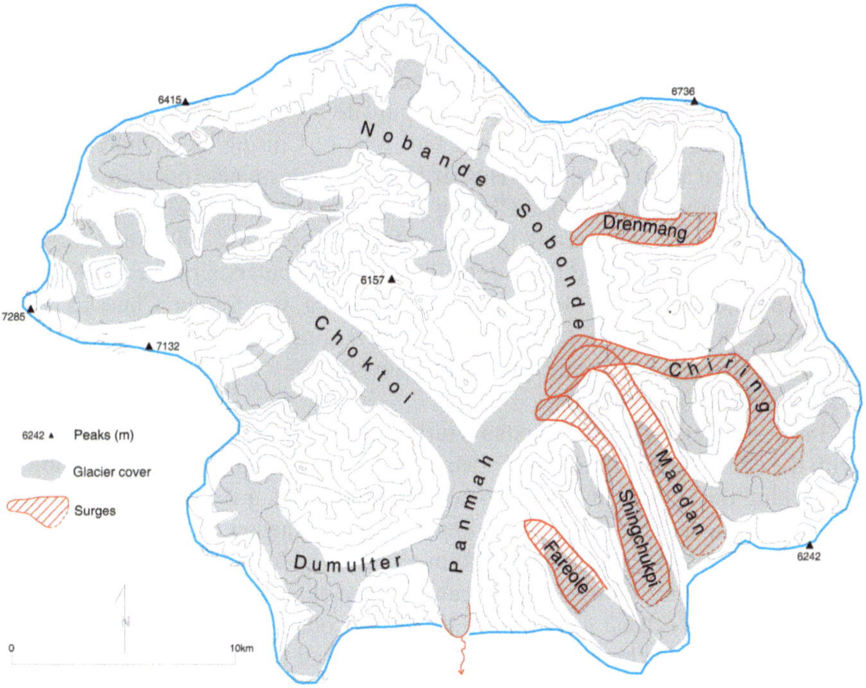


Fig. 9.1 Tributary surges in Panmah Glacier basin between 1994 and 2011 (After Hewitt 2007a)

**Plate 9.6** Post-surge conditions in the lower Chiring Valley. Sheared-off ice on opposite wall shows the height of the surge, 150 m above the declining glacier surface (Hewitt 1996)

century. In October 1992, I camped on the ice-free floor below Chiring snout, which was about 1.5 km above the Nobande Sobande. The 1994–1995 surge had occurred when I returned in 1996. Our campsite lay beneath 250–300 m of ice. The Nobande Sobande contained a 3 $km^2$ lobe of Chiring ice compressing the main glacier to a narrow ribbon (Hewitt 1998a). Steep arcuate pressure ridges up to 100 m high formed the rim of Chiring ice. Along the valley walls of the Chiring were trim-lines of sheared-off ice up to 150 m above the 1996 surface. They marked the surface during the surge (Plate 9.6). Tributaries of the Chiring had been sheared off from Tsoka upwards, recalling the Sultan Chussku surge described above. Massive subsidence and crevassing of the ice was observed to about 5,300 m elevation in the firn basin below the West Mustagh Pass (5,600 m), 12 km from the main glacier.

In September 1992, the Maedan's terminus was some 3 km from its junction with the Chiring. By 1996, it had advanced 1.5 km and satellite imagery showed it reaching the Chiring in 2000. Some time after 2002, a surge occurred involving thickening and break-up of the lower 5 km of Maedan, with masses of dirty ice thrust up in places more than 100 m along the side-walls (Hewitt 2007a). A lobe of Maedan ice entered the Nobande Sobande, marked by a series of concentric pressure ridges. The highest was at the outer rim, and Chiring Glacier was now overridden by Maedan ice and compressed to a narrow thread opposite the junction.

**Plate 9.7** Surge of the Shingchukpi Glacier viewed from Nobande Sobande, showing thrust up dirty ice, heavy crevassing and seracs and pressure ridges in main glacier ice (Hewitt 2005)

Shingchukpi Glacier[2] lies parallel to the Maedan and has a similar morphology. In 1996, it was 2.5 km from the main glacier but then made a steady advance of almost 2 km before a surge commenced in winter 2004–2005. By June 2005, the lower two-thirds of the glacier had become severely broken, and dirty marginal ice was thrust up tens of metres against the valley walls. Where they met, Nobande Sobande ice remained quite distinct in a series of concentric pressure ridges (Plate 9.7).

The latest Drenmang surge came from the east branch, the larger of its tributaries, and probably began in 2004. It created vertical ice cliffs along both flanks where broken, thrust up masses of dirty ice overrode the old lateral moraines.

Recent events aid in recognising past surges. Published and archival materials were found to document a Maedan surge between 1857 and 1861, Chiring in 1886–1887 and Drenmang in 1930–1931 and 1977–1978 (Hewitt 2007b). Recurrence intervals are indicated of some 155, 120 and 50 years, respectively. No report of previous Shingchukpi surges was found but local shepherds identify it with glaciers that surge.

Each of these tributaries is completely independent or, at least, had no connection to each other when they surged. Nevertheless, they have some common features. Entry into the main glacier absorbed much of their energy, stalling them and greatly reducing surge run-out. All the surges lasted a few months with the typical rapid movement, thickening and heavy crevassing. However, major

[2]The name given to me by local shepherds, also called 'Second Fareole' ('Karakoram Orographical Sketch Map', Swiss Foundation for Alpine Research, 1990, Sheet 1). 'Fareole', however, means 'opposite' and seems used to point out the glacier from the west-flank path, not to name it.

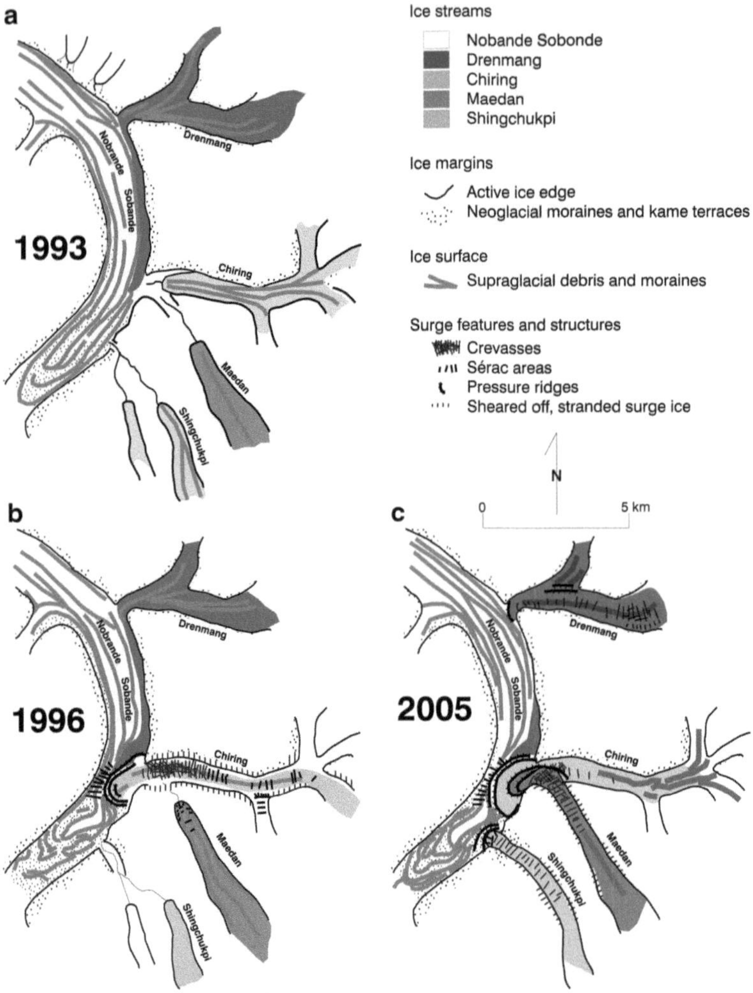


**Fig. 9.2** Surface changes at Panmah Glacier associated with recent surges: (**a**) in 1993 prior to the Chiring surge, showing ice of the 1977–1978 surge of Drenmang having travelled 0.5 km to reach the mouth of Chiring by this time, (**b**) 1996, following the 1994–1995 Chiring surge and pre-surge advance of Maedan, (**c**) 2005, following the surges of Maedan, Shingchukpi and Drenmang, showing the complex interactions of surge ice and the main ice stream

pre- and post-surge developments continued for years or decades (Fig. 9.2). Adjustments of surge ice in the main glacier have gone on for 18 years in the case of the Chiring lobe and seem likely to continue for much longer. The surges have not yet affected the main glacier terminus (2010), but a delayed, non-surge advance is likely to occur later, something that may explain Panmah's curious 150-year record (see Chap. 11).

## 9.7 Relations of Surges to Climate and the Glacial Record

The 55 surges reported in 1860–2010 could mean one in every 2.7 years. Since many surges must have been missed, the incidence in the past 35 years may be more accurate, that is, close to one event per year. If at all representative, these two sets suggest between at least 3,700 and possibly over 10,000 surges in the Holocene. There is, as yet, no compelling reason to say surge-type numbers or conditions have changed although they may have (see Chap. 11).

At present, it is not known whether the major Pleistocene glaciations included as many, more, fewer or – as is implied by a literature that fails to mention them – no surge-type ice masses. Given that a common strategy in the Karakoram, as elsewhere, is to reconstruct glacial sequences from terminus positions and moraines, caution seems in order. Some of the most intensively studied glacial deposits involve several of the surge-type glaciers identified above, apparently without any awareness of what effect that might have (Li Jijun et al. 1984; Kalvoda 1992). Yet, as shown above, surges can drastically reschedule glacial sequences, and further complications can be expected with tributary surges (Chap. 12).

## References


Barrard N, Murray T (2006) Multivariate controls on the incidence of glacier surging in the Karakoram Himalaya. Arct Antarct Alp Res 38(4):489–498

Clarke GKC, Blake EW (1991) Geometric and thermal evolution of a surge-type glacier in its quiescent state, Trapridge Glacier, Yukon Territory, Canada, 1969–89. J Glaciol 37(125):158–169

Clarke GKC, Schmok JP, Ommanney CSL, Collins SG (1986) Characteristics of surge-type glaciers. J Geophys Res 91:7165–7180

Cogley JG (2010) A more complete version of the World Glacier Inventory. Ann Glaciol 50(53):32–38

Conway WM (1894) Climbing and exploration in the Karakoram-Himalaya. 3 vols. Appleton, New York

Copland L, Pope S, Bishop MP, Shroder JF, Clendon P, Bush A, Kamp U, Seong YB, Owen LA (2009) Glacier velocities across the central Karakoram. Ann Glaciol 50(52):41–49

Croot DG (ed) (1988) Glaciotectonics: forms and processes. Balkema, Rotterdam

Dainelli G, Marinelli O (1928) Le Condizioni Fische Attuali. Spedizione Italiana de Filippi 4(series 2):1–68

Desio A (1954) An exceptional glacier advance in the Karakoram-Ladakh region. J Glaciol 2(16):383–385

Desio A, Marussi A, Caputo M (1961) Glaciological research of the Italian Karakoram Expedition, 1953–1955. International Association Hydrological Sciences. IAHS Pub 54, pp 224–232

di Savoia-Aosta D (1936) La Spedizione Geografica Italiana al Karakoram 1929. S.A. Arti Grafiche Bertarelli, Milano

Eisen O, Harrison WD, Raymond CF, Echelmeyer KA, Bender GA, Gordeev VV (2005) Variegated Glacier, Alaska, USA: a century of surges. J Glaciol 51(174):399–422

Fowler AC, Murray T, Ng FSL (2001) Thermally controlled glacier surging. J Glaciol 47(159):527–538

Gardner JS, Hewitt K (1990) A surge of Bualtar Glacier, Kararoram Range, Pakistan: a possible landslide trigger. J Glaciol 36(123):159–162
Godwin-Austen HH (1864) On the Glaciers of the Mustagh Range (Trans-Indus). Proc R Geogr Soc 34:19–56
Goudie AS (1984) The Geomorphology of the Hunza Valley, Karakoram Mountains, Pakistan. In: Miller KJ (ed) International Karakoram Project, vol 1. Royal Geographical Society, London, pp 359–410
Harrison WD, Post AS (2003) How much do we really know about glacier surging? Ann Glaciol 36:1–6
Hayden HH (1907) Notes on certain glaciers in Northwest Kashmir. Rec Geol Surv India 35:127–137
Hayward GW (1870) Journey from Leh to Yarkand and Kashgar. J R Geogr Soc 40:33–166
Hedin SA (1910) The Kumdan Glaciers in 1902. Geogr J 36:184–194
Hewitt K (1969) Glacier surges in the Karakoram Himalaya (Central Asia). Can J Earth Sci 6:1009–1018
Hewitt K (1988) Catastrophic landslide deposits in the Karakoram Himalaya. Science 242:64–67
Hewitt K (1998a) Glaciers receive a surge of attention in the Karakoram Himalaya. EOS Trans Am Geophys Union 79(8):104–106
Hewitt K (1998b) Recent glacier surges in the Karakoram Himalaya, northern Pakistan. Supplementary Materials, EOS, Amer Geophys Union http//www.agu.org/eos_elecas97106e.html
Hewitt K (1998c) Himalayan Indus streams in the Holocene: Glacier-, and Landslide-'Interrupted' fluvial systems Chap 1. In: Stellrecht, I (ed) Karakorum-Hindu Kush-Himalaya: dynamics of change part I. Rudiger Koppe Verlag, Koln
Hewitt K (2006) Rock avalanche dams on the transHimalayan Upper Indus streams: a survey and assessment of hazards-related characteristics. Ital J Eng Geol Environ (Spec Issue I):61–66
Hewitt K (2007a) Tributary glacier surges: an exceptional concentration at Panmah Glacier, Karakoram Himalaya. J Glaciol 53(181):181–188
Hewitt K (2007b) Rediscovering colonized landscapes: the first Europeans at the Mustagh Pass, Karakoram Himalaya, Inner Asia. In: Gervers M, Bulag U, Long G (eds) The exploitation of the landscape of Central and Inner Asia, Toronto studies in Central and Inner Asia, 9. Asian Institute, University of Toronto, Toronto, pp 41–67
Hewitt K (2009) Rock avalanches that travel onto glaciers: disturbance regime landscapes, Karakoram Himalaya, Inner Asia. In: Crosta G (ed) Geomorphology. Int J Pure Appl Geomorphol 103:66–79
Hewitt K, Liu J (2010) Ice-dammed lakes and outburst floods, Karakoram Himalaya: historical perspectives and emerging threats. Phys Geogr 31(6):528–551
SIHP (Snow and Ice Hydrology Project) (1989) Annual Report 1988. Cold Regions Research Centre, Wilfrid Laurier University, Waterloo, Ontario
Iturrizaga L (2011) Glacier lake outburst floods. In: Singh VP, Singh P, Haritashya UK (eds) Encyclopaedia of snow and glaciers. Springer, Dordrecht, pp 381–399
Jiskoot H (2011) Glacier surging. In: Singh VP, Singh P, Haritashya UK (eds) Encyclopaedia of snow and glaciers. Springer, Dordrecht, pp 415–428
Jiskoot H, Murray T, Boyle P (2000) Controls on the distribution of surge-type glaciers in Svalbard. J Glaciol 46(154):412–422
Kalvoda J (1992) Geomorphological record of the Quaternary orogeny in the Himalaya and the Karakoram, vol 3, Developments in earth surface processes. Elsevier, Amsterdam
Kamb B (1987) Glacier surge mechanism based on a linked cavity configuration of the basal water conduit system. J Geophys Res 92:9083–9100
Kamb B, Engelhardt H (1987) Waves of accelerated motion in a glacier approaching surge: the mini-surges of variegated Glacier, Alaska, USA. J Glaciol 5:159–172
Kavanaugh JL, Clarke GKC (2006) Discrimination of the flow law for subglacial sediment using in situ measurements and an interpretation model. J Geophys Res 111, F01002. doi:10.1029/2005JF000346

Kick W (1989) The decline of the last Little Ice Age in High Asia compared with that in the Alps. In: Oerlemans J (ed) Glacier fluctuations and climate change. Academic, Dordrecht, pp 129–142
Kotlyakov VM (ed) (1997) Atlas of snow-and-ice resources of the world. Institute of Geography, Moscow University and UNESCO, Plate 406
Kotlyakov VM et al (ed) (2010) Glaciers of the former Soviet Union. In: Williams RS, Ferrigno JG Jr (eds) Satellite image atlas of glaciers: Asia. United States Geological Survey, Denver, Professional Paper 1386-F, Washington, F-1, pp 1–93
Kreutzmann H (1994) Habitat conditions and settlement processes in the Hindukush Karakoram. Petermanns Geogr Mitt 138:337–356
Li Jijun, Derbyshire E, Shuying X (1984) Glacial and paraglacial sediments of the Hunza Valley, North–west Karakoram, Pakistan: a preliminary analysis. In: Miller KJ (ed) International Karakoram project, vol 1. Cambridge University Press, Cambridge UK, pp 496–535
Longstaff TG (1910) Glacier exploration in the Eastern Karakoram. Geogr J 35(6):622–658
Longstaff TG (1920) The Karambar Glaciers. Alp J 33:159
Lyall Grant IH, Mason (1940) The upper Shyok glaciers in 1939. Himal J 12:52–63
Karakoram Orographical Sketch Map (1990) 2 sheets, Swiss Foundation of Alpine Research, Zurich
Mason K (1930) The glaciers of the Karakoram and neighbourhood. Rec Geol Surv India 63(2):214–278
Mason K (1931) Expedition notes: tours in the Gilgit Agency. Himal J 3:110–115
Mason K (1935) The study of the threatening glaciers. Geogr J 85:24–41
Meier MF, Post AS (1969) What are glacier surges? Can J Earth Sci 6:807–819
Murray T, Strozzi T, Luckman A, Jiskoor H, Christakos P (2003) Is there a single surge mechanism? Contrasts in dynamics between glacier surges in Svalbard and other regions. J Geophys Res 108(B5):2237. doi:10.1029/2002JB001906
Neve A (1907) Rapid glacial advance in the Hindu Kush Alp. J 50:55–62
Paffen KJ, Pillewizer W, Schneider HJ (1956) Forschungen im Hunza-Karakorum. Erdkd 10:1–33
Paterson WSB (1994) The physics of glaciers, 3rd edn. Elsevier, New York
Pecci M, Smiraglia C (2000) Advance and retreat phases of the Karakorum glaciers during the 20th century: case studies in Braldo Valley (Pakistan). Geogr Fı's Din Quat 23(1):73–85
Quincey DJ, Copland L, Mayer C, Bishop M, Luckman A, Belo M (2009) Ice velocity and climate variations for the Baltoro Glacier, Pakistan. J Glaciol 5(194):1–11
Quincey DJ, Braun M, Glasser NF, Bishop MP, Hewitt K, Luckman A (2011) Karakoram glacier surge dynamics quantified by cross-correlation feature tracking. Geophys Res Lett 38(18):L18504. doi:10.1029/2011GL049004
Schlagintweit HA, Schlagintweit R (1861/1866) Results of a scientific mission to India and High Asia undertaken between the years 1854 and 1858, etc., vol 4. Truebner, London
Schneider HJ (1967) Minapin-Gletscher und Menschen im NW-Karakoram. Erdkd 100(2–4):226–286
Sharp M (1988) Surging glaciers: behaviour and mechanisms. Progr Phys Geogr 12(3):349–370
Shipton E (1938) The Shaksgam expedition. Geogr J 91:313–39
Shaw R (1871) Visits to high tartary, Yarkand and Kashgar (formerly Chinese Tartary), and return journey over the Karakorum pass. John Murray, London
Shroder J, Bishop M (2010) Glaciers of Afghanistan. In: Williams RS, Ferrigno JG Jr (eds) Satellite image atlas of glaciers: Asia. United States Geological Survey, Denver, Professional Paper 1386-F, Washington F-3, pp 167–199
Smiraglia C, Mayer C, Mihalcea C, Diolaiuti G, Belo M, Vassena G (2008) Himalayan-Karakoram glaciers: results and problems in the study of recent variations of major non-polar glaciers. In: Bonardi L (ed) Terra glacialis, special issue, mountain glaciers and climate changes in the last century. Servizio Glaciologico Lombardo, Milan, pp 149–164
Todd HJ (1930) The Shyok flood in the Gilgit Agency. Himal J 2:37–39
Visser C (1926) Explorations in the Karakoram. Geogr J 68:457–473
Visser C (1932) Gletscherüberschiebungen im Nubra- und Shyok-Gebiet des Karakorum. Z. Gletsch Glazialgeol 20

Visser PhC, Visser-Hooft J (1935–1938) Wissenschaftliche Ergebnisse der niederländischen Expeditionen in den Karakorum und die angrenzenden Gebiete in den Jahren 1922, 1925 und 1929–30. E.J. Brill, Leiden

Wake CP (1989) Glaciochemical investigations as a tool to determine the spatial variation of snow accumulation in the Central Karakoram, Northern Pakistan. Ann Glaciol 13:279–284

Wake CP, Searle MP (1993) Rapid advance of Pumarikish Glacier, Hispar Glacier basin, Karakoram Himalaya, correspondence. J Glaciol 39(131):204–206

Workman FB (1908) Further explorations in the Hunza-Nagar and the Hispar Glacier. Geogr J 32:495–6

Workman FB (1910) The Hispar Glacier: its tributaries and mountains. Geogr J 35:23–29

Workman WH, Workman FB (1911) Call of the snowy Hispar: a narrative of exploration and mountaineering on the Northern Frontier of India. Constable, London

Younghusband FE (1892) Journeys in the Pamirs and Adjacent countries. Proc R Geogr Soc 14:205–234

Zhang X (1984) Recent variations of some glaciers in the Karakoram mountains. In: Miller KJ (ed) International Karakoram project, vol 1. Cambridge University Press, Cambridge, UK, pp 39–50

# Chapter 10
# Glacial Impoundments and Outburst Floods

**Abstract** Countless small and many larger lakes exist in Karakoram glacier basins. The region has a long history of outburst floods from them. Impoundments may be on the glacier, beside, in front of and even beneath the ice. The actual dams may be formed of ice, moraines or a combination of the two. Smaller glacial lake outburst floods (GLOFs) seem to occur somewhere in every year, the commonest glacier hazard. More rarely, large lakes occur and threaten much greater damage. Most have involved ice dams of a single type, where a substantial tributary glacier advances across and impounds a main river valley. Over the last two centuries, more than 100 glaciers of over 10 km in length have interfered with upper Indus and Yarkand streams. Large reservoirs have only been definitely identified with 23 glaciers, but other evidence shows many more at some time in the past. Large Karakoram ice dams develop quickly and rarely last more than a few months. The most dangerous cases, in particular at Chong Khumdan and Kyagar Glaciers, have involved two or more major outburst floods in episodes lasting several years. These GLOF hazards differ from smaller ones in the Karakoram and those receiving attention recently in the rest of the Himalaya. Specifically they require glacier advances. In the past decade or so, some of the glaciers associated with large ice dams have advanced and caused, or threatened to cause, GLOFs.

**Keywords** Glacial hazards • Glacier impoundments • GLOFs • Chong Khumdan Glacier • Kyagar Glacier

## 10.1 Introduction

Ponds and small lakes are present in great numbers on and around the ablation zones of Karakoram glaciers in a wide variety of conditions and types of impoundment. The great majority of glacial ponds and lakes are not identified with damaging events (Plate 10.1). They may drain slowly or survive for decades without being

K. Hewitt, *Glaciers of the Karakoram Himalaya: Glacial Environments, Processes, Hazards and Resources*, Advances in Asian Human-Environmental Research,
DOI 10.1007/978-94-007-6311-1_10, 

**Plate 10.1** Small glacial lakes along the middle Panmah, Nobande Sobande branch, viewed from Drenmang. They include some on the ice, between the ice and lateral moraines and behind lateral moraines (Hewitt 1998)

hazardous. Many are important resources for mountain pastoralists, they are favourite places for their temporary summer villages or for wildlife and they attract tourists (Chap. 13). However, some pose dangers due to inundation, closing access routes along valleys or across glaciers, sometimes loss of land in settlements near the ice. The more destructive problems involve glacial lake outburst floods (GLOFs). Small GLOFs occur in most years from short-lived lakes at lateral ice margins. Their size means they mainly threaten communities in the mountains (Kreutzmann 1994; Hewitt 2004). Given the abundance of unconsolidated sediments, floodwaters can quickly become debris flows which are the more serious hazards (Santi et al. 2010).

There is also a history of infrequent but large glacial lakes, mainly associated with the higher, most heavily glacierised Karakoram. These may cause GLOFs bringing destruction far through the mountains, even into the Indus or Yarkand Basin lowlands. Several have occurred in the past decade on the upper Yarkand River. The last large events on the Indus were over 50 years ago although, recently, glaciers associated with them have advanced and become more threatening (Hewitt 1989; Hewitt and Liu 2010).

An assessment of GLOFs cannot ignore the considerable concern over dangerous glacial lakes in other parts of the Himalaya and High Asia (Richardson and Reynolds 2000; Iturrizaga 2011). In Nepal, some 25 outburst floods have been recorded since the 1930s, with especially destructive events in 1985 and 1991 (Yamada 1998; Mool et al. 2001; Kattelmann 2003; Bajracharya and Mool 2009). Bhutan also has a number of dangerous lakes, of which one outburst had disastrous consequences in

1994 (Iwata 2010). Others are reported in the Tien Shan and Pamirs (Liu 1992; Ng et al. 2007). A variety of impoundments and outburst types have been reported, but the lakes and triggers of the GLOFs have been attributed to glacier retreat and climate warming (Mool et al. 2001; WWF 2005; Armstrong 2010). Developments in the Karakoram raise some of the same problems. However, the larger and most dangerous lakes involve distinctively different conditions that will be the main focus of this chapter.

## 10.2 Types and Characteristics of Glacial Lakes

Water may be ponded by glacial processes in a variety of ways and contexts. Ice dynamics and sedimentation can be important factors. Weather conditions affect lake size, and ablation rates can also influence whether outbursts occur and their timing. Dams may be composed of ice, moraine or mixed ice and moraine. There are also compound barriers, in which other earth surface processes play a role, notably mass movements. The composition and state of the barrier can be critical factors of risk. Of special interest is where water lies in relation to the ice mass, whether on the surface, within, under, beside or in front of the glacier and whether active or stagnant ice are involved (Maag 1969). Terrain, the catchment for meltwater, glacier surface and marginal and subsurface drainage systems are also important considerations. Identifying the different types of impoundment provides a perspective in relation to glacial environments and hazardous developments.

Most existing classifications focus on barrier composition, morphology in relation to the ice mass and surrounding terrain and size (Blachut and Ballantyne 1976; Costa and Schuster 1987; Tweed and Russell 1999; Clague and Evans 2000). These criteria help to identify the diversity of situations found in the Karakoram. Examples of virtually every class of impoundment occur in the region, except those associated with volcanic activity.

### *10.2.1 Supraglacial Lakes*

These are the most frequent of lakes associated with active ice. Most are more like pools or ponds, rarely growing beyond a few tens of metres in diameter (Plate 10.2). Their importance arises because of sheer numbers, with hundreds on the larger glaciers. They occur mainly in areas of heavy supraglacial debris or among the penitentes in higher elevation ablation zones of the eastern Karakoram.

In other regions, supraglacial lakes are said to be serious hazards, but are not known to pose much danger in the Karakoram, even locally. No reports have identified damaging events with this class of impoundment. The more widespread examples, amid heavy supraglacial debris, have turbid water, and are usually quite shallow. Sediment washing in from the sides reduces penetration of sunlight that can cause deepening in clear water ponds on cleaner ice.

**Plate 10.2** A small supraglacial lake on the lower Baltoro, surrounded by debris-covered ice and facets, relatively shallow and ponding turbid water (Hewitt 2005)

### *10.2.2 Lateral Ice-Margin Lakes*

Ice-margin lakes of intermediate size, 0.5–2 km in length, are quite common and the most frequent sources of locally destructive GLOFs. The type of barrier is critical for the stability and longevity of lakes. They can be impounded by ice or lateral moraines, against active ice or in stagnant ice. Those between active ice and lateral moraines are more likely to form and drain suddenly (Plate 10.3 a, b). Lakes behind or within lateral moraines tend to be more stable, and many persist for decades. Lakes in ice-margin troughs can last decades, coming to an end by infilling with sediment or vegetation. An example that did cause a destructive GLOF was at Aling Glacier where a long-lived lake was suddenly released in 1989 or 1990 by the Lokpar tributary surge (Chap. 9).

### *10.2.3 Lakes at Glacier Termini*

Impoundments may occur before glacier termini where advancing ice encounters reverse slopes or older terminal moraines and in depressions left behind frontal moraines as glaciers retreat. In the southeastern part of Sasir Mustagh, for example, in the upper Shyok basin, satellite images show dozens of lakes clustered around existing and former glacier termini. The same is common immediately west of the Karakoram in the Hindu Raj and Hindu Kush, much less so in the Mustagh Karakoram.

**Plate 10.3** Ice-margin lake beside Panmah glacier near the junction of Nobande Sobande and Choktoi branches. (**a**) Lake in 2005 is full, (**b**) Lake in 2009 is drained (Hewitt 2005, 2009)

Dams at termini may be of ice or moraine or both, at the active glacier margin, in stagnant ice or among the moraines of former terminal positions. Some have led to local GLOFs but many persist for decades without incident. Even the larger ones have not been associated with major GLOFs like those reported from such lakes in Bhutan and Nepal.

## 10.3 Large Glacier Dams and Lakes

Rarer but much larger impoundments can generate GLOFs that travel hundreds of kilometres through the mountains causing major devastation and have occurred on the Yarkand and the Indus. The lakes generally exceed 2.5 km in length and some over 25 km. All reported examples involved *ice dams* and nearly all were of the class of dam that Hutchinson's (1957) general classification identifies as Type 24b, '... a main valley [lake] dammed by ice from a lateral valley....' (p. 159).

Unlike the most dangerous events reported elsewhere in the Himalaya, these GLOFs do not involve moraine barriers, mixed ice and moraine or dead ice. And they only occur with *advancing glaciers*. The impoundments develop quickly and are short lived, usually for a few months, never more than 2 years. No Karakoram examples are known triggered by avalanches or landslides into the lakes, although the possibility certainly exists. They are 'self-dumping' lakes (Plate 10.4). The critical interactions are between the ice barrier and impounded water (Matthews 1964). It means that a useful tool in other areas, the glacial lake inventory, is of little use (Mool et al. 2001). Only regular monitoring of glaciers known to create dams offers a chance to be prepared (Hewitt and Liu 2010).

**Plate 10.4** An ice-dammed lake on the Karambar River, Chhateboi Glacier (Not to be confused with the Chhateboi Glacier 60 km away on the upper Yarkand, although the name, in both cases, refers to them being 'lake-prone'). The lake and stranded icebergs reflect recent emptying beneath the ice, which happens two or three times in most years. It is an example of the 'self-dumping' mechanism. Old strand lines indicate much larger lakes from a much thicker glacier in the past (Hewitt 1993)

The sites of concern are concentrated within the highest Karakoram valleys across the northern rim (Fig. 10.1). On the Yarkand flank, all known examples came from a 30 km section of the Shaksgam River and five glaciers that descend immediately north and east of K2 (8,611 m). On the Indus, most of these GLOFs have also come from glaciers across the northern rim, from far western parts in the Karambar, Chapursan and Shimshal River Basins to far eastern parts along the upper Shyok. Most of the glaciers have a northerly orientation, including all in the Yarkand Basin (Hewitt 1982). On the upper Karambar and upper Shyok Rivers, some flow in southerly, westerly and easterly directions and all appear to be surge type (Chap. 9). In the Central Karakoram, two Karakoram glaciers known to have caused GLOFs, Batura and Biafo, flow east or southeast, are large and not known to surge. Otherwise, the glaciers involved are of intermediate size.

The greater number of destructive events on record occurred in the Little Ice Age, followed by a marked reduction from the early 1930s. On the Yarkand, some major disasters occurred into the final decade of the twentieth century and in the present one. None have been reported on the Indus since the mid-twentieth century, but recently some of the Indus glaciers concerned have had advances.

### *10.3.1 Historical Records of Large Ice Dams and Outburst Floods*

In the upper Indus and Yarkand Basins, more than 100 glaciers of 10 km in length or more have interfered with rivers of which they are tributaries. Most are in the Karakoram. For completeness, those causing larger GLOFs from the upper Yarkhun River in the Hindu Raj are included that affect the Chitral and Kabul tributaries of the Indus. The Nanga Parbat also involves some examples that drain to the Indus just south of the Gilgit junction. In the past two centuries, 93 cases of destructive floods are on record from dams of more than 25 glaciers (Mason 1935; Hewitt 1982; Zhang 1992; Iturrizaga 2007). Other glaciers show indirect evidence for impoundments such as lake strandlines, spillways and deposits of large floods downstream, for which GLOF dates are not known.

The upper Yarkand is uninhabited. Before satellite coverage, few dams were observed and outbursts identified only from floods reaching the Tarim Basin (Zhang et al. 1990). Records compiled for a 120-year period show at least 23 GLOFs (Table 9.1). Published reports of the glaciers concerned began with the second, 1889, exploration of Younghusband (1904, pp. 237–245). Desio (1930) and Mason (1930, 1935) visited the glaciers and determined that they had blocked the Shaksgam River. In 1978 and 1986, satellite images showed the Teram Kangri and Kyagar Glaciers across the river, leading to renewed concern, and some destructive floods followed (Hewitt 1982; Ma and Li 1990; Yafeng et al. 2010).

On the upper Indus, ice dams were reported frequently in the early nineteenth century. Oral accounts recalled GLOFs in the eighteenth century and much older traditions relating to them. Inventories were compiled by Hedin (1917/1922) and

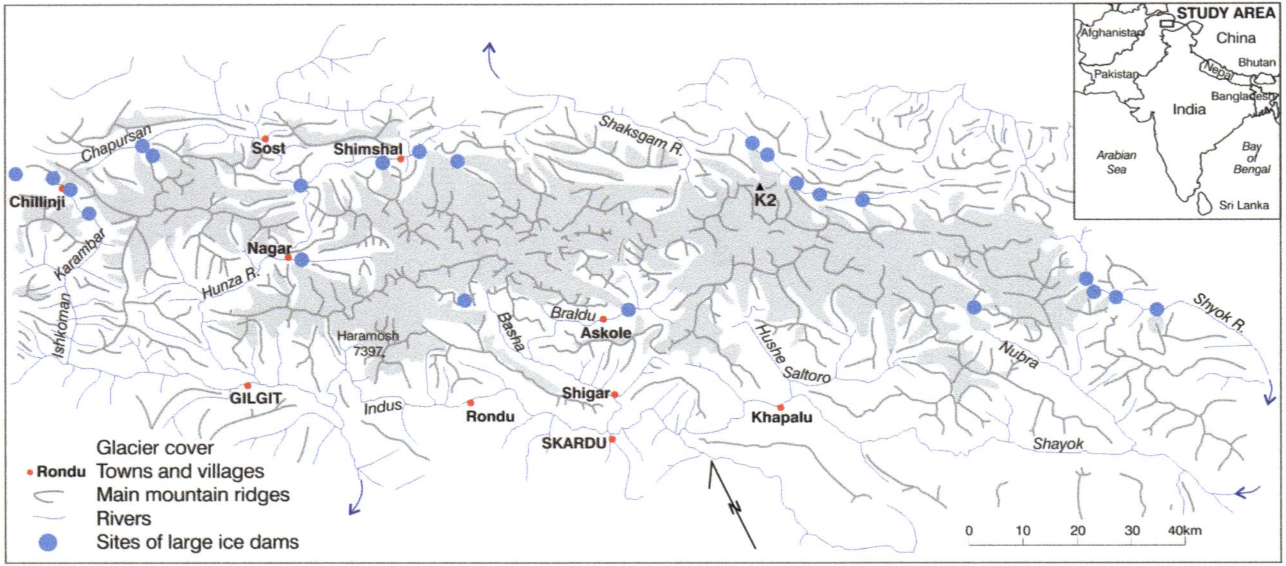


**Fig. 10.1** Sites of large ice dams identified on the upper Indus and upper Yarkand Rivers. It will be seen that they are highly concentrated in certain sub-basins and across the main axis of the Karakoram

**Table 10.1** The Upper Yarkand Basin inventory: 24 reports of GLOFs and 11 (*main set*) that generated major flood waves and damages downstream. Peak discharges are estimated at Kaqun gauging station about 500 km from the Shaksgam ice dams (After Zhang et al. 1990; Zhang 1992; Hewitt and Liu 2010)

| Kaqun # | Date (Y/M/D) | Peak discharge ($m^3/s$) | Glacier | Comment (main set in bold) |
|---|---|---|---|---|
| 1 | 1880.9 | 9,140 | ? | **Largest known flood** |
| 2 | 1892.6 | 6,000 | ? | **Major flood** |
| 3 | 1897.6 | 5,120 | ? | **Large flood** |
| 4 | 1917.6 | 3,000 | ? | – |
| 5 | 1937.7 | 5,250 | ? | **Large flood** |
| 6 | 1959.08.29 | 2,460 | Kyagar | – |
| 7 | 1961.09.03 | 6,270 | Kyagar | **Large flood** |
| 8 | 1963.06.24 | 1,690 | Kyagar | Minor |
| 9 | 1964.09.06 | 2,450 | Kyagar | Some damages |
| 10 | 1965.08.21 | 1,770 | Kyagar | Minor |
| 11 | 1968.08.10 | 3,150 | Kyagar | Some damages |
| 12 | 1971.08.02 | 4,570 | Kyagar | **Destructive flood** |
| 13 | 1978.09.06 | 4,700 | Kyagar | **Destructive flood** |
| 14 | 1980.10.21 | 802 | Kyagar | Minor flood |
| 15 | 1982.11.16 | 856 | Kyagar | Minor flood |
| 16 | 1983.10.28 | 854 | Kyagar | Minor flood |
| 17 | 1984.08.30 | 4,570 | Kyagar | **Major flood** |
| 18 | 1986.08.14 | 1,980 | Kyagar | Minor flood |
| 19 | 1997.08.03 | 4,040 | Kyagar | **Major damages, $12.5 million** |
| 20 | 1998.11.05 | 1,850 | Kyagar | Minor damages, roads, land |
| 21 | 1999.08.11 | 6,070 | Kyagar | **Major damages, $25 million** |
| 22 | 2002 | 4,610 | Kyagar | **Destructive flood** |
| 23 | 2005.09.02 | 1,880 | Kyagar | No damage |
| 24 | 2009.08 | 1,470 | Kyagar | No damage |

Mason (1929) as detailed in Hewitt (1968, 1982). The available records identify 71 events (Table 9.2). Nearly all occurred before the 1930s and, apparently, none with serious impacts since the 1960s (Hewitt 1982). GLOFs are identified in some cases where the actual glacier responsible is not known.

GLOFs large enough to persist as large flood waves at gauging stations far down the rivers are of singular interest and are treated as main sets here. For the Yarkand, 11 such events are established and 21 on the Indus (Tables 10.1 and 10.2). The Yarkand set comes from the Kaqun hydrometric station, the Indus ones from Attock gauging station, just below the junction with Kabul River. Attock is 800 km below the Hunza–Gilgit dam locations and 1,400 km from the Khumdan glaciers. Kaqun is 500 km below the Shaksgam dams. The distances show the remarkable persistence or, more exactly, recuperative power of these floods as they have travelled through the mountains. The flood waves have a recognisable shape and timing.

The 1929 Indus flood is the largest GLOF in these records and the only upper Indus case with some detailed observations (see below). The greatest number of upper Shyok GLOFs floods with known source has come from the same glacier. It has

**Table 10.2** The Upper Indus Basin inventory of 71 reports of GLOFs from ice dams where a glacier dammed a trunk stream and glacier where known. The main set of flood waves that appear on Attock gauge record are marked (*). The (?) refers to uncertainty about glacier responsible, not the GLOF (Sources: 1. Hewitt 1982; 2. Kreutzmann 1994 and Personal communication 2009; 3. Hewitt and Liu 2010)

| Date | Glacier | River basin | Comment | Source |
|---|---|---|---|---|
| WADPDA 1533 | ? Khumdan | Shyok | No details | 1. |
| 1780 (appr.) | ? Khumdan | Shyok | No details | 1 |
| Late eighteenth century | Biafo | Braldu | Severe damages | 1 |
| 1826 | ? Khumdan | Shyok | Serious flooding | 1 |
| 1833a | ? Khumdan | Shyok | Mountains only | 1 |
| 1833b | Yashkuk Yaz | Chapursan | Major flood | 1, 2 (gives 1830)* |
| 1835 | Sultan Chussku | Shyok | Extensive floods | 1 |
| 1839 | ? Khumdan | Shyok | Less than 1835 | 1 |
| 1842 | ? Chong Khumdan | Shyok | Small | 1 |
| 1844 | ? U. Karambar R. | Ishkoman | Damage at Gilgit | 1 |
| 1850 (appr.) | Tarshing | Astor | Local damages | 1 |
| 1855a | ? Khumdan | Shyok | Damage at Gol | 1 |
| 1855b | ? Khumdan | Shyok | Major flood | 1* |
| 1865 | ? U. Karambar R. | Ishkoman | Damage to Gilgit | 1 |
| 1871 | ? Khumdan | Shyok | Large flood (Nov.) | 1* |
| 1873 | Batura | Hunza | Damage at Pasu | 1,2 |
| 1877 | ? | ? | Great flood (Nov.) | 1* |
| 1879(?) | ? Khumdan | Shyok | Great flood(Aug.) | 1*3 |
| 1882a | ? Khumdan | Shyok | Largest (July) | 1* |
| 1882b | ? U. Shimshal | Shimshal | Large flood(Sept.) | 1* |
| 1884 | ? U. Shimshal | Shimshal | Damage in Hunza | 1, 2* |
| 1890 | ? | ?Shimshal | Nomal, Gilgit flood | 2 |
| 1892a | ? | ? | Major flood(Aug.) | 1* |
| 1892b | ? | ? | Nomal, Gilgit flood | 2 |
| 1893a | ? U. Shimshal | Shimshal | Hunza damages | 1, 2 |
| 1893b | ? U. Karambar R. | Ishkoman | Major flood | 1* |
| 1894 | Hassanabad | Hunza | Bridge destroyed | 2 |
| 1895a | ? U. Karambar | Ishkoman | Large flood (July) | 1* |
| 1895b | ? | Hunza | Damages, Ganesh | 2 |
| 1898 | ? Khumdan | Shyok | Large flood (July) | 1* |
| 1899 | Hassanabad4 | Hunza | Bridge destroyed | 2 |
| 1900 | ? | Hunza | Bridge destr. | 2 |
| 1901a | ? Khumdan | Shyok | Major flood (May) | 1* |
| 1901b | ? Shimshal | Shimshal | Damages Hunza | 1,2 |
| 1902 | Hassanabad | Hunza | Repeated flooding | 1 |
| 1903a | Hassanabad | Hunza | Destr. canal | 1 |
| 1903b | Kichik Khumdan | Shyok | Local flooding | 1 |
| 1904a | Hassanabad | Hunza | Destr. canals | 1 |
| 1904b | Khurdopin | Shimshal | Local damages | 2 |
| 1905a | Kichik Khumdan | Shyok | Small flood | 1 |

(continued)

**Table 10.2** (continued)

| Date | Glacier | River basin | Comment | Source |
|---|---|---|---|---|
| 1905b | Karambar Gl. | Ishkoman | Big flood to Gilgit | 1* |
| 1905c | Khurdopin | Shimshal | Great flood (Aug.) | 1, 2* |
| 1905d | Malangutti | Shimshal | Local damages | 1, 2 |
| 1906a | Khurdopin | Shimshal | Major losses(Aug) | 1, 2* |
| 1906b | 'Khurdo' or Malangutti | Shimshal | Small flood | 2 |
| 1907 | Khurdopin | Shimshal | Small flood | 1, 2 |
| 1910 | ? | Hunza | Bridge destr. | 2 |
| 1914a | ? | ? | Major flood (Apr.) | 1* |
| 1914b | ? | Hunza | Bridge destr. | 2 |
| 1922 | ? | Shimshal | Flood damage | 2 |
| 1923 | Khurdopin | Shimshal | Flood damages | 2 |
| 1926 | Chong Khumdan | Shyok | Great flood (Oct.) | 1* |
| 1927a | Beskiyeng | Chapursan | Local damage, Hunza | 1 |
| 1927b | Khurdopin | Shimshal | Small flood | 1,2 |
| 1928 | Kilik Valley | Hunza | Local flood | 1 |
| 1929 | Chong Khumdan | Shyok | Great flood (July) | 1* |
| 1932 | Chong Khumdan | Shyok | Major flood (Oct.) | 1* |
| 1933 | Chong Khumdan | Shyok | Major flood (Aug.) | 1* |
| 1934 | ? | Hunza | Local damage Ganesh | 2 |
| 1944 | ? Khurdopin | Shimshal | Damage at Pasu | 2 |
| 1959 | ? Shimshal | Shimshal | Pasu destr. (Aug.) | 1, 2 |
| 1960 | Khurdopin | Shimshal | Local damage | 2 |

had episodes of repeated blockages over a decade or more (Mason 1935). The 1929 flood was part of the last series from 1925 to 1933 in which four GLOFs show up as large rises at Attock. The others were reported in 1926, 1932 and 1933. Since then, however, the glacier posed no threat until 2009 when a massive forward push was observed on satellite imagery (Chap. 9). For the Yarkand, in only one case are there observations at the glacier as well as Kaqun, the August 5, 1987, outburst from a Kyagar ice dam (Zhang 1992). The outburst of August 1999 was the most destructive in recent years. The only event known to have occurred in winter on the Yarkand was on November 5, 1998, as observed at Kaqun station.

### *10.3.2 Characteristics of the Glaciers*

The majority of large impoundments are by glaciers of intermediate size for the region (15–40 km long) and just two of large size (Table 10.3). Some surge types are small although their length is greatly extended in advances that create impoundments. The glaciers generally have exceptional elevation ranges (Fig. 10.2). Average gradient tends to be inversely proportional to length. Most have precipitous icefall sections in upper parts, but the lobes that form ice dams are of gentle slope.

**Table 10.3** Dimensions of selected Karakoram glaciers with a history of large ice dams and outburst floods (Fig. 10.1)

| Sub-basin Glacier | Terminus (m.asl) | Highest point (m) | Range (m) | Gradient (m per km) | Length (km) | Area ($km^2$) |
|---|---|---|---|---|---|---|
| U. Shyok (35.07N; 77.44E) | | | | | | |
| Chong Khumdan | 4,700 | 7,530 | 2,830 | 135 | 21 | 152 |
| Kichik Khumdan | 4,680 | 7,530 | 2,850 | 143 | 20 | 86 |
| Aqtash[a] | 4,590 | 6,736 | 2,150 | 165 | 13 | 35 |
| Sultan Chussku[a] | 4,550 | 6,690 | 2,140 | 153 | 14 | 33 |
| Braldu (35.41N; 74.21E) | | | | | | |
| Biafo | 3,070 | 7,285 | 4,215 | 62 | 68 | 627 |
| Hunza (36.30N; 74.53E) | | | | | | |
| Batura[b] | 2,700 | 7,785 | 5,085 | 86 | 59 | 285 |
| Hispar (36.15N; 74.45E) | | | | | | |
| Bualtar–Barpu | 2,450 | 7,458 | 5,008 | 152 | 33 | 241 |
| Shimshal (38.20N; 76.39E) | | | | | | |
| Khurdopin[a] (!) | 3,200 | 7,760 | 4,560 | 111 | 41.0 | 280 |
| Yazghil | 3,055 | 7,880 | 4,825 | 170 | 30.0 | 145 |
| Malangutti | 2,900 | 7,900 | 5,000 | 219 | 23.0 | 105 |
| Karambar (36.46N; 74.01E) | | | | | | |
| Karambar[a] | 2,850 | 7,160 | 4,310 | 183 | 23.5 | 202 |
| Chillinji(?[a]) | 3,300 | 6,680 | 3,380 | 282 | 12.0 | 54 |
| Chhateboi | 3,530 | 6,400 | 2,870 | 169 | 17.0 | 62 |
| Chapursan (36.49N; 74.21E) | | | | | | |
| Yashkuk Yaz | 3,370 | 7,150 | 3,780 | 158 | 24.0 | 252 |
| Yarkhun (36.49N; 73.17E) | | | | | | |
| Pechus–C. | 3,340 | 6,871 | 3,531 | 196 | 18 | 45 |
| Shaksgam (35.44E; 76.55N) | | | | | | |
| Gasherbrum | 4,250 | 8,047 | 3,800 | 158 | 24.0 | 160 |
| Urdok | 4,260 | 8,068 | 3,810 | 149 | 25.5 | 162 |
| Staghar | 4,410 | 7,202 | 2,790 | 100 | 28.0 | 115 |
| Singhi | 4,520 | 7,245 | 2,725 | 114 | 24.0 | 169 |
| Kyagar | 4,740 | 7,245 | 2,505 | 134 | 19.0 | 106 |

[a] Known surging glaciers

[b] The Batura snout is heavily debris covered and moraine was likely a factor in dams

(!) These involve lakes dammed in ice-free tributaries by the main glacier named

Ice dams are controlled by specific interrelations of topography, climate, glacier dynamics and hydrology. Mason (1929) was, perhaps, the first to recognise a type of 'threatening glacier' characteristic of the Karakoram involving 'Transverse glaciers whose snouts project unchecked into main valleys [where they] are always liable to cause obstructions'. (p.10). Thick tongues of ice must descend well into the fluvial zone and across stream valleys gentle enough for large lakes to develop. Favourable topographic relations arise from ubiquitous high relief, and large tributary glaciers transverse to it so they enter the river valleys more or less at right angles. Equally important for reservoir size are valleys deeply incised in geologically recent time but with deep valley fill helping create broad, flat floors.

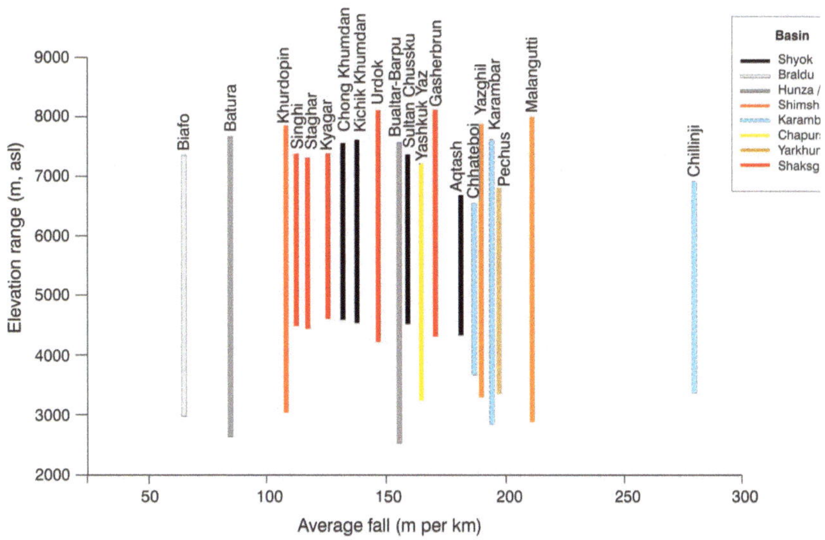


**Fig. 10.2** Elevation ranges of glaciers forming large ice dams (After Hewitt and Liu (2010))

## 10.4 Phases of Impoundment and Outburst Events

GLOFs involve a set of phases, each dependent upon what has come before and controlling conditions that are critical for a particular phase (Table 10.4). The list is a guide to what needs to be monitored but only rarely has been. A problem in explaining and predicting the ice dams is that many more advances are known across river channels than impoundments. In these cases, rivers maintain or keep opening up channels around or beneath the ice. It suggests that differences in ice dynamics may be important and perhaps related to thermal, substrate and drainage systems and debris in the ice. However, information about such conditions is generally lacking for most examples. No existing monitoring system addresses them. Satellite coverage could help with some dimensions but requires systematic observing and should likely serve mainly to alert communities and initiate ground-based investigations.

### *10.4.1 Glacier Advance and Sealing of an Ice Dam*

Flow rates of main ice streams are comparatively low in the outer tongue, commonly less than 50 m/year except during surges (BIG 1979; Hewitt et al. 1989). Mason (1935) believed the upper Indus streams were too powerful to be blocked by ice except during exceptional advances. However, most dams are sealed in winter when, although ice movement is slower, river flow is low or absent. At the elevations

**Table 10.4** Phases of ice dam and outburst flood hazards. A checklist of properties and attendant variables that can control the development, dimensions and risk of a GLOF. 'Worst case' refers to conditions that have been critical in at least some GLOFs that reached the lowlands

| Phase | Processes | Key variables | Critical for 'worst case'? |
|---|---|---|---|
| 1. Glacier advance? | Cross valley? | Speed, | Yes |
| | | Site geometry, | Yes |
| | | Timing (winter/summer) | Yes |
| | Individual/general? | Glacier geometry, dynamics | |
| | Surge? | Glacier geometry, dynamics | Yes |
| 2. Ice dam? | Barrier profile | Height, width | Yes |
| | Post-seal dynamics | Continuing ice inflow | |
| | | Thermal properties | |
| 3. Reservoir? | Actual and potential size | Length, width, volume | Yes |
| | Basin hydrology | Discharges into lake | Yes |
| | Weather conditions | Snow cover, ablation | Yes |
| 4. Stability? | Ice/lake relations | Lake depth at dam | Yes |
| | Dam-break form | Overtopping? | |
| | | Subglacial ('tunnelling?') | Yes |
| | | Full height? | Yes |
| | Speed of breaching | Rapid/gradual | Yes |
| | Complete/partial? | | Yes |
| 5. Outburst flood? | Size | Volume | Yes |
| | | Max. height | Yes |
| | | Max. discharge | Yes |
| | Flood wave | Dynamic? | Yes |
| | | Kinematic? | |
| | | Rate of rise | Yes |
| | Downstream routing and modifications? | Attenuation? | |
| | | Regrowth and dynamics? | Yes |
| | | Changing channel geometries | Yes |
| | Erosion | Bank erosion? | Yes |
| | | Bed scouring? | Yes |
| | Sediment transfers | Concentration/deposition | Yes |

of the dams, sub-zero average temperatures last 90–170 days. Rivers freeze over from October or November through March, in some cases into April. Snow cover can persist from October to May. Marked fluctuations in movement rates have been documented for Karakoram glaciers and may help seal and maintain ice dams despite low average rates (Chap. 7).

## *10.4.2 Reservoir Size and Dam Breaching*

Lake size depends upon barrier height, valley geometry and inflows from up valley. Summer weather is a major factor in the timing and rates of lake infilling. Direct measurements were made only at Chong Khumdan in 1929 and Kyagar in 1987 as described below. The timing of outbursts has been related to weather conditions, especially summer temperatures affecting ablation and meltwater input to the lake (Rabot 1905, p. 545). Those reported for the days preceding the 1929 Chong Khumdan outburst were said to exceed previous records (Todd 1930, p. 39). Exceptionally warm conditions were associated with the Kyagar 1999 GLOF (see below).

Flood size also depends upon how, and how rapidly, the barrier is breached and water evacuated. Reports suggest most catastrophic outbursts involve sudden drainage *under* the ice dam rather than by overtopping or drainage around the margin (Zhang et al. 1990, p. 120). The more extreme events and highest peak discharges from Chong Khumdan, Khurdopin and Kyagar involved subglacial and complete lake drainage. If the lake is deep enough, there may be hydraulic tunnelling as water pressure exceeds the yield stress of glacier ice (Glen 1954), or the barrier may be floated off when depths are over 90 % of ice thickness.

The Kyagar 1987 outburst seems to have drained through subglacial tunnels possibly exploiting the pre-existing drainage system within the ice or aided by rapid increases in crevassing (Ma and Li 1990, p. 107). The lake level was not more than 60 m at the dam, much less than the 200 m necessary for hydraulic tunnelling according to Glen's (1954) hypothesis. The enormous quantity and sizes of ice blocks carried down the Shyok in 1929, and other large outbursts do suggest mechanical erosion of the ice barrier is important.

### 10.4.3 Outburst Floods

Lake size and the speed and completeness of lake drainage are critical to the dimensions and behaviour of outburst floods. Once released, valley geometry and pre-existing river discharges affect the behaviour of the floodwaters. Reports of flood waves usually show a very steep rise to peak and more gradual recessional limb. This is the opposite of the classic 'jokulhlaup' GLOF shape, described from Iceland (Björnsson 2004, p. 169). It has steadily increasing outflows followed by an almost vertical recessional limb due to closure of the outlet. The absence of observations close to the ice dams in the Karakoram may be a factor. Most reports are from downstream areas where passage through narrow, steep gorge sections may have modified the flood waves. One case that was observed close to the source, the August 5, 1987, outburst from Kyagar Glacier dam, had a wave shape more like the classic profile. At the gauging stations far downstream, however, a sudden rise and gradual falling limb were observed (Hewitt and Liu 2010).

Outburst timing is also important because of the role of pre-existing water discharges downstream. Since most GLOFs occur in summer and warmer weather, they tend to coincide with higher stream flows. In glacial regime rivers, winter floods encounter very low stream flows. Seepage and storage losses along valley floors are much greater. Thus, few seem likely to reach distant gauging stations and those that do seem relatively small. However, even the latter must reflect very large impoundments to explain far travel. The floods on the Indus in November 1877 and October 1926 and the Kyagar 1998 event are examples.

## 10.5 The Chong Khumdan Events 1926–1931

When the Chong Khumdan 1929 outburst occurred, the lake had reached 18 km in length and an estimated volume of 1.5 $km^3$. Lake level at the barrier reached some 150 m, 'probably 150 ft [46 m]' below the top (Mason et al. 1930, p. 46). Failure occurred in a catastrophic breach, and complete drainage of the lake took 2 days. Shortly afterwards a 'canal' was found cutting directly across the centre of the ice lobe from the Shyok right bank at the upstream end terminating at the left bank (Gunn 1930, p. 35). It was 122 m wide and appears to have begun through a subglacial tunnel, before becoming a full height breach.

Flood behaviour was observed at various sites above Attock (Fig. 10.3). A sharp rise to peak occurs at most of the sites, especially those closest to the dam. The greatest height was at Saser Brangsa, the most extreme concentration of the wave at Khalsar. Subsequently, in narrow valley sections, it was regenerated as a steep-fronted, dynamic wave. The recuperative power in the gorges below the open Skardu Basin is notable, indicative of the complexity of such events and difficulties of extrapolating to other sites and when seasonal flows differ.

Another danger was shown by the 1929 flood in which reverse flood waves travelled up the more gentle tributary valleys (Mason et al. 1930). These floods occurred all the way from the upper Shyok to the plains. The most destructive was, in fact, on the Kabul River just above Attock (Hewitt 1982), an extraordinary threat from a flood originating 1,400 km away. As shown in the figure, within a few days, the danger was compounded by the largest known rainstorm flood on the Kabul!

## 10.6 Recent Kyagar Outbursts

The largest recent outburst flood on the Yarkand was in August 1999 and brought enormous damage to southern Kashgar District (Table 10.1). The November 5, 1998, event, observed at Kaqun station, is the only occurrence known in winter. The hydrograph shows the stream discharge started to rise from a base flow of 87 $m^3/s$ at about 20:00 h and reached a peak discharge of 1,850 $m^3/s$ within 2 h. It subsided slowly back to base flow by November 10 after about 98 h. Here again is a typical rapid onset-slow decline flood event. It was estimated from previous experience that the released water volume must have been at least $88.1 \times 10^6$ $m^3$, much larger than the surveyed capacity of Kyagar Lake in 1987. Yet the Kyagar Lake had not drained completely and, in fact, retained a contribution to the 1999 flood. It implies either growth of a much higher ice dam or combined drainage of the Kyagar and Teram Kangri lakes – or that the model, based on summer events, may not be accurate for winter outbursts.

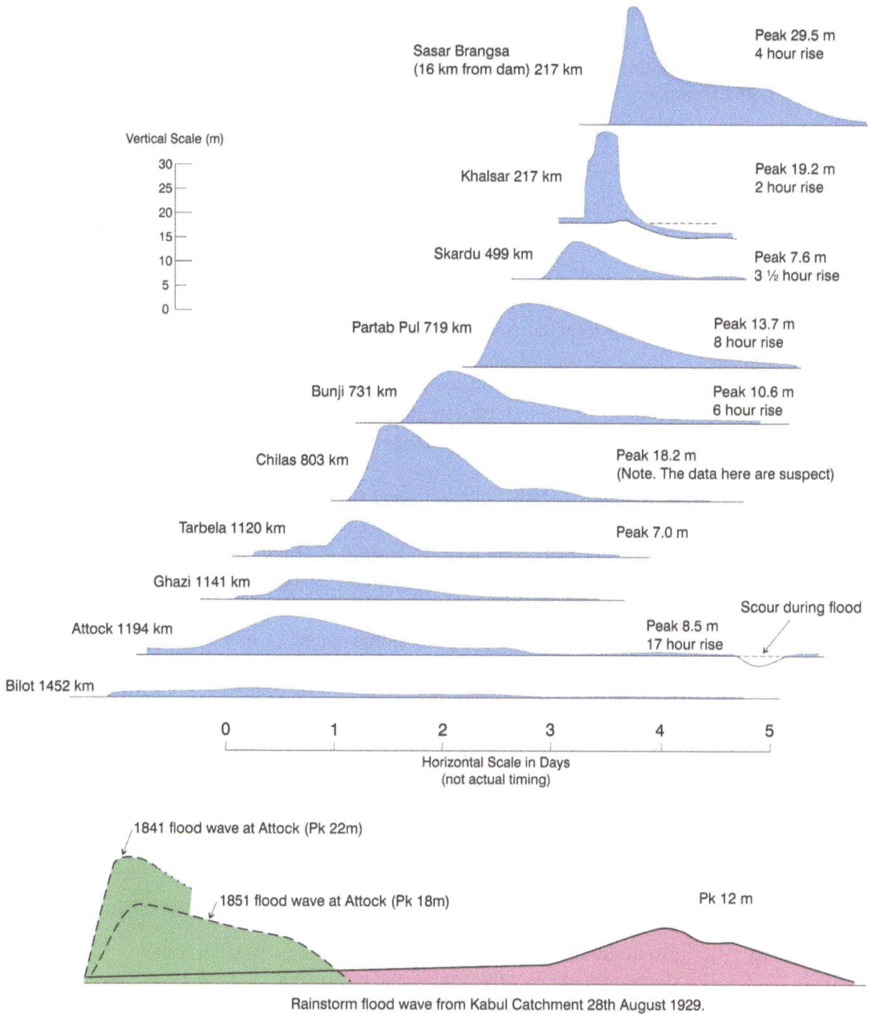


**Fig. 10.3** Progress of the 1929 Chong Khumdan GLOF (After Gunn 1930). Profiles for the largest historical floods at Attock the 1841 and 1958 landslide dam-break floods and the August 1929 rainstorm flood from the Kabul River offer comparisons. The latter's shape is conspicuously different from the dam-break flood waves, and its occurrence within a few days of the GLOF suggests some 'worst case' possibilities (Hewitt 1982; Hewitt and Liu 2010)

Unusually warm climatic conditions were involved. The only high-elevation meteorological site in the region, at Tashkurgan (3,091 m), recorded an average annual air temperature in 1998 of more than 5.5 °C, which was 2.1 °C higher than the previous 50-year mean. In the cold months of 1998, from January to March and again in November, air temperatures were anomalously high, at −3.6 °C and −1.7 °C, respectively. The former was the highest for the previous 50 years, the latter close to the maximum of −1.6 °C for any November between 1958 and 2007. It suggests how climate change effects on winter temperatures may alter outburst flood risks, not just the ablation season as usually assumed.

**Table 10.5** Present status of glaciers across the northern rim of the Karakoram with histories of GLOFs. Qualitative risk status (danger rating) and monitoring requirements are identified from visual inspection of satellite imagery with respect to formation of a large dam. Ranks 1 (imminent high risk) and 2 (imminent possible risk) require constant monitoring, ranks 3 (growing risk) and 4 (uncertain risk) require some intermittent checks each year and rank 5 (no immediate risk) an annual check in summer

| Glacier | Recent behaviour | Danger rating |
|---|---|---|
| *Upper Shyok* | | |
| Chong Khumdan | Major thickening and advance of NE branch underway 2008–2009. Frontal lobe advancing across Shyok in mid-2009, but had not at opposite wall | 1 |
| Kichik Khumdan | Thickened and advanced across Shyok in the early 2000s with steep frontal cliff. River maintained flow beneath terminal lobe. Began to retreat in 2008 | 4 |
| Aqtash | Advanced across Shyok, 2009, possibly surging as in 1902–1903 (Hewitt 1969) | 2 |
| Sultan Chhusku | Advancing in 2009 but 5 km away from Shyok | 5 |
| *Shimshal* | | |
| Khurdopin | Surged in 1998–1999 formed small dams 1998, 1999, 2000 Small outburst floods, with minor damage. Terminus presently advancing | 4 |
| Yazghil | Advancing slowly since 1998. Some danger of impoundment | 3 |
| Malangutti | Advancing into river and undercut by it since the late 1990s | 3 |
| *Karambar* | | |
| Karambar | Surged in 1993 reached Karambar R. but did not form dam | 5 |
| Chillinji | Advancing since the 1990s, partly as result of 1991 rock avalanche. Completely across valley in 2009, river in tunnel beneath ice | 1 |
| Chhateboi | Glacier thickening since the late 1990s. Permanently across Karambar, but small outburst floods every year. Larger lakes may form as the ice thickens | 4 |
| *Chapursan* | | |
| Yashkuk Yaz | No discernable change in ice front since the 1990s but some nearby glaciers thickening and advancing | 4 |
| *Yarkhun* | | |
| Pekhin | Slowly advancing since the mid-1990s, ice front undercut by river | 3 |
| *Shaksgam* | | |
| Gasherbrum | Advancing in the 2000s, valley blocked 2008, river maintained under ice | 1 |
| Urdok | No discernable change in terminus in recent years | 4 |
| Staghar | Advancing in the 2000s, not yet blocking river | 2 |
| Singhi | Advancing in the late 2000s, completely blocks Shaksgam Valley but river maintaining channel underneath the ice | 1 |
| Kyagar | Advancing in the late 2000s, several completely blockages of Shaksgam Valley, last 4 km long in 2009 drained slowly | 1 |

## 10.7 Status of Glaciers of Interest

After the interest generated by GLOFs on the Indus in the early years of the twentieth century, they largely faded from popular and scientific concern. Renewed interest in GLOFs in the eastern Himalaya concerns glacial lakes associated with waning glaciers. By the 1990s, climate warming and reports of the so-called disappearing glaciers made it seem that glacier retreat would be the norm, and these kinds of large upper Indus ice dams would cease to occur. However, the problem has returned in the past two decades. Large ice dams have formed on the Shaksgam River, and some serious floods have already occurred. Some Indus glaciers with a history of large GLOFs have been advancing, and it is an open question of whether they represent a temporary anomaly or consequences of climate change that may lead to a recurrence of large GLOFs (Hewitt 2005, 2007). The status of the glaciers assumes urgent interest and underlines the need to rate the dangers (Table 10.5).

There are singular problems for emergency planning and the scientific advice, given the sudden re-emergence of these GLOF risks. Many of the dam sites have not been visited by glaciologists for decades, if at all. Not one has been subject to regular monitoring or research to determine characteristics critical in lake formation and outburst floods. As noted, on the upper Indus, these phenomena were monitored in 1929 and 1932, not since. There is renewed interest in the Shaksgam glaciers but little had been done to 2012.

Since most events on record were associated with the Little Ice Age, a period of global cooling, a pressing issue is whether the present period of global warming alters critical properties and potential. Changed thermal properties of the ice could be significant, including the extent of frozen and unfrozen bed in these thermally mixed ice masses. Summer ablation and inflows may have changed. Since the 1960s, in the valleys below the glaciers, winter temperatures seem to have increased, but summers cooled (Archer and Fowler 2004). Finally, what has changed risks the most since the 1930s, say, are much larger or more concentrated populations exposed in the valleys downstream and far more extensive infrastructure. There are more development activities in the valleys of the upper Indus and where the Yarkand enters the Tarim Basin than ever before. It complicates attempts to compare changing risks now with those a century ago or more (Chap. 13).

## References


Archer DR, Fowler HA (2004) Spatial and temporal variations in precipitation in the upper Indus River: global teleconnections and hydrological implications. Hydrol Earth Sci Syst 8:47–61

Armstrong RL (2010) The glaciers of the Himalayan-Hindu-Kush region. Technical paper, The International Centre for Integrated Mountain Development (ICIMOD), Kathmandu, 16p

Bajracharya SR, Mool GP (2009) Glaciers, glacial lakes, and glacial lake outburst floods in the Mount Everest region, Nepal. Ann Glaciol 50:81–85

BIG (Batura Investigations Group) (1979) The Batura Glacier in the Karakoram Mountains and its variations. Sci Sin 22:958–974
Björnsson H (2004) Glacial lake outburst floods in mountain environments. In: Owens PN, Slaymaker O (eds) Mountain geomorphology. Arnold, London, pp 165–186
Blachut SP, Ballantyne CK (1976) Ice-dammed Lakes: a critical review of their nature and behaviour. Department of Geography discussion paper #6, McMaster University, Hamilton
Clague JJ, Evans SG (2000) A review of catastrophic drainage of moraine-dammed lakes in British Columbia. Quat Sci Rev 19:1963–1983
Costa JE, Schuster RL (1987) The formation and failure of natural dams, Open-file report. U.S. Geological Survey, Vancouver, pp 87–392
Desio A (1930) Geological work of Italian expedition to the Karakoram. Geogr J 75:402–411
Glen JW (1954) The stability of ice-dammed lakes and other water-filled holes in glaciers. J Glaciol 2:316–318
Gunn JP (1930) Hydraulic observations on the Shyok flood of 1929. Government of Punjab, Irrigation Branch Paper #32, Lahore
Hedin S (1917–1922) Southern Tibet: discoveries in former times compared with my own researches in 1906–1908, vol 7, History and exploration in the Karakoram Mountains. Lithographic Institute, Swedish Army, Stockholm
Hewitt K (1968) Studies in the geomorphology of the mountain regions of the upper Indus Basin. 2 vols. Unpublished dissertation, University of London
Hewitt K (1969) Glacier surges in the Karakoram Himalaya (Central Asia). Can J Earth Sci 6:1009–1018
Hewitt K (1982) Natural dams and outburst floods of the Karakoram Himalaya. In: Glen J (ed) Hydrological aspects of Alpine and high mountain areas. International Hydrological Association. (I.A.H.S.) Publication No. 138, Exeter, pp 259–269
Hewitt K (1989) Hazards to water resources development in high mountain regions; the Himalayan sources of the Indus. In: Starosolszky O, Melder OM (eds) Hydrology of disasters: proceedings of the technical conference, Geneva, November 1988. James and James, London, pp 294–312
Hewitt K (1998) Himalayan Indus streams in the Holocene: glacier-, and landslide-'Interrupted' fluvial systems. In: Stellrecht I (ed) Karakorum-Hindu Kush-Himalaya: dynamics of change part I. Rudiger Koppe Verlag, Koln, pp 1–28
Hewitt K (2004) Geomorphic hazards in mountain environments. In: Owens P, Slaymaker O (eds) Mountain geomorphology. Hodder Scientific, London, pp 187–218
Hewitt K (2005) The Karakoram anomaly? Glacier expansion and the "elevation effect", Karakoram Himalaya. Mt Res Dev 25:332–340
Hewitt K (2007) Tributary glacier surges: an exceptional concentration at Panmah glacier, Karakoram Himalaya. J Glaciol 53:181–188
Hewitt K, Liu J (2010) Ice-dammed lakes and outburst floods, Karakoram Himalaya: historical perspectives and emerging threats. Phys Geogr 31(6):528–551
Hewitt K, Wake CP, Young GJ, David C (1989) Hydrological investigations at Biafo Glacier, Karakoram Himalaya: an important source of water for the Indus River. Ann Glaciol 13:103–108
Hutchinson GE (1957) A treatise on limnology. Pt 1: geography and physics of lakes. Wiley, New York
Iturrizaga L (2007) Historical and recent glacier variations in the Karakoram Mountains. Eur Geosci Union Geophys Res Abstr 9:05470
Iturrizaga L (2011) Glacier lake outburst floods. In: Singh VP, Singh P, Haritashaya UK (eds) Encyclopaedia of snow, ice and glaciers. Springer, Dordrecht, pp 381–399
Iwata S (2010) Glaciers of Bhutan: an overview. In: Williams RS, Ferrigno JG Jr (eds) Satellite image atlas of glaciers of the world: Asia, vol 1386-F, U.S. geological survey professional paper. U.S. Government Printing Office, Washington, DC, pp F201–F257
Kattelmann R (2003) Glacial lake outburst floods in the Nepal Himalaya: a manageable hazard? Nat Hazards 28:145–154

Kreutzmann H (1994) Habitat conditions and settlement processes in the Hindukush Karakoram. Petermann Geogr Mitt 138:337–356
Liu JS (1992) Jokulhlaups in Kunmalike River, the Southern Tien Shan Mts., China. Ann Glaciol 16:85–88
Ma QH, Li NJ (1990) The Kyagar Thso Lake and its geomorphological characteristics. In: Study on glacier-dammed lake outburst floods of Yarkant River, Karakorum Mountains, Lanzhou. Institute of Glaciology and Geocryology. Science Press, Beijing, pp 101–108
Maag H (1969) Ice-dammed lakes and marginal glacial drainage on Axel Heiberg Island, Canadian Arctic Archipelago, Axel Heiberg Island research reports. McGill University, Montreal
Mason K (1929) Indus floods and Shyok glaciers. Himal J 1:10–29
Mason K (1930) The glaciers of the Karakoram and neighbourhood. Rec Geol Surv India 63(2):214–278
Mason K (1935) The study of threatening glaciers. Geogr J 85:24–35
Mason K, Gunn JP, Todd HJ (1930) The Shyok flood in 1929. Himal J 2:35–47
Matthews WH (1964) Water pressure under a glacier. J Glaciol 5(38):235–240
Mool PK, Bajracharya SR, Joshi SP (2001) Inventory of glaciers, glacial lakes, and glacial lake outburst floods: monitoring and early warning systems in the Hindu Kush Himalayan Region, Nepal. International Centre for Integrated Mountain Development and United Nations Environment Programme, Regional Resource Centre for Asia and the Pacific, Kathmandu
Ng F, Liu S, Mavlyudov B, Wang Y (2007) Climatic control of the peak discharge of glacier outburst floods. Geophys Res Lett 34, L21503. doi:10.1029/2007GL031426
Rabot C (1905) Glacial reservoirs and their outbursts. Geogr J 25:534–548
Richardson SD, Reynolds JM (2000) An overview of glacial hazards in the Himalayas. Quat Int 65(66):31–47
Santi PM, Hewitt K, VanDine DF, Barillas Cruz E (2010) Debris-flow impact, vulnerability, and response. Nat Hazards 56(1):371–402. doi:10.1007/s11069-010-9576-8
Todd HJ (1930) The Shyok flood in the Gilgit agency. Himal J 2:37–39
Tweed FS, Russell AJ (1999) Controls on the formation and sudden drainage of glacier-impounded lakes: implication for jokulhlaup characteristics. Prog Phys Geog 23:79–110
WWF (World Wildlife Fund) (2005) An overview of glaciers, glacier retreat, and subsequent impacts in Nepal, India and China. World Wildlife Fund, Nepal Program, March, 70p
Yafeng S, Desheng M, Tandong Y, Qunzhu Z, Chaohai L (2010) Glaciers of China. In: Williams RS Jr, Ferrigno JG (eds) satellite image atlas of glaciers: ASIA, United States Geological Survey, Denver, Professional Paper 1386-F, Washington, F-2, pp 127–166
Yamada T (1998) Monitoring of Glacier Lake and its outburst floods in Nepal Himalaya. Japanese Society of Snow and Ice, Tokyo, Monograph 1
Younghusband FE (1904) The heart of a continent: a narrative of travels in Manchuria, across the Gobi Desert, through the Himalayas, and Chitral, 1884–1894. John Murray, London (reprinted 1994 by Book Faith, New Delhi)
Zhang XS (1992) Investigation of glacier bursts of the Yarkant River in Xinjiang, China. Ann Glaciol 16:135–139
Zhang XS, Wang WT, Li NJ, Wang YL (1990) The dewatering way of the Kyagar Thso Lake and the simulated calculation of the GLOF. In: Lanzhou Institute of glaciology and geocryology, study on glacier-dammed lake outburst floods of Yarkant River, Karakorum Mountains. Science Press, Beijing, pp 109–121

# Chapter 11
# Rock Glaciers and Related Phenomena

**Abstract** Rock glaciers are ice-cored lobes or tongues of coarse, angular debris that form below steep rock walls and move slowly across or down valleys. Typically they are associated with glaciated mountain terrain and are transitional forms between glacial and periglacial process regimes. Largely neglected in work on the Karakoram, they are present in great numbers and a wide variety of forms and sizes. They occur in a well-defined elevation band, rarely of more than 1,400 m vertically, varying with slope orientation and across the region. Watershed or interfluve elevations appear as primary constraints. Rock glaciers are absent below the lowest *and* the highest Karakoram interfluves. They are fairly uncommon in most heavily glacierised parts of the Mustagh Karakoram or Nanga Parbat–Haramosh Massif. The heaviest concentrations are in lower offshoots or sub-ranges of the Karakoram and in surrounding lesser mountains. Where they occur their dimensions, morphology and behaviour reflect relations of seasonal temperatures, snowfall, avalanching, debris sources and delivery. The larger ones tend to be glacier-derived and originate in the upper parts of the rock glacier zone. Some glacier basins like Batura have considerable numbers in their extensive periglacial areas. The genesis of many rock glaciers is hardly separated from the history of glacial advance and retreats. This reinforces the sense that transitional as well as transglacial forms and processes are involved. Rock glaciers are also associated with formerly but not presently glaciated areas, including cirques, glacially sculpted valley walls and floors with abundant glacial deposits. It suggests there are also paraglacial relations. They include numerous and distinctive avalanche- and talus-derived forms, generally concentrated towards the middle and lower parts of the zone, again underscoring their transitional nature. In such cases, an important gradient is that between relatively humid source areas and semiarid, even arid, valley floors into which the lowest lobes spread. Finally, there are a great many inactive or relict rock glacier forms, usually continuous with the zone where active ones occur. This too suggests involvement of the history of deglaciation and Holocene climate change. Regrettably, there are no data to explore these relations or the interior of Karakoram rock glaciers and their

K. Hewitt, *Glaciers of the Karakoram Himalaya: Glacial Environments, Processes, Hazards and Resources*, Advances in Asian Human-Environmental Research,
DOI 10.1007/978-94-007-6311-1_11, 

mechanics. This chapter can only highlight the scope of this huge but neglected topic and something of what visible morphologies, dimensions and environmental relations can reveal.

**Keywords** Periglacial • Land systems • Debris sources • Talus • Rock glacier thresholds • Transglacial • Paraglacial • Fragmented drainage systems

## 11.1 Introduction

Rock glaciers are visible as lobes or tongues of coarse, angular debris, unsorted on the surface, with a steep frontal slope and, in many cases, steep lateral margins. They are usually raised up a few tens of metres above the surrounding valley floor or above troughs at the margins. Many larger examples have distinctive sets of ridges and furrows at right angles to movement, usually curved and convex downslope (Plate 11.1). Up valley, longitudinal ridges and furrows may occur, especially towards the lateral margins. When active, the rock glacier mass moves or deforms downslope. There are no measurements for Karakoram examples, but

**Plate 11.1** Rock glacier country, upper Thalle Valley, Baltistan, descending northwards to about 4,100 m from watersheds up to 5,500 m. Ridge-and-furrow features are evident in the near ground. The rock glaciers include active tongues, some advancing and inactive, vegetated ones (Hewitt 1998)

recorded velocities in other regions are between a few centimetres and tens of centimetres a year, rarely more than a metre (Kääb 2008; Röer 2007; Delaloye et al. 2010). This is much slower than most valley glaciers, excepting some terminal areas, but gives rise to certain similarities of form. The movement and characteristic morphologies are attributed to a core of ice and/or ice-cemented debris (Giardino et al. 2011).

The debris stream is also critical in rock glacier formation. Most are in formerly or presently glaciated mountains where there is an abundance of moraine, talus or avalanched debris. They are generally found at the base of rock walls. Debris falling from free faces is the dominant factor (Gardner 1980; Barsch and Jacob 1998). There is usually a link to permafrost conditions, and some are in areas of well-developed periglacial patterned ground (Dixon and Abrahams 1992; French 2011). In other cases, the more evident link is to existing glaciers. A great many Karakoram rock glaciers are linked to existing glaciers or where they have diminished or been lost since the Little Ice Age. However, equally large periglacial areas support rock glaciers in the absence of present-day glaciers. Some extend beyond the permafrost zone but originate on and below cliff faces where it is present.

Difficulties of identification and typology arise in the many cases where rock glaciers are found in close proximity to other coarse, rock debris features, including moraines, talus slopes and cones, avalanche run-out zones and small glaciers. A variety of other coarse debris forms have been mistaken for rock glaciers in the past and the reverse. They include rock or rubble streams, talus cones and ramparts, avalanche boulder tongues, rock avalanche deposits and debris-mantled glaciers. In this respect too, after a period of intense debate, somewhat more eclectic views have emerged, partly to accommodate the diversity of conditions becoming known around the world (Giardino and Vitek 1988; Whalley and Martin 1992; Humlum 1998; Kääb et al. 2002; Millar and Westfall 2008).

If rock glaciers can arise from a range of conditions and processes, they do seem to converge towards common sets of features and behaviour (Corte 1999). It has been argued they should be treated as *transitional* forms and processes, within and between glacial and periglacial environments (Giardino and Vitek 1988). A continuum with certain types of 'slope failures' has also been suggested (Shroder 1987), a reminder of the critical roles slope processes play, and that the debris stream is as important as the presence of ice.

In the Karakoram, firm identifications of rock glacier types and sources are limited by a lack of definitive measurements of movement or of internal structure and composition, including ice content. There is evidently an extreme diversity of forms, often in close proximity, perhaps more than in most better-known regions. The well-defined individual rock glaciers that have contributed so much to understanding elsewhere are relatively uncommon. A sense of rock glaciers existing within a continuum of landscape forms and linked to the range of cryospheric conditions seems particularly relevant. They are also among the most significant, if neglected, transglacial forms and processes, and many occur where paraglacial conditions affect contemporary developments (Chap. 8; Iturrizaga 2006; Hewitt 2009).

### 11.1.1 Some Karakoram Background

The world literature on rock glaciers is vast and has increased exponentially in recent years, especially their place in cryosphere responses to climate change (Haeberli and Gruber 2009; Giardino et al. 2011). However, the Karakoram has an even smaller place in this literature than its glaciers in world glaciology. Short or incidental references to the rock glaciers occur in reports from the region (Hewitt 1968, 1989; Owen and Derbyshire 1989; Shroder and Bishop 2010). Only one paper of any length is available, Owen and England's (1998) valuable survey of examples in a transect between Shimshal and Khunjerab Pass. This leaves no doubt of the high concentrations of rock glaciers that can occur. Nevertheless, they have been absent from the interests that draw scientists to the region and also, perhaps, because they are rare on the paths we usually follow. These are either along the arid valley floors giving access to and through the region or into the highest mountains. The former are too low down or arid. Rock glaciers are generally precluded from much of the Mustagh Karakoram, the Haramosh and the Nanga Parbat Massifs by steepness and the sheer extent of glacier ice. Travelling the Karakoram Highway, for example, the only rock glaciers encountered, and a small handful, are well north of the Karakoram Range as the road approaches Khunjerab Pass on the China border.

Thus, it would be easy to pass quickly over the topic. A whole chapter is devoted to it here mainly to establish its importance from a regional cryosphere standpoint, and for a close relative of the glaciers, rather than my own or any other track record of research. First and foremost, rock glaciers are widespread and characteristic of large parts of the Karakoram. Some valleys are choked with countless active and inactive rock glaciers where they cover tens of square kilometres. It is likely there are thousands of individual rock glaciers in the Karakoram itself, as numerous as glacier masses if not more so, though generally smaller individually, and less in total cover. They typify the considerably larger areas of periglacial or seasonally thawed environments in and around glacier basins. Rock glaciers are even more common in surrounding ranges, where glaciers are fewer and smaller, and probably amount to tens of thousands of cases (see below).

It also seems likely that rock glaciers in the greater Karakoram Region comprise the most diverse and complete spectrum of such forms known anywhere. This applies to the interrelations of morphology and size range, likely genesis, states of development and countless relict forms. They are linked to the range of cryospheric conditions in the region, glacial, periglacial and slope processes, including avalanche and rockfall, and will be shown to have an important place in local meltwater drainage systems and stream hydrology. It is clear that, over time, rock glaciers have waxed and waned, while behaving and responding to climate changes somewhat differently from glaciers. Today, many rock glaciers seem to be inactive, dying or remnants of past conditions, but others are expanding, even where the surrounding glaciers are retreating (Plate 11.2).

Finally, for mountain communities, rock glaciers often provide or influence critical water resources (cf. Azocar and Brenning 2010). Their role in the Indus Basin

**Plate 11.2** A diversity of rock glaciers terminal lobes in upper Thalle Valley, Baltistan, including active and apparently inactive and well-vegetated relict examples. In such locations, relations are complex, and active lobes are seen overriding inactive ones, possibly responsible for starving or rejuvenating them (Hewitt 1998)

waters as a whole is much less than the glaciers, but occurring below the perennial snow zone, they affect mountain pastures, tree cover, timber lines and headwater streams. They could be useful in tracking cryosphere changes, including those to which people can, and need to, adapt. Such are the reasons for trying to establish the, hitherto unknown, geography of rock glaciers in the region and, to the extent possible, what controls their distribution and behaviours.

## 11.2 Rock Glacier Distribution

The main concentrations of rock glaciers are in lesser offshoots of the Mustagh Karakoram and ranges of intermediate elevations surrounding it (Fig. 11.1). In addition to the Ghujerab Mountains of northern Shimshal described by Owen and England (1998), they occur in great numbers in valleys of the Deosai Mountains and margins of the Deosai Plateau; in the Shimshak Mountains, for example, at the heads of Baumaharel and Thalle Valleys; and in the Naltar and Lukpar Mountains. As noted, they are rare in the highest most heavily glacierised parts of the Karakoram.

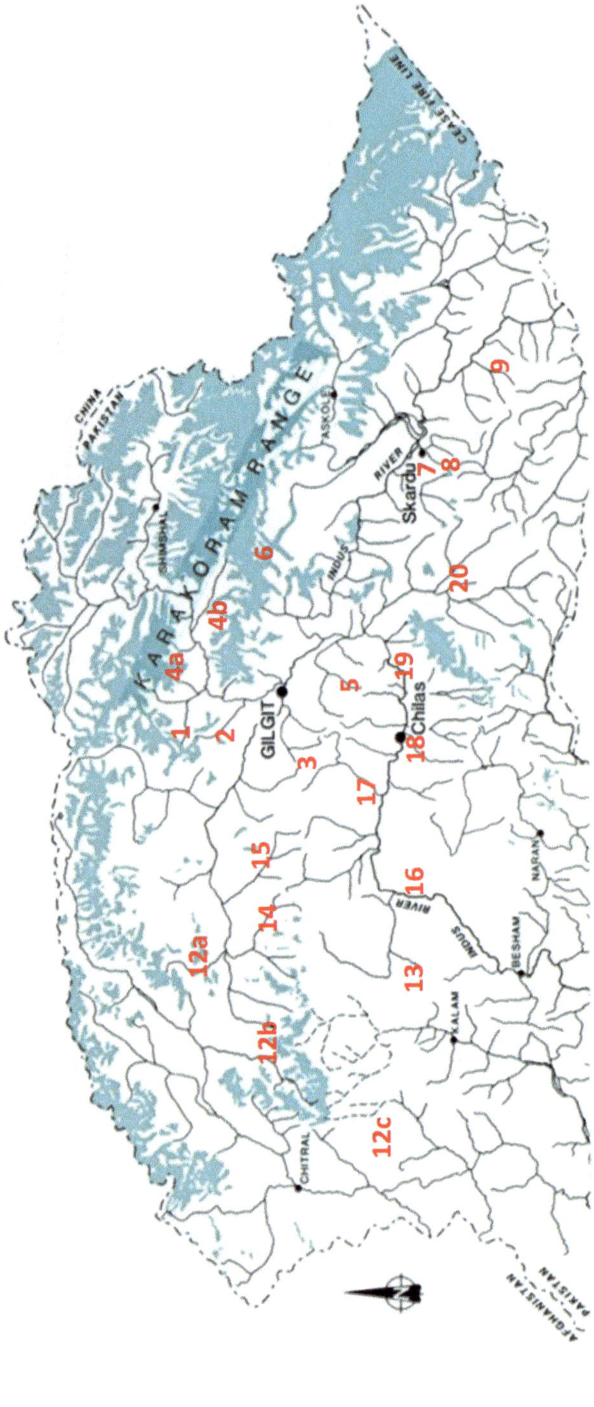


**Fig. 11.1** Mountain ranges in the upper Indus Basin of Northern Pakistan, with major concentrations of rock glaciers: 1. Lupghar, 2. W. Batura Mustagh, 3. Naltar Mts, 4a. Ghujerab Mts (North), 4b. Ghujerab Mts (South), 5. Rakaposhi-Haramosh, 6. Wesm Mts (Lower N. Flanks), 7. Mango Gusor Mts, 8. Shimshak, 9. South Saltoro Mts, 10. Ne. Rimo Mts, 11. S. Sasir Mustagh 12a. Hindu Kush, Rich Gol (NE), 12b. Hindu Kush, Tirich Mir (Central), 12c. Hindu Kush, Nuristan (SW), 13 Kalam Hr, 14. Yasin Hr, 15. Darkot Hr 16. Swat Kohistan, 17. Gilgit Mts, 18. Indus Kohistan, 19. NW Kohistan, 20. Deosai Mts (see Table 11.1)

**Plate 11.3** Lobe of a well-known glacier-derived rock glacier fed from a relatively low interfluve of Mitre Peak (6,025 m) entering Baltoro Glacier near Concordia (Hewitt 2005)

Strong gradients and differences in interfluve elevations, or between north and south slopes, mean rock glaciers can be clustered close to other areas where they are absent. Some isolated, 'eccentric' examples, like a much-photographed one on Mitre Peak in Baltoro Basin, are exceptions that support the general rule (Plate 11.3). There are none on the heavily glacierised south flanks of Batura Glacier, but its north slope tributaries, with lower watersheds, contain dozens.

Conditions where rock glaciers are concentrated include the driest and the most humid parts of the region. Those in the Hindu Raj and Hindu Kush valleys in Pakistan and Afghanistan can resemble arid land examples described for the Andes of Argentina and Chile (Shroder and Giardino 1978; Azocar and Brenning 2010). In the Deosai and Shimshak Mountains, they resemble those of the humid Scandinavian or Alaska–Yukon Mountains (Johnson 1978). Where a generous part of their source is in the zone of maximum precipitation, they can resemble the 'maritime' types described from Svalbard, Iceland and western Greenland (Harris and Murton 2005; Humlum et al. 2007). In all, available elevation range, orographic, topoclimatic and lithological diversities help underpin an exceptional variety of rock glaciers and related forms.

## *11.2.1 Rock Glacier Verticality*

The indications that rock glaciers are rare or absent in the higher parts and the lower valleys of the Karakoram suggest elevation and verticality relations are important.

Thresholds and zones for rock glacier formation were derived from surveys along the same transects as for glaciation levels (Chap. 2, Table 2.5). Satellite images from between 2003 and 2009 were used and for late summer (mid-August to mid-October). Along the four transects, the limits of apparently active and fresh rock glaciers were determined, and apparently inactive or relict ones, where the surface is subdued, strongly weathered or vegetated (Table 11.1). The specific measures were for the following:

(i) *Rock glacier level or threshold*: the lowest interfluves which support an active rock glacier below or descending from them.
(ii) *Highest rock glacier source*: the highest interfluve with active rock glaciers below it. Unlike glaciers, rock glaciers do have well-defined upper as well as lower limits on their development. This is, equally, another 'glaciation limit' since rock glaciers are replaced by glaciers above these levels, evidently due to a relative excess of (avalanched) snow rather than the absence of debris.
(iii) *Rock glacier zones*: As with glaciers, many rock glaciers commence well below the interfluves concerned, fed by avalanches descending over rock walls and talus above. According to inputs and local terrain, they may spread well below the elevations which determine whether they exist at all.
(iv) *Rock glacier spans*: examples of the larger elevation ranges of individual rock avalanches identified along the transect.
(v) *Rock glacier free zones*: the highest and lowest interfluves without active rock glaciers below.
(vi) *Lowest inactive and relict rock glaciers*: lowest identified deposits and features in valleys along the transects, indicative of Holocene fluctuations in rock glacier activity.

In the absence of any sustained monitoring of rock glaciers, it is not possible to assign 'rock glacier initiation altitudes' or RILA (Humlum 2000, p. 42). However, they are probably complicated by similar problems to ELAs and snowlines (Chaps. 4 and 7). Local terrain, lithologies, topoclimates, wind and avalanche dynamics may well override or preclude well-defined contour or isothermal thresholds.

The data show that rock glaciers are found within broad elevation ranges. For the whole region, they exist in a total vertical range of some 2,000 m but varying in span across the mountains. A 'rock glacier zone' is rarely reported from other regions except the high Andes, and it stands out in the Karakoram partly because of the total range of elevations, over most of which *no* active rock glaciers are seen.

No rock glaciers were observed where interfluves are less than 4,300 m or where they are higher than about 6,000 m. In three of the transects, the range between upper and interfluve elevations is a quite narrow, a few hundred metres at most. There are differences in the limits for northerly versus southerly exposures of between 200 and 450 m. In the Central Karakoram, if an interfluve does not reach above 4,900 m, it does not support a contemporary rock glacier, regardless of steepness, orientation or rock type. If the crest is somewhere between 4,900 and 5,100 m, it is most likely that a small one will develop below it or related talus

**Table 11.1** Elevation limits and relations of rock glaciers (RG) at four transects across the Greater Karakoram Range. Along the transects, interfluve heights were determined at 50 points for rock glaciers. At least five points were sought for each of the parameters listed. Satellite images were used from between 2003 and 2009 and for late summer (mid-August to mid-October)

| Transect | Long. 73° 31′ (Darkot–Karambar Pass) | | Long. 74° 32′ (Hunza–Bagrot–Sost) | | Long. 76° 15′ (K2–Shimshak Mts.) | | Long. 77° 30 (Rimo–Chang Chenmo) | |
|---|---|---|---|---|---|---|---|---|
| *Glaciation level* | Average | Extreme | Average | Extreme | Average | Extreme | Average | Extreme |
| North faces | 4,600 | 4,500 | 4,900 | 4,550 | 5,200 | 5,050 | 5,800 | 4,750 |
| South faces | 5,200 | 5,150 | 5,350 | 5,300 | 5,500 | 5,350 | 6,100 | 5,930 |
| Rock glaciers | | | | | | | | |
| *RG level or limit* | | | | | | | | |
| N facing | 4,400 | | 4,900 | | 4,450 | | 5,650 | |
| S facing | 4,900 | | 5,100 | | 4,850 | | 5,500 | |
| *Highest with RG* | | | | | | | | |
| N facing | 4,700 | | 5,300 | | 5,300 | | 5,800 | |
| S facing | 5,350 | | 5,350 | | 5450 | | 5900 | |
| *Highest without* | | | | | | | | |
| N facing | 4,300 | | 5,200 | | 4250 | | 5600 | |
| S facing | 4,800 | | 4,400 | | 4,500 | | 6,100 | |
| *RG span (e.g.)* | | | | | | | | |
| N facing | 4,470–3,850 | | 5,300–3,650 | | 5,200–3,865 | | 5,600–4,750 | |
| | 4,790–3,850 | | | | 5,100–4,260 | | 5,900–4,800 | |
| | 4,900–3,900 | | | | 5,300–4,450 | | | |
| S facing | 4,900–4,400 | | 5,350–4,575 | | 4,850–4,345 | | 5,950–5,100 | |
| | 5,100–4,150 | | 5,375–4,520 | | 5,700–4,105 | | 5,950–5,050 | |
| | 5,350–3,990 | | | | 5,200–4,145 | | | |
| *Lowest inactive/relict* | 3,350–3,720 | | 3,430–3,660 | | 3,650–3,900 | | 3,810–5,150 | |

features. As the interfluve height increases, more and larger rock glaciers are likely below it. Above 5,500 m, glaciers will be generated. Rock glaciers are excluded unless glacier ice grades into them lower down. If the interfluve is above 7,000 m, it will support glaciers that overwhelm any possibility of rock glacier growth at lower elevations. The highest examples yet identified commence around 6,100 m in the eastern Karakoram. The lowest reach is 4,200 m in the western part – lower, if the base of some tip-heap deposits from rock glaciers higher up are included. The Central Karakoram offers the broadest span within the region of about 1,400 m. Conditions vary considerably along the transect from around K2 (8,611 m) to the Indus where the rock glacier zone is much narrower.

It is important to distinguish between interfluve thresholds, the actual elevations where rock glaciers form and their elevation ranges. As with glaciers, many rock glaciers commence well below the interfluves concerned, fed by avalanches and rockfall from above (Fig. 11.2 and 11.3). Their size and spread depends upon inputs of snow and debris carried to them by other processes, in turn influenced by rock type, orientation, topoclimate and valley morphology.

The exact lower limits can be difficult or impossible to determine visually. Active rock glaciers may grade into or override seemingly stalled or inactive ones. Sharp or fresh features are employed to identify active features, but there are no data on how quickly these fade or are lost in a stalled or 'dead' rock glacier. Seemingly stable, weathered and vegetated lobes indicate past rock glaciers that extended hundreds of metres vertically below today's lower limits and as much as 1,000 m in some cases. The chances of preservation likely declines with age, but more detailed ground inspection may well extend the record downwards and back in time.

## 11.3 Rock Glacier Morphology and Associated Features

Virtually every rock glacier form described in the literature seems to be present somewhere in the Karakoram, involving a full range of morphological relations to valleys and cliffs, to glaciers and periglacial features such as patterned ground (Davies 1969, pp. 42–47; French 2011). In given areas, particular types or characteristics are found, but it is common for a variety to occur in close proximity or overlapping. There are marked differences between north- and south-facing rock glaciers in the same valleys and elevation ranges. Perhaps the much greater available relief and steepness are responsible, and pathways for delivery of snow and debris presumably increase proportionately.

There is huge variety in sizes, including many that seem much larger than the most intensively studied cases in the European Alps, Scandinavia or the Western USA (Haeberli 1985; Barsch 1988, 1992). There are lobate- and tongue-shaped rock glaciers filling valleys from side to side, many exceeding 2 km in length, some as much as 4 km. Some continue for 6 km if the glacier ice in source areas or seemingly inactive sections into which they grade are included. It is, however, unusual for width to exceed 0.5–0.7 km except for some terminal lobes.

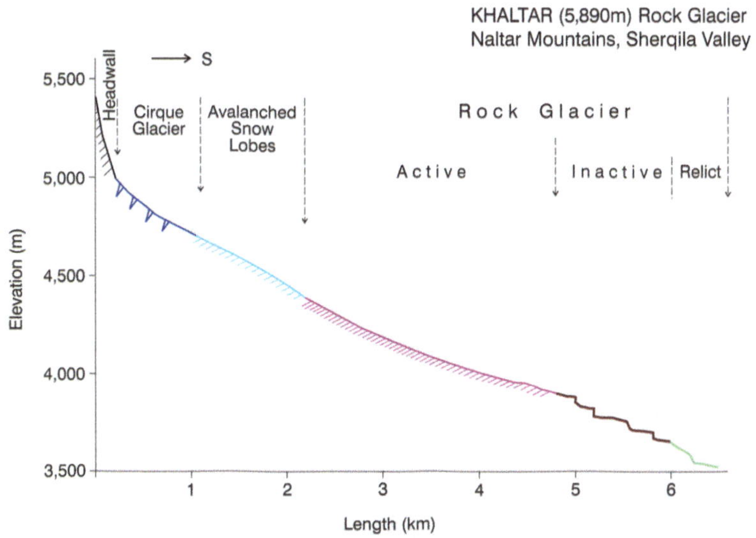


**Fig. 11.2** Long profile of Khaltar Peak (5,890 m) SE rock glacier complex, Naltar mountains, Sher Qila Valley (see Plate 11.7), with tongue-shaped morphology, glacier-, avalanche- and talus-derived. The active rock glacier is 2.8 km long, ending at a smaller inactive lobe, with a relict lobe below this. Basin area above the active terminus is ten times larger than the rock glacier

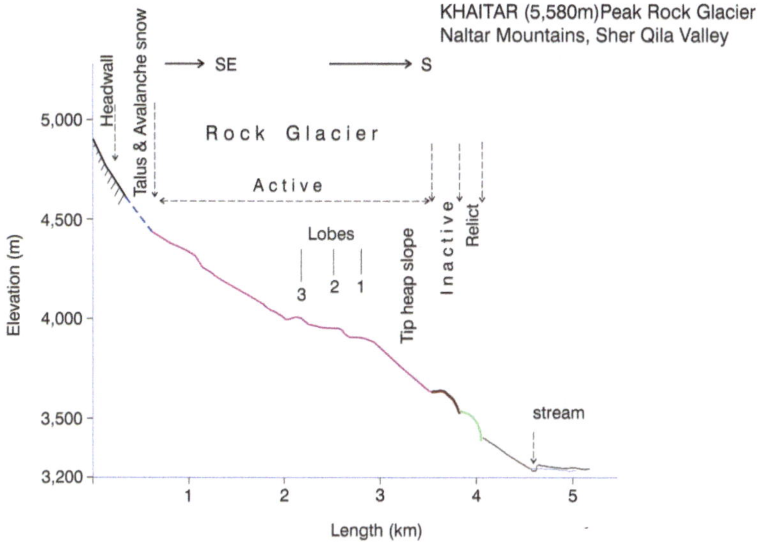


**Fig. 11.3** Long profile of the Khaitar Peak (5,580 m) SW rock glacier complex, Naltar mountains, Sher Qila Valley (see Plate 11.8), a tongue-shaped, predominantly avalanche- and talus-derived form, terminating in a long tip-heap slope with smaller inactive and relict RGs below. Basin area above the active terminus is five times larger than the rock glacier

**Plate 11.4** Rock glacier complex with combined and interacting lobes, upper Thalle Valley, Baltistan (Hewitt 1998)

Rock glaciers reported elsewhere seem to be mostly single lobes. In the Karakoram and its more extensive developments, they comprise drainage networks or systems. Two or more tributaries enter a main valley complex and run parallel or coalesce into a single lobe that may be 2–3 km across. They usually have particularly well-developed and complex surface relief of subparallel furrows and ridges. In such environments, rock glacier lobes may be advancing with steep, tip-heap fronts over other slower or inactive ones (Plate 11.4). These networks may not be active throughout or continuously so. The longest complexes identified combine more active and less active sections, apparently inactive ones, and relict lobes, usually in a downslope sequence but sometimes arranged parallel. Complications arise where more vigorous tributaries enter further down valley. The connected systems may be affected by small rock glaciers that reach and deform their margins without joining the main network.

In the higher and more humid areas, it is difficult to separate rock glaciers from debris-covered glacier tongues. Most of the longer rock glaciers emerge from glacier ice higher up, yet they exhibit classic ridge-and-furrow forms, lobes with tip-heap fronts and no sign of ice along most of the margins. Rock glaciers associated with or close beside contemporary glaciers exhibit considerable diversity and at least three distinct forms. Two types could be termed *glacier-head* types (Plate 11.5). In one of these, the glacier surface merges into the rock glacier tongue with no clear boundary. There may be a section between the upper glacier and lower rock glacier portions disturbed by crevasses and surface ponds (Plate 11.6). In the second case, the glacier lobe or ice stream has a well-defined, steep snout, immediately below which a rock glacier fills the same valley. A third type or set,

**Plate 11.5** Source zone of a 'glacier-head-type' rock glacier in Naltar Valley WC Karakoram (lat. 36° 11′N; long. 74° 06′E). As is usual, it combines ice from a wasting cirque glacier with heavy avalanche nourishment. Massive talus cones spread over the margins and reach well up the headwall, still hidden under snow in this mid-June scene. The active feature continues 2 km below the photograph and the whole is 3.3 km long from the cliff base to a tip-heap terminal slope where debris falls into a steep canyon. The feature descends from about 4,300 to 3,600 m, progressively acquires more 'classic' rock glacier features including well-developed ridge-and-furrow forms. Its terminal lobe combines with two other rock glaciers, one 2.3 km long and another 1.5 km that is inactive where they meet (Hewitt 2010)

the most common, is quite detached from glaciers, but glacier ice exists in the upper basin, and there is evidence of recent glaciation, or glacier ice may exist in the rock glacier and is occasionally revealed in crevasses or depressions. In other cases, rock glaciers and glaciers are in close proximity, but the rock glaciers commence independently below heavily avalanched or talus slopes, and the two move parallel along the same valley (Plate 11.7).

There are valley-head rock glaciers that originate under the headwall of a former glacier and occupy part or all of its former valley but without visible contemporary glacier ice. A case may be made for calling these *paraglacial*, in that their genesis and development reflect a continuing influence of former glaciation (Jijun et al. 1984; Ballantyne and Benn 1996; Hewitt 2009).

The ridge-and-furrow forms on the surface seem to be good visual indicators of a rock glacier, features with a relief of a few metres and spaced some metres or tens of metres apart. There are countless Karakoram examples and an astonishing number on the surface of some examples. Conversely, there are many others with poorly developed or absent ridge-and-furrow forms but with steep, fresh tip-heap slopes at the front and lateral margins and other signs of rock glacier behaviour.

**Plate 11.6** Intermediate zone between a wasting glacier and full-blown rock glacier, in the same valley as Plate 11.5. Strong surface relief, heavy debris covers, surface ponds and occasional facets of glacier ice are typical of such cases (Hewitt 2010)

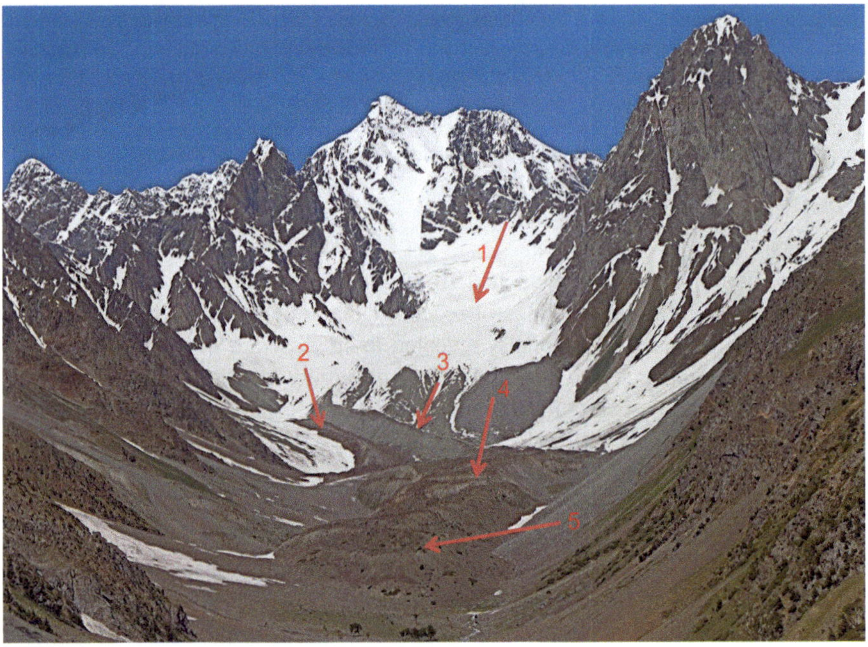


**Plate 11.7** Rock glacier complex, Sher Qila Valley, Naltar Mountains, descending from the south-east flank of Khaltar Peak (5,890 m). 1. Active small glacier, 2. avalanche-derived active rock glaciers, 3. eroded lateral moraine of former glacier overlooking active rock glacier, 4. inactive or barely active rock glacier, 5. relict rock glacier (Plate 11.9). Parts of the apparent talus along the valley sides descend from tip-heap fronts of small rock glaciers in tributary valleys (Hewitt 2010)

Among the largest rock glaciers are curious ones resembling the ‘bird’s-foot’ delta. In satellite images, their flow forms suggest a viscous fluid, like treacle or warm toffee. The comparison sometimes made with lava flows seems apt in these cases (Bodvarsson 1983). There are multiple overlapping small lobes in, on or diverging from a main valley rock glacier. Segments that move(d) at an angle to the general direction may reach to the edges and protrude from them as rounded lobes or small distributaries. Some other examples combine the viscous lobes with complex ridge-and-furrow forms.

## 11.4 Active, Inactive and Relict Forms

Just as there is a highly variable and complex transition to glaciers around the upper limits of the rock glaciers zone, similar complex transitions apply between presently active, waning or growing rock glaciers and apparently inactive and relict or ‘fossil’ forms. Many are continuous from active to relict, and it is not always clear where the transition from active to inactive occurs. Relations are often complicated by talus and avalanche activity. In other cases, visible differences between active, inactive and relict forms can be obvious. Even in short visits, active rock glaciers may be seen to advance over the ground in front, with boulders repeatedly tumbling down the bounding slopes (Plate 11.8). Rock glaciers were described as advancing

**Plate 11.8** Rock glacier with huge tip-heap terminus and steep flanks, Sher Qila Valley flowing west from Khaltar Peak (5,454 m). Relict rock glacier lobes lie below the fresh material falling down (Hewitt 2010)

and overriding vegetated areas and young river terraces in north of Shimshal, when most glaciers were retreating (Owen and England 1998, p. 206). If actual movement is not observed, activity is indicated by unweathered material and fines exposed in steep lateral and terminal slopes, evidently at or close to the angle of repose. More rarely, wet material and even ice may be exposed (Barsch 1988). The absence of such features and more rounded and weathered surfaces point to inactive rock glaciers, although where slow or intermittent motion ceases may not be obvious without measurement. At the lowest levels in given valleys are old forms completely vegetated, even wooded, or buried and smoothed by wind-blown dust. Unlike active rock glaciers, which strongly affect streams, force them against the valley side and even cause ponding, the inactive ones are eroded by the streams.

Some feel that rock glaciers should not be described as 'active' or 'inactive', only in terms of the processes involved (Giardino et al. 2011). That may make sense from theoretical perspectives and in regions where processes are well established and monitored. It is hard to require or justify in the Karakoram. The evidence available may suggest process, but is not based on it. Diverse and overlapping forms defy easy identification of process.

Differentiating forms as, say, relict, inactive and active seems a valuable notion for a geomorphologist. Process may be key, but the forms of interest are in solids, many of which hold their shape more or less long after the processes imparting it have ceased to operate. This makes 'memory' an integral part of the solid Earth surface and how cold region landforms record varying cold environment histories (Haeberli and Gruber 2009). Landscapes of the cryosphere are replete with features made by processes no longer present. Visual differences that distinguish between active, inactive or relict forms in a poorly known context offer a preliminary way to explore changing glacial–periglacial conditions or, at least, what needs monitoring to confirm apparent differences.

In the Karakoram region, there are clearly inactive, vegetated lobes indicating that rock glaciers once extended hundreds of metres lower than the limits of today's activity, a 1,000 m lower in some cases (Plate 11.9). The chances of preservation likely declines with age, but more detailed ground inspection may extend the record downwards and back in time.

## 11.5 Rock Glacier Genesis and Development

Formerly there was debate about whether rock glaciers have an ice content, its characteristics and how it originates, is sustained or is depleted. It seemed the most likely explanation of their dynamics but is rarely evident at the surface. Eventually, massive ice cores and ice-cemented debris were established by excavation and geotechnical methods. They support a general explanation of motion due largely or wholly to creep in the ice, possibly with some shearing. Unlike the coarse material comprising their surface, a range of sediments occurs in the interiors of rock glaciers, referred to as 'soil' in many studies and partly segregated into silty, sandy

**Plate 11.9** The relict, lowest main lobe of the Khaltar Peak (5,790 m) rock glacier complex in Sher Qila Valley, Naltar Mountains (see Plate 11.7). The base of this main valley lobe is at 3,650 m, a source of firewood and intensively grazed in summer. Difficulties of reconstructing environmental change are indicated by nearby *active* lobes, from the steeper chutes on valley walls, that reach somewhat lower elevations (Hewitt 2010)

or gravelly units (Whalley and Martin 1992). Finer material can emerge in steep, unstable margins of rock glacier lobes. These materials seem likely to affect strength and ice-debris relations, possibly movement, as in other active layer processes in permafrost (Nickling and Bennett 1984). In most Karakoram examples, such conditions can only be inferred, but there is no reason to think ice cores are not the main source of movement, their depletion leading to inactive forms.

The origin of the ice cores has also been subject to much debate. Permafrost, glaciers and avalanches have been supported as critical or allowable sources and rejected! It is thought that cold air drainage through the openwork boulders may cause evaporative cooling, sublimation or refrigeration of moisture infiltrating from the surface (Delaloye and Lambiel 2005). Questions and disagreements remain, but an emerging consensus accepts a diversity of origins.

The dominant processes are thought to generate three main types of rock glacier: *talus-*, *glacial-* and *avalanche-derived types* (Giardino and Vitek 1988; Corte 1999; Humlum et al. 2007). They are commonly identified with particular environments or climates. However, all exist in the Karakoram. In some parts, one or the other type predominates, but examples of all types can occur in single valleys, even adjacent and overlapping.

Talus-derived rock glaciers depend upon periglacial processes, are associated with permafrost and develop through build-up of ice-cored talus cones or lobes. They are sometimes called *protalus* or *valley wall* rock glaciers (Whalley and Martin 1992). In the Karakoram, as elsewhere, there is difficulty in deciding which of the many protalus forms that occur has, or should be ascribed, rock glacier characteristics. The deposits at the base of talus slopes include countless minor lobes, bulging and rampart-like features, protruding a few tens of metres across the valley involved. There are apronlike forms that spread across most of the valley with steep fronts and many that cross and dam high valleys. There is some doubt about which ones should be included as rock glaciers. Rather, it often seems there is a continuum from purely talus cones and tongues, lobes with bulging ramparts suggestive of creep or incipient rock glacier processes, to well-developed rock glacier lobes with ridge-and-furrow forms.

Avalanche types are widely present, possibly the most common, especially at higher locations and interspersed with glacier-derived forms (Humlum et al. 2007). Rubble or rock streams, buried by avalanches annually and for most of the year, are found beside and deforming the edges of well-developed rock glaciers.

Glacial rock glaciers develop from the burial of glacier ice where it is starved of sufficient snow and overwhelmed by the quantities of debris. Sometimes they are called *moraine* or *debris rock glaciers*. In most Karakoram examples, the debris arrives directly from rockfalls and avalanches, much as it does or did with the glaciers they are derived from. It can be difficult or impossible to distinguish rock glaciers and debris-covered glacier tongues or to determine whether or how far the cores consist of glacier ice. The near-complete protection from ablation beneath thick debris and slowness of movement mean that rock glacier ice may be centuries, even millennia, old and, hence, from a distant glacier ancestor or even predating glacier expansions in the Little Ice Age.

A key criterion may be that moraine-like embankments do not develop with rock glaciers. They have *tip-heap margins* and what have been termed 'arched rolls' (Corte 1999, his Fig. 16). This contrasts with the moraine-banked margins typical of debris-mantled glaciers (Benn and Evans 1998, p. 244). It separates rock glaciers from their closest glacier relatives, 'moraine-dammed' glaciers (Müller 1958; Benn et al. 2005). Different processes operate at an ice margin, as opposed to a debris lobe. In glacier-derived Karakoram rock glaciers, a switch from moraine-banked to tip-heap may be observed in mid-valley areas. It is not unusual for rock glaciers to be flanked and overlooked by old lateral moraines from larger, former glaciers, but separated from them by a trough and their own tip-heap margins (Plate 11.7).

Wherever they occur, rock glaciers reflect and vary with local terrain and thermal conditions, supplies of debris and lithology, winter snow avalanching and the timing and patterns of summer melting. Karakoram examples generally originate in the zone of maximum (orographic) precipitation for the valleys concerned. The rock glacier zone can be snow-covered for 8–11 months of the year, depending on elevation, orientation and location within the region. Near the upper limits, they may be only briefly snow-free in summer. Avalanches and wind-drifted snow tend to survive from year to year over the upper parts. Remnants of winter snow in furrows persist after most of the rock glacier is exposed, indicating wind action can be important in both moisture supply and thermal conditions. The porous, openwork nature of the surface means there are important but potentially complex relations between snowfall, freeze–thaw in the snowpack, moisture infiltration, ice accretion or depletion and thermal conditions at depth.

## 11.6 Rock Glacier-Interrupted Drainage

Within some alpine watersheds, rock glaciers have a major impact on the hydrology of streams.... (Giardino et al. 1992)

The hydrological roles of rock glaciers are substantial wherever they are concentrated. They act as stores of ice considerably buffered against ablation and, to a varying extent, of liquid water in internal aquifers and surface ponds. The relation to periglacial environments and mountain resources as well as climate change warrants consideration, hitherto missing (Giardino et al. 2011). The internal hydrology of Karakoram examples is not known, but they can be important regulators in the vertical cascade of moisture and of stream hydrographs. Research elsewhere and the clear water streams commonly associated with Karakoram rock glaciers show an influence in water quality that is distinctly different from glaciers (Corte 1999).

In the absence of any work to quantify such matters just how important they are cannot be established. Yet, they seem to be more directly and extensively part of human settlement and land uses in the mountains than the glaciers are. Failure to recognise this goes back to comments made earlier about the preoccupations and locations of most scientific activity.

In terms of their hydrological and geomorphic roles, striking features involve the countless rock glaciers that interrupt streams in headwater basins and pond water there. This refers to the headwaters of most streams in, say, the Naltar, Deosai and Shimshak Mountains; the Hindu Raj; and north slope of the Kohistan–Gilgit Himalaya. The Shigarthang River is an example (lat. 35° 28′ N; long. 75° 29′ E). It drains a central part of the Deosai Mountains and joins the upper Indus near its exit from Skardu Basin at Katzarah. The lowest rock glacier identified on satellite imagery is at 3,600 m elevation, a few kilometres above Shigarthang village. It once entered and blocked its Dari Lungma tributary but is now seemingly inactive and being eroded. It does still disturb the river's course. Above this, along the 21 km upper reach of the Dari Lungma, there are 61 places where the main stream is disrupted by well-defined rock glaciers, or by large talus cones, with some signs of mobility and, therefore, possibly ice-cored rock glaciers. Similar interference is seen along the main river and its tributaries comprising some 75 km of channel length. The whole upper basin above Shigarthang Village covers about 610 km$^2$. It contains more than 150 rock glaciers larger than 0.5 km in diameter and possibly twice as many smaller than this. The largest ones identified are around 2 km in diameter, mostly glacier-derived.

Drainage interruption is exaggerated by other processes; notably many more talus cones and boulder lobes that are too small to rate as rock glaciers or lack any indications of creep. There are at least equal numbers of avalanche cones across the rivers in early summer, at the same places and others. The generally greater mass and height of rock glacier tongues and lobes, especially where they are active, helps protect and magnify stream interruptions by other processes. Meanwhile, a number of large rock avalanches block the Shigarthang, starting with the Katzarah megaslide at its mouth and four others confirmed in field surveys of the basin (Hewitt et al. 2011; Hewitt

2011). Those discovered are below the lowest rock glaciers, but it is possible others exist higher up. Interference with fluvial drainage by geomorphic processes comprises a hierarchy in which larger, usually rarer, events tend to facilitate the action of smaller ones and partially protect them from stream erosion. While many of the lower rock glacier barriers are inactive or relict and undergoing trenching and removal by streams, they remain morphological influences over their valleys and hydrology.

Similar developments are found in the other mountains listed as having concentrations of rock glaciers. In all, several thousand square kilometres of mountain terrain have drainage partly and, in places, severely fragmented by rock glaciers and related processes. There are rock avalanche barriers on many of the same rivers. Towards the higher elevations, rock glacier interruptions give way to glacier-interrupted fluvial systems, including those where glacial lakes and outburst floods are reported (Chap. 10). Again, hierarchies of interruption exist that interact to create chronically fragmented drainage, but also most of the land suitable for human use (Hewitt 2010). What appears as disturbance or fragmentation from a fluvial point of view amounts to a major, systemic regulation of fresh water hydrology variously exploited by mountain peoples.

## 11.7 Rock Glaciers as Resources

In the whole Indus or Yarkand Basins, the hydrological contribution of rock glaciers may seem insignificant. Even for the trans-Himalayan headwaters of the upper Indus, they are tiny contributors compared to snowfall and glaciers, even rainfall.

Even where rock glacier research is most advanced, there is an impression of their limited importance. Humans, other than scientists, perhaps mountain walkers, seem largely absent. Except for a few rare, catastrophic debris flows triggered in degraded rock glacier areas, they are seen as indicators rather than critical parts of broader environmental change (Haeberli and Gruber 2009). This is far from being the case where rock glaciers are concentrated in and around the Karakoram. They tend to be near or in the headwaters of the more populous valleys of the intermediate mountains and help support a broad spectrum of high mountain economies and cultures.

The surfaces of active rock glaciers are largely avoided but are closely associated with a variety of critical resources. Streams issuing from their basins are tapped for irrigation agriculture and other uses in permanent settlements. In these valleys, rock glaciers may be the main sources of water and much prized because the water is clear and more reliable than glaciers or snowmelt. Rock glaciers are surrounded by or overlook extensive mountain pastures (Chap. 13). The ponds on and beside them, or the streams issuing from them, are integral parts of summer activities. The surfaces of hundreds of inactive rock glaciers, especially older relict features, provide grazing, sites for summer villages and wood for fuel and construction and are important for wildlife and hunting.

The hydrological phenomena are largely unknown, but they occur in more sensitive elevations relative to climate warming than most glacier ice. Rock glaciers should help conserve or slow losses of water supplies, at least for mountain communities. However, adverse effects may arise from other conditions described. Stream fragmentation and ponding may lead to greater evaporative losses, places favourable for insect infestations under warmer conditions or increased water contamination by domestic and wild animals. Unfortunately, even in other regions where rock glaciers and their response to climate change are being carefully researched, there is rarely recognition of these issues, and there are few investigations that might help guide future work in the Karakoram.

## References


Azocar GF, Brenning A (2010) Hydrological and geomorphological significance of rock glaciers in the dry Andes, Chile (27°–33°S). Permafr Periglac Process 21:42–53

Ballantyne CK, Benn DI (1996) Paraglacial slope adjustment during recent deglaciation and its implications for slope evolution in former glaciated environments. In: Anderson MG, Brooks SM (eds) Advances in hillslope processes, vol 2. Wiley, Chichester, pp 1173–1195

Barsch D (1988) Rock glaciers. In: Clark MJ (ed) Advances in periglacial geomorphology. Willey, Chichester, pp 69–90, Chap 4

Barsch D (1992) Studies and measurements on rock glaciers at Macun, Lower Engadine. In: Evans DJA (ed) Cold climate landforms. Willey, Chichester, pp 457–473, Chap 21

Barsch D, Jacob M (1998) Mass transport by active rock glaciers in the Khumbu Himalaya. Geomorphol 26:215–222

Benn DI, Evans DJA (1998) Glaciers and glaciation. Arnold, London

Benn DI, Kirkbride MP, Owen LA, Brazier V (2005) Glaciated valley landsystems. In: Evans DJA (ed) Glacial landsystems. Hodder Arnold, London, pp 370–406, Chap 15

Bodvarsson G (1983) Lava flows and forms. Jokull 33:57–60

Corte AE (1999) Rock glaciers. U.S. Geological Survey, U.S. Department of the Interior. http://pubs.usgs.gov/pp/p1386i/chile-arg/dry/rock.html#4

Davies JL (1969) Landforms of cold climates: introduction to systematic geomorphology III. MIT Press, Cambridge, USA

Delaloye R, Lambiel C (2005) Evidence of winter ascending air circulation through talus slopes and rock glaciers situated in discontinuous permafrost (Swiss Alps). Nor Geogr Tidsskr 59:194–201

Delaloye R, Lambiel C, Gärtner-Roer I (2010) Overview of rock glacier kinematics research in the Swiss Alps. Seasonal rhythm, interannual variations and trends over several decades. Geogr Helv 65(2):135–145

Dixon JC, Abrahams AD (1992) Periglacial geomorphology: proceedings of the 22nd annual binghamton symposium in geomorphology. Wiley, New York

French HM (2011) Periglacial. In: Singh VP, Singh P, Haritashaya UK (eds) Encyclopaedia of snow, ice and glaciers. Springer, Dordrecht, pp 827–841

Gardner JS (1980) Frequency, magnitude and spatial distribution of mountain rockfalls and rockslides in the Highwood Pass area, Alberta, Canada. In: Coates DR, Vitek JD (eds) Thresholds in geomorphology. Allen and Unwin, London, pp 267–293

Giardino JR, Vitek JD (1988) The significance of rock glaciers in the glacial-periglacial landscape continuum. J Quat Sci 3(1):97–103

Giardino JR, Vitek JD, DeMorett JL (1992) A model of water movement in rock glaciers and associated water characteristics. In: Dixon JC, Abrahams AD (eds) Periglacial geomorphology: proceedings of the 22nd annual Binghampton symposium in geomorphology, John, New York, pp 159–184

Giardino JR, Regmi NR, Vitek JD (2011) Rock glaciers. In: Singh VP, Singh P, Haritashaya UK (eds) Encyclopaedia of snow, ice and glaciers. Springer, Dordrecht, pp 943–948

Haeberli W (1985) Creep of mountain permafrost: internal structure and flow of alpine rock glaciers. Mitteiliungen Versuchsanstalt fur Vasserbau. Hydrol Glaziologie 77:1–142

Haeberli W, Gruber S (2009) Global warming and mountain permafrost. In: Margesin R (ed) Permafrost soils, vol 6, Soil biology. Springer, Berlin, pp 370–406, Chap 14

Harris C, Murton JB (2005) (eds) Cryospheric systems: Glaciers and permafrost. Geological Society, London, Special Publications 242, pp 11–27

Hewitt K (1968) Studies in the geomorphology of the mountain regions of the Upper Indus Basin. 2 vols. Unpublished dissertation, University of London

Hewitt K (1989) Hazards to water resources development in high mountain regions; the Himalayan sources of the Indus. In: Starosolszky O, Melder OM (eds) Hydrology of disasters: proceedings of the technical conference, Geneva, November 1988. James and James, London, pp 294–312

Hewitt K (2009) Paraglacial rock slope failures, disturbance regimes and transitional landscapes, Upper Indus Basin, northern Pakistan. In: Knight J, Harrison S (eds) Periglacial and paraglacial processes and environments. The Geological Society, London, Special Publications, vol 320, pp 235–255

Hewitt K (2010) Gifts and perils of landslides: Catastrophic rockslides and related landscape developments are an integral part of human settlement along upper Indus streams. Am Sci 48:410–419

Hewitt K (2011) Rock avalanche dams on the trans-Himalayan Upper Indus Streams: a survey of late Quaternary events and hazard-related characteristics. In: Evans SG, Hermanns RL, Strom A, Scarascia-Mugnozza G (eds) Natural and artificial rockslide dams. Lecture Notes in Earth Science. Series 133, Springer, Chap 6, pp 177–204. doi:10.1007/978-3-642-04764-0_6

Hewitt K, Clague JJ, Gosse J (2011) Rock avalanches and the pace of late Quaternary development of river valleys in the Karakoram Himalaya. Geol Soc Am Bull doi:10.1130/B30341.1. 16 p, Accessed 14 June 2011

Humlum O (1998) The climatic significance of rock glaciers. Permafr Periglac Process 9:375–395

Humlum O (2000) The geomorphic significance of rock glaciers: estimates of rock glacier debris volumes and headwall recession rates in West Greenland. Geomorphology 35:41–67

Humlum O, Christinasen HH, Juliussen H (2007) Avalanche-derived rock glaciers in Svalbard. Permafr Periglac Process 18:75–88

Iturrizaga L (2006) Transglacial landforms in the Karakoram (Pakistan): a case study from Shimshal Valley. In: Kreutzmann H (ed) Karakoram in transition: culture, development and ecology in the Hunza Valley. Oxford University Press, Oxford, pp 49–72, Chap 5

Jijun L, Derbyshire E, Shuying X (1984) Glacial and paraglacial sediments of the Hunza Valley, North-West Karakoram, Pakistan: a preliminary analysis. In: Miller KJ (ed) International Karakoram project, vol 1. Cambridge University Press, Cambridge, UK, pp 496–535

Johnson PG (1978) Rock glacier types and their drainage systems, Grizzly Creek, Yukon territory. Can J Earth Sci 15:1496–1507

Kääb A (2008) Remote sensing of permafrost-related problems and hazards. Permafr Periglac Process 19:107–136

Kääb A, Paul F, Maisch M, Hoelzle M, Haeberli W (2002) The new remote sensing derived Swiss glacier inventory: II. First results. Ann Glaciol 34:362–366

Millar C, Westfall RD (2008) Rock glaciers and periglacial rock-ice features in the Sierra Nevada: classifications, distribution, and climate relationships. Quat Int 188:90–104

Müller F (1958) Eight months of glacier and soil research in the Everest region. In: The Mountain world 1958/59. Swiss Foundation for Alpine research, Harper, New York, pp 191–208

Nickling WG, Bennett L (1984) The shear strength characteristics of frozen, granular debris. J Glaciol 40:348–357

Owen LA, Derbyshire E (1989) The Karakoram glacial depositional system. Zeit Geomorph 76:33–73

Owen LA, England J (1998) Observations on rock glaciers in the Himalayas and Karakoram Mountains of northern Pakistan and India. Geomorph 26(1–2):199–214

Röer I (2007) Rock glacier kinematics in a high mountain geosystem, Bonner Geographische Abhandlungen. Asgard, Sankt Augustin, 117

Shroder JF Jr (1987) Rock glaciers and slope failures; high plateaux and La Sal Mountains, Colorado. In: Giardino JR, Shroder JF, Vitek JD (eds) Rock glaciers. George Allen and Unwin, London, pp 193–238

Shroder J, Bishop M (2010) Glaciers of Afghanistan. In: Williams RS, Ferrigno JG Jr (eds) Satellite image atlas of glaciers: Asia, vol 1386-F, United States Geological Survey, Denver, professional paper. U.S. G.P.O, Washington F-3, pp 167–199

Shroder JF Jr, Giardino JR (1978) Progress of rock glacier research. Transactions. Neb Acad Sci 6:51–54

Whalley WB, Martin HE (1992) Rock glaciers: II models and mechanisms. Prog Phys Geogr 16(2):127–186

# Chapter 12
# Karakoram Glaciers and Climate Change

**Abstract** This chapter looks at fluctuations in Karakoram glaciers, mainly in the last 200 years, their consequences and implications for future responses to climate change. Most of the evidence available concerns terminus changes. Improvements in satellite coverage and analytical techniques have increased the range and quality of information but results can raise as many problems as answers. Records are reviewed back to the mid-nineteenth century for some of the larger and more frequently visited glaciers, including Baltoro, Biafo, Batura, Chogo Lungma and Hispar. This information broadly confirms an ice cover decline since the Little Ice Age (LIA), although it does not seem to exceed 5 % of the greatest LIA extent. Large glaciers with high elevation watersheds appear less reduced than small and minor ice masses, although evidence from the latter is very limited. The timing of the greatest advances in the LIA varies by decades for different glaciers, in some cases by centuries. Between the 1920s and 1980s, most of the larger glaciers had a net retreat, but an almost chaotic situation emerges from the late LIA through the early twentieth century, and again since the mid-1990s. This is only partly accounted for by unsystematic fluctuations in surge-type glaciers. There has been no great loss of ice, and more than 40 high Karakoram glaciers have undergone advances of varying extent. At the latest time frame to 2010, few if any glaciers were at their most advanced positions of the past 150 years but also, no case was at its greatest reported retreat. Advances and retreats have been more or less out of phase even in neighbouring glaciers. Few of the large glaciers have retreated in proportion to the amount of thinning, and some have even advanced while appearing to thin. The recent picture differs from other parts of the Himalaya and common global trends. This may be due to the distinctive climatic regime or how it responds to global climate change as discussed in earlier chapters. Out-of-phase relations of terminus fluctuations may also follow from different styles of nourishment, thermal regimes and shifts causing movement instability. As yet unrecognised surge-type glaciers are another source of complexity. Confusion has also arisen in relating Quaternary glaciations to the state of present-day glaciers, notably debris covers, and ice

K. Hewitt, *Glaciers of the Karakoram Himalaya: Glacial Environments, Processes, Hazards and Resources*, Advances in Asian Human-Environmental Research,
DOI 10.1007/978-94-007-6311-1_12, © Springer Science+Business Media Dordrecht 2014

margin and pro-glacial deposits near them. This is illustrated by former and emerging interpretations of the so-called Great Lateral Moraine, the nearly ubiquitous, relatively well-preserved and massive lateral margin deposits overlooking present ice levels. They create an impression of glacier decline that may be misleading. Originally viewed as equivalent to the '1850' Little Ice Age moraines of the European Alps, the deposits turn out to be more diverse in age and origin. Post-glacial geomorphic developments along the Indus streams are at least as important as climate change and neoglaciation.

**Keywords** Terminus fluctuations • Climate change • Biafo Glacier • Hispar Glacier • Chogo Lungma Glacier • The Great lateral Moraine (GLM) • Intermontane sedimentation

## 12.1 Introduction: 'Disappearing' Glaciers

Glacier mass and behaviour can be sensitive indicators of climate variability. In recent years, this has driven interest in Himalayan glaciers to an overwhelming extent. Rapid and possibly catastrophic reduction of the ice cover has dominated reports from the region. Global climate change has had major effects on the weather systems of High Asia, and a direct link is commonly made between global warming and 'the melting Himalayas' (ICIMOD 2007, 2011; Armstrong 2010). The phrase 'disappearing glaciers' came into widespread use early in the twenty-first century (Table 12.1). Threatened water resources and increased risks from glacier hazards such as glacier lake outburst floods are widely described and a potentially large role of Himalayan glacier melting in rising sea levels (Bajracharya and Mool 2009; Immerzeel et al. 2010; Jacob et al. 2012). Greater ice losses have been predicted in the foreseeable future in relation to global warming, which seems likely to go on increasing (Kotlyakov and Severskiy 2009). These notions of rapid, if not catastrophic, change have been applied to Karakoram glaciers and colour discussions of them, despite the otherwise very limited coverage of them in the glaciological literature. In fact, the notion that they have been 'disappearing' seems to be based on assumption not observation.

Marked reductions in ice cover have affected nearly all glacierised areas of the world since early in the twentieth century. Most glaciers with reliable monitoring records show an accelerated loss in the last two of three decades (Haeberli et al. 1998; Bonardi 2008; Zemp et al. 2009). The same seems true in most of High Asia (Tsevetkov et al. 1998; Yafeng et al. 2006; Williams and Ferrigno 2010). However, more recent investigations suggest that changes in the Hindu Kush–Karakoram–Himalaya (HKH) are not happening as fast or consistently as many reports had claimed (Bolch et al. 2012; Kääb et al. 2012; Schmidt and Nüsser 2012). The use of terminus variations, possibly misread in critical cases, seems to have exaggerated the rates of change (Raina 2009; Scherler et al. 2011). Large differences occur across the region, and losses seem to have been least or less

**Table 12.1** Some recent statements about Himalayan glaciers and the Indus waters

| |
|---|
| IPCC Climate Change Asia, 2005: |
| ' – *Glacier melt in the Himalayas is projected to increase flooding... and affect water resources within the next two to three decades. This will be followed by decreased river flows as the glaciers recede'.* |
| World Bank Report No. 34081-PK, *Pakistan ...Water Economy: Running Dry.* 2005: |
| '...*The Indus basin depends heavily on the glaciers of the western Himalayas ...climate change is already affecting these western glaciers in a dramatic fashion (far more seriously, for example, than in the damper Eastern Himalayas)... best estimates ....are that there will be fifty years of glacial retreat, during which time river flows will increase...But then the glacial reservoirs will be empty...'.* |
| WWF Nepal Program March, 2005: |
| '*In 1999, a report by the Working Group on Himalayan Glaciology (WGHG) of the International Commission for Snow and Ice (ICSI) stated: 'glaciers in the Himalayas are receding faster than in any other region".* (p. 29)[a] |
| OXFAM, November 2007: |
| '...*the retreat of glaciers ...in the Himalayan highlands in north Asia has accelerated. Supplying seven of Asia's great rivers – [including the] Indus – the glaciers... ensure a year-round supply of water to billions of people. The impacts ...increased risk of flooding in the short-term, reduced river flow ...long-term...'.* |
| The Economist, June 2008: |
| '...*Mr Hasnain estimates that Himalayan glaciers will be gone in 20–30 years.... According to the IPCC this may be the fate of the Indus, Ganges and Brahmaputra by 2035'.* |
| Worldwatch Institute, Washington, DC 2012: |
| '*Accelerated melting of glaciers in the Himalayas is...posing a catastrophic threat to the 1.3 billion people in [the region's] river basins'*, said Uday Sharma, secretary of Nepal's Ministry of Environment, who attended the meeting in Kathmandu in early September. |

[a]WWF Nepal program March 2005. However, in subsequent editions the following appears: 'This statement was used in good faith but it is now clear that this was erroneous and should be disregarded.... WWF regret any confusion this may have caused'. An overview of glaciers, glacier retreat, and subsequent impacts in Nepal, India and China

consistently negative in the Karakoram (Hewitt 2005, 2011a; Gardelle et al. 2012). There is no cause for complacency, but the 'disappearing glaciers' image seems exaggerated if not misleading. Major problems emerge from the tendency to oversimplify regional diversity and developments that are complex in time and space, not least in their human implications (Chap. 13).

For the Karakoram and upper Indus Basin as a whole, available information does confirm substantial losses since the Little Ice Age (LIA), losses associated with periods of higher flows in the Indus streams. These glacier changes rarely seem to equal 5 % of the area of larger glaciers, and their length reductions are an even smaller percentage (see below). More challenging is the lack of consistent directions and rates of change. Substantial fluctuations occurred during as well as since the nineteenth and early twentieth centuries. In the late twentieth and early twenty-first centuries, total ice cover has remained relatively stable. Between the 1960s and 2010, ice mass seems to have been changed little or not at all in area, although thicknesses show a bewildering range of variations (Gardelle et al. 2012; Bhambri et al. 2012). Since the mid-1990s, while retreat or stability has been most common, dozens of terminus advances have been observed in the high Karakoram (Hewitt 2005; Liu et al. 2006). Then again, many glaciers used as evidence for unusual behaviour turn out to be surge type (Chap. 9). Their advances and retreats are not expected to be in-phase with climate fluctuations (Hewitt 2007a; Jiskoot 2011).

Glacier change is a prevailing fact of the Karakoram and climate is a major factor. However, it hardly conforms at all closely to the rest of the Himalaya or, in the past decade at least, to the world at large (Kick 1962; Goudie et al. 1984; Hewitt 2005; Scherler et al. 2011). Perhaps it is because the regional climate is distinctive or that global climate change affects it differently.

Fowler and Archer (2006), investigating temperature data at seven weather stations in the region for 1961–2000, found strong contrasts between the behaviour of winter and summer temperatures. Mean and maximum temperatures showed significant increases in winter and consistent decline in summer. An increase in diurnal temperature range was found in all seasons and annually. There was an observed 1°C fall in mean summer temperature since 1961, and even greater reductions in spring months. They suggest that this explains, or relates to, a ~20 % decline in summer runoff from the Hunza and Shyok Rivers. These are among the most heavily glacierised Indus tributaries in the high Karakoram and seem to reflect greater mass storage by the glaciers rather than signs of depletion. It is not clear how this relates to terminus fluctuations (cf. Rees and Collins 2006).

Perhaps Karakoram glaciers, with their exceptional elevations, elevation range and concentration, respond differently to climate change. Perhaps there is just a longer, but still temporary, lag in response to global warming. However, since the earliest modern studies, there have been suggestions of different or anomalous timings and scales of glacier adjustments here. Some even viewed the major Quaternary glaciations as out of phase, with events in the Northern Hemisphere generally or varying in magnitudes (Li Jijun et al. 1984; Shroder et al. 1993; Owen 2006; Seong et al. 2007). Little Ice Age (LIA) advances do not coincide with better-known ones in Europe (Kick 1989). Such views do involve some contested evidence (Fort 1995; Kuhle 2004; Hewitt 1999; 2009a). There has been a general failure to recognise the numbers of surge-type glaciers and their implications. The need to look more closely is clear. In the absence of extensive monitoring evidence, this must be done in terms of conditions in Karakoram glacial environments and processes discussed in previous chapters.

## 12.2 Glacier Change in the Last 200 Years

> ...Ultimately, the movements are due to climate and snowfall, but the factors are so varied that the snout movements appear to be peculiar to each glacier. There may be little resemblance between... neighbouring glaciers... even if they have the same exposure; sometimes there is no similarity between... two branches of the same compound glacier; occasionally one side of a glacier tongue may be advancing while the other retreats.... (Mason 1930, pp. 218–219)

Terminus changes are usually the only information available to track glacier fluctuations before the late twentieth century, and much effort has been devoted to reconstructing them (Hewitt 1968; Mercer 1975; Goudie et al. 1984; Shroder and Bishop 2010). Reasons will emerge for caution about how accurately or consistently terminus change follows climatic and mass balance changes. As a rule the results of mass balance change will tend to accumulate progressively down glacier and be greatest towards the terminus, immediately suggesting the risk of exaggerated fluctuations (Paterson 1994, p. 318). It has also been shown that morphological, debris-cover, and thermal complexities in Karakoram glaciers are associated with widespread flow instabilities. Observations on termini and terminal moraines have not only been the main evidence there is to work with. Until very recently they dominated discussion of glacier change in the Karakoram, and assumptions based on that work continue to do so. It is important to review what has been found.

Mason (1930) assembled evidence for terminus changes in 34 Karakoram glaciers, going back to the 1880 s for most of the region and to the 1860 s in the Braldu and Shyok Basins. He arrived at two compelling results. For the 1920s he showed 85 % of his set were in retreat. Only five glaciers were advancing: Batura, Yazghil and Khurdopin in his 'Hunza–Nagar' group, Chong Khumdan on the upper Shyok and Kyagar on the Yarkand side (Mason 1930, his Plates 6 and 7). However, all the glaciers had been advancing at some time between the 1880 s and 1920. Most had their greatest advance between 1890 and 1915, strong support for Kick's (1989) view that the LIA culminated later in the Karakoram than the European Alps (cf. Mayewski and Jeschke 1979). Mason's overall results present a fairly chaotic picture in any given decade and an apparent lack of synchrony between the glaciers involved.

To explain these differences, Mason (1930) proposed three classes of behaviour: 'secular variations', responding fairly directly to climate change; 'periodicity', or cycles attributed to unique conditions in the given basin; and 'accidental variations'. The latter identify most of what are now classed as surge types. Mason attributed their advances to external triggers such as earthquakes, heavy avalanching or, in those creating ice dams, 'variable river erosion'. His was a first, impressive attempt to address the singular problems of the Karakoram record. The quotation above needs little qualification in terms of what has occurred since. There are, however, problems in detail and innovations in glaciology that came after Mason's time[1] notably the whole new domain of internal dynamics related to surges. It now emerges that all of his 'secular' types have surging tributaries, likely to complicate their behaviour. To his credit he suggested an overlap may be present such that most if

[1]It now seems odd to find Hispar, Biafo and Pasu in the same 'secular variation' category (p. 221), and the latter two also with 'some periodicity' and in the same group as Aqtash and Minapin (p. 223). Recent events appear to confirm his elaborate figure comparing Chong and Kitchik Khumdan Glaciers with diametrically opposed rhythms of advance and retreat (Mason 1930, his Plate 7). However, the 45-year 'Periodicity Curves' he deduced, partly from Indus floods he attributed to ice dams not actually observed, do not withstand close scrutiny (Hewitt 1969, p 1015).

not all glaciers have elements of all three 'styles' of response. He found evidence to suggest that some in the first two classes also underwent 'accidental' advances, and he believed climate change had some effect on all types. The period since his work could be said mainly, though not entirely, to reinforce his impressions.

Goudie et al. (1984, p 452) found that 'from the 1930s to the 1970s the glaciers nearly all displayed patterns dominated by retreat', but '...during the years 1920 to 1940', more glaciers were advancing or stationary. Their survey of nine Hunza glaciers showed some undergoing a short-term advance and retreat in the 1960s. All except two were retreating in 1980.

From the late 1980s through the early 1990s, very few glaciers had advances anywhere in the Karakoram, excluding rapid advances of nine surge-type glaciers (see Chap. 9). The region seemed, indeed, to have entered a phase of general retreat and thinning, much as in the rest of the Himalaya. It was easy to infer this was part of the worldwide depletion identified with global warming. However, in the years 1995 through 2011, evidence was again very mixed. Some expansion took place in the highest parts of the Karakoram. To date there has been nothing to remotely support the 'disappearing glaciers' scenario.

Recent patterns recall Mason's from 1890 to 1915. Through the late 1990s, I encountered an increased number of high elevation tributaries with much steepened and more active snouts and, in some cases, advancing. This was not readily observed with the tools for satellite analysis at that time. By 2002, I had observed 13 glaciers of intermediate size between the Hunza and Kondus valleys that were advancing. Along eight larger glaciers, including the mid- and upper ablation zone of Biafo and Chogo Lungma, there were sections with marked thickening. In some places ice was overriding lateral moraine that had been ice free for some decades (Hewitt 2005, p 332–333). Further developments of the same kind occurred through the rest of the decade (Fig. 12.1).

As a growing number of glaciers showed undeniable advances of their termini, the matter began to be taken more seriously (Pecci and Smiraglia 2000; Yafeng et al. 2006; Cogley 2012). With more sophisticated tools, satellite imagery revealed ice-margin expansion, thickening or accelerated movement associated with them but also irregular negative as well as positive shifts in glacier thickness (Bishop et al. 2008; Scherler et al. 2011; Kääb et al. 2012). From a range of sources, there is evidence that more than 60 glaciers have undergone advances since the late twentieth century, some continuing through the period, some short-term, followed by no apparent change or again retreating. Comparing remote sensing images of some 40 glaciers from the Baltura/Hispar and Baltoro regions and the mid-1970s to 2003 or 2004, Haritashya et al. (2007) reported that 45 % of them were either advancing or no change could be discerned. Some of the glaciers are surge type and several larger ones have surge-type tributaries, but no rates of movement were observed comparable with major surge events. As at Panmah, there is increasing evidence of surge-type glaciers also undergoing periods of seemingly 'normal' advance before or after a surge event. Shroder and Bishop (pers. comm.) concluded that as much as 60 % of Karakoram glaciers were expanding in 2008. To delve further, a closer look at glaciers with the more extensive records is possible.

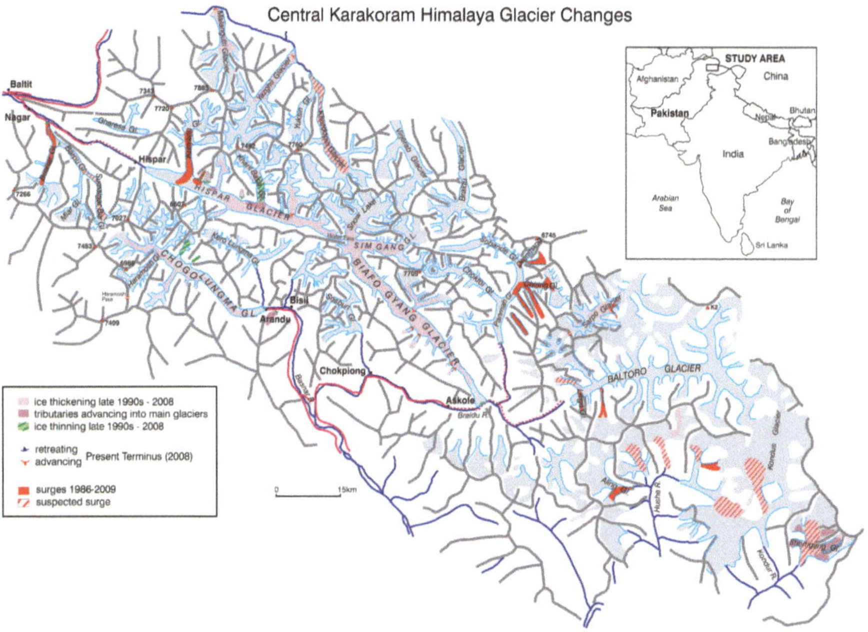


**Fig. 12.1** Glacier change in the Central Karakoram early in the twenty-first century (After Hewitt 2005)

## 12.3 Histories of Some Large Karakoram Glaciers

The largest glaciers show a general tendency to reduce over the whole period of record but, as Mason found, no other common pattern emerges (Fig. 12.2). Net reductions seem much less than reported elsewhere in the same period (UNESCO 1998; Oerlemans 2001; Gallucio and Scotti 2008, p 119). They share something of the diversity of, say, glaciers of the more arid – but not the humid – Tien Shan (Savoskol 1997), of Iceland (Sigurdsson 2008, p 57), and the New Zealand Alps (Fitzharris et al. 1999; Kerr and Owens 2008). It seems that diverse basin conditions and glacier types intervene to reconfigure the relation to climate.

### *12.3.1 Biafo Gyang Glacier*

> …This glacier is peculiar: its basin is complicated by a vast accumulation of névé in the snow-lake at its head…. (Mason 1930, p. 221)

Biafo Glacier lies on a much-travelled route to the Baltoro Glacier, K2 and the other highest Karakoram peaks and, formerly, to the West and East Mustagh Passes

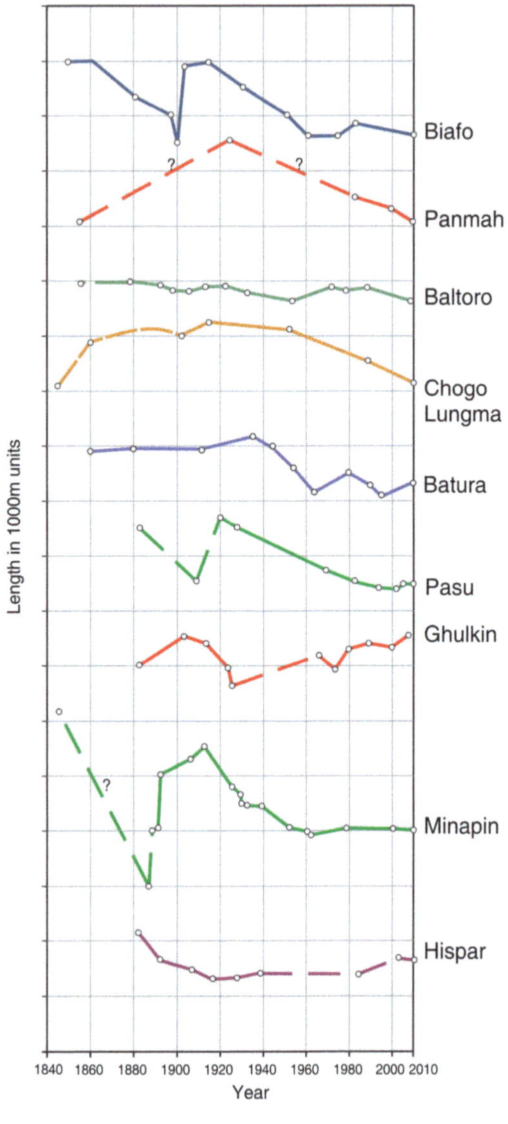


**Fig. 12.2** Fluctuations of selected main glacier termini since the nineteenth century from source outlined in text, illustrating the considerable differences in their modern behaviour (With modifications updating Hewitt 2010)

(Hewitt 2007). A relatively long and detailed modern record can be reconstructed starting in 1856 (Table 12.1). It includes several sketch maps and photographs to help track changes (Fig. 12.3).

Contrary to Mason's suggestion, one might expect an Alpine-type glacier like Biafo to track climate changes fairly well but, instead, as he implies, the record is quite confusing. The first 60 years of the twentieth century seem to trace a typical post-LIA, 'global warming' response, with a steady retreat of up to 1.7 km. However, huge complications appear before this and return in the late twentieth century. An LIA maximum occurs in

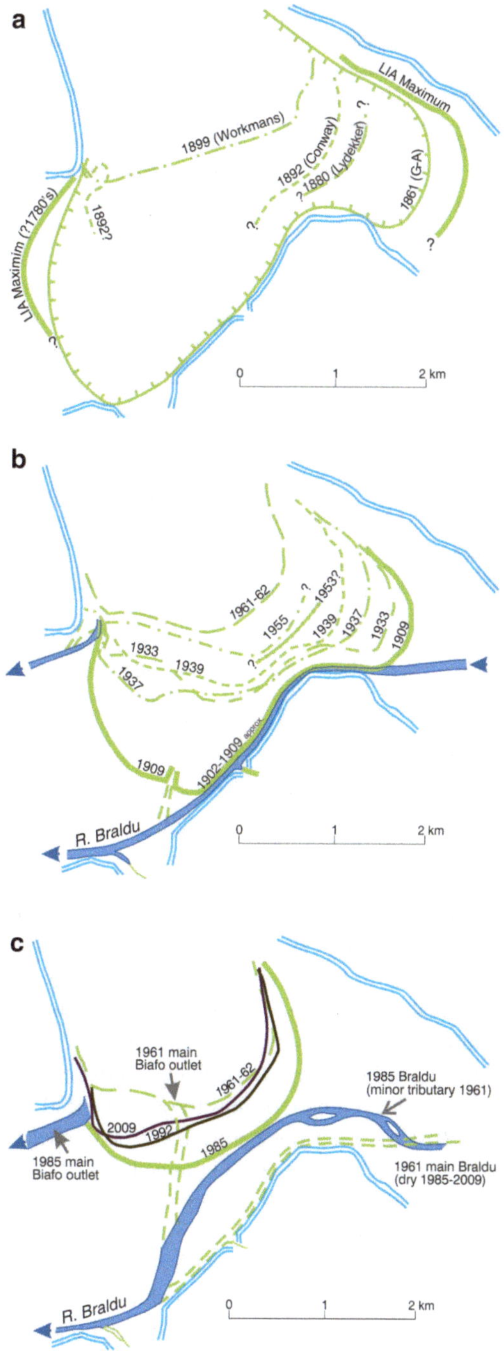


**Fig. 12.3** Fluctuations of Biafo Glacier terminus for the past 160 years (see Table 12.2 and text, Hewitt 1998a, 2005; Hewitt et al. 1989; Shroder and Bishop 2010). (**a**) Nineteenth century, 1861–1899. LIA maximum after Seong et al. (2007). (**b**) Twentieth century, 1909–1962. (**c**) Selected positions, 1962–2009

**Table 12.2** Biafo Glacier: an example of the sources of information for reconstructing historical terminus fluctuations (Auden 1935; Hewitt 1998a, 2005; Hewitt et al. 1989; Seong et al. 2007; Shroder and Bishop 2010)

| Date | Type of information |
|---|---|
| 1857 | Thickness and position in Braldu valley |
| 1861 | Map and description |
| 1880 | Map |
| 1892 | Map, description, watercolour from Skoro La Path |
| 1899 | Map, description, photo Skoro La Path |
| 1902 | Description and position relative to Braldu R. |
| 1905 | Description and position relative to Braldu R. |
| 1909 | Description and photo from Skoro La Path |
| 1913 | Map and description |
| 1922 | Description |
| 1923 | Position relative to Braldu |
| 1929 | Map and relation to Bakhor Das |
| 1933 | Detailed map, description, photos from Laskam |
| 1937 | Map |
| 1939 | Map |
| 1953 | Position relative to Bakhor Das |
| 1954 | Map |
| 1955 | Sketch map |
| 1959 | Position relative to Bakhor Das |
| 1961–1962 | Map, description |
| 1985 | Description, photos from Laskam and Hum Bluk and Skoro La Path |
| 2005 | Description and photos from Laskam |
| 2005 | Photos and satellite image |
| 2009 | Photo from Skoro La Path |

the nineteenth century, but also a large recession. The greatest retreat in the record was in 1899. The rapid changes between 1892, 1899 and 1909 might lead to scepticism but are derived from reports of several expeditions (Dainelli 1959). Meanwhile, most of the remarkable nineteenth-century recession came earlier, between 1861 and 1892 when many other Karakoram glaciers had their LIA maxima. Turning to more recent decades, after negligible change through the 1960s and early 1970s, a strong advance affected the western and central terminus through the mid-1980s. A steady recession has followed. By 2010, the front was barely 100–200 m ahead of the 1960s. The ice front and active lobe in the Braldu valley had thinned tens of metres compared to the 1980s.

Further complications involve differences between the eastern and western Biafo ice front. They may advance or recede independently. The former was more advanced in the early twentieth century, the latter since the late 1970s advance. The main portal of the Biafo also exhibits marked shifts. Before the 1950s, and again since the 1970s, it emerged on the right/west flank of the terminus, and along the north side of Braldu valley. Through the mid-twentieth century, it issued from near the centre of the snout and flowed south into the Braldu. Since the 1960s the Braldu itself has shifted northwards and removed large swaths of old Biafo moraines.

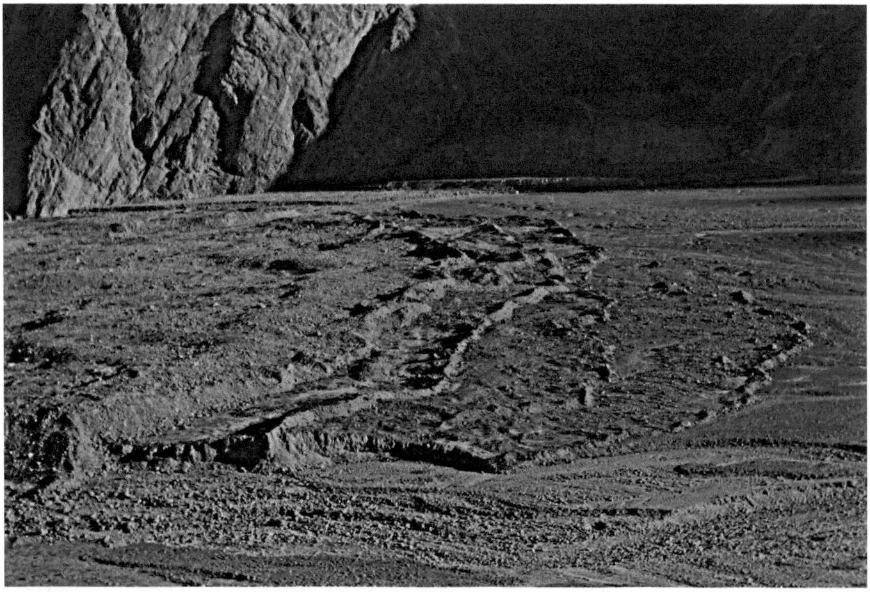

**Plate 12.1** Well-preserved old, mainly LIA terminal moraines of Biafo looking south across the west flank (Hewitt 2005)

The oldest terminal moraines, preserved on the Braldu River terraces, probably represent greatest expansions of the Little Ice Age, which were pre-nineteenth century (Plate 12.1). Godwin-Austen (1864) reported a local story of a Biafo ice dam and outburst flood in the late eighteenth century. Such a GLOF is recorded in major flood deposits across terraces west of Biafo and on down the Braldu valley (Plate 12.2). This suggests the LIA maximum came before the nineteenth century (Hewitt 1964; 1982; 2008). Seong et al. (2007, p 3399) dated boulders on the east flank moraines as between 0.4 and 1.2 ka. Finally, there have been large landslide dams on the Braldu above and below Biafo (see Chap. 8). These have affected ice levels through episodes of aggradation and, as at present, trenching, when the barriers have been breached (Hewitt 1998b, 2011b; Seong et al. 2007).

Finally, near the end of the twentieth and through the twenty-first centuries, thickening of 2–15 m has been observed in ice-margin levels and mid-glacier ice heights along the midsections of the Biafo ablation zone (see Plate 7.3). These seem to represent zones moving down the glacier. Areas that were thickest in the late 1990s had subsided again when revisited in 2010. In the latter year, by contrast, the Baintha profile (4,050 m) was much flatter than in the 1980s, yet the ice at the margins was higher (Plate 12.3). This may relate to the apparent thinning of this section shown by Gardelle et al. (2012). It seems that changing ice dynamics is involved, not only ablation. Ice-thermal changes due to high-altitude warming may be a factor and, if so, further complicate efforts to track mass balance changes or their relation to climate.

**Plate 12.2** Kettle holes in coarse GLOF deposits from the eighteenth-century Biafo ice dam, just beyond the outermost terminal moraines on the southwest side (Hewitt 2005)

**Plate 12.3** Ice-margin thickening at Ho Bluk on the Biafo Glacier *left* flank where ice was crowding and, in places, overriding old lateral moraines. This is the Baintha section and, despite marginal advance and thickening, the whole cross-glacier profile has lost 15 m or more of height and relief compared to 1985 (Hewitt 2010)

### 12.3.2 *Baltoro Glacier*

It might be expected that Mustagh-type glaciers, given the large role of avalanches, would be more erratic, perhaps less sensitive to climate change. This does not seem so of Baltoro Glacier, a near neighbour of Biafo. Its terminus has retreated since the mid-nineteenth century, but to a very small extent compared to Biafo, and as a proportion of its total length. Photograph comparisons indicate minor fluctuations between 1913 and 2010 (Smiraglia et al. 2008). The greatest recession was in the 1950s. However, since then it *advanced* slowly in most years to the late 1990s. A slight recession was observed in the present century but in 2004 it was tens of metres in advance of its 1954 position (ibid, p 158).

Today, ablation zone ice appears generally lower than in the extensive photographic coverage of 1909 (De Filippi 1912; Associazione Macromicro 2010). At the beginning of the twentieth century, the ice generally overtopped the lateral moraines but in 2010 was generally lower by 15–20 m along the lower ablation zone. A famous photograph site between Liligo and Rhobutse (Sella 1987, p 33) shows much greater thinning, perhaps 40–50 m but exceeds other parts in 1909 photographs. Rather than general thickening of the glacier, it could represent something similar to the 1990s Liligo surge and the more recent (2005–2010) Uli Biaho tributary on the north flank. A sudden advance or surge of the nearby Yermanendu tributary might have been involved. Indeed, the small oscillations in the lower Baltoro may well be responses to tributary glacier changes or surging more than secular climate change (Smiraglia et al. 2008, p 159).

Mayer et al. (2006) suggest Baltoro is protected from large or sudden retreat by its extensive debris cover. However, the ablation zone *without* heavy debris cover has a greater area and is likely more critical to mass balance (Chap. 7). Other evidence is beginning to show that thickness changes in Himalayan glaciers, including the Karakoram, differ little between clear and debris-covered ice (Kääb et al. 2012). What is clear is that *in the second largest basin and ice mass in the region, no substantial or sudden changes in frontal position have occurred during the past 150 years, and only a modest decrease in mass since the LIA.*

### 12.3.3 *Hispar Glacier*

Hispar Glacier shares headwaters with Biafo, but terminus behaviour is more like Baltoro's.[2] In 2005 it was thinner but almost 200 m in advance of Calciati' and Koncza's (1910) map of 1908, a negligible change in 97 years. It was about 0.5 km further back than in Conway's (1894) map for 1892. He did write that the terminus had retreated 'three-quarters of a mile… comparatively recently' (1894, p 331). In 1906 Hayden estimated a further retreat of 'a few hundred yards' from the 1892 position, much as Calciati observed. If so, it exceeded anything since that time.

[2]Reports on Hispar terminus; Conway 1894; Hayden 1907; Calciati and Koncza 1910; Mott 1950

Mason (1930, p 227) remarked on the relatively minor and slow changes in the Hispar and believed that '...fluctuations of the lowest tributaries are 'out of time' with each other and have little apparent effect [on the main terminus]...'.

A re-advance occurred through the 1930s as indicated in the Shipton, 1939, survey (Mott 1950) but there was no discernible change from then to when I visited in 1985. Local guides did report considerable erosion of the ice front due to a change in the portal from right to left flank. Satellite imagery for the Hispar's snout in 2005 showed it to be advanced by 150 m or so compared to the 1908 survey (Calciati and Koncza), yet the portal was 1.5 km further back and along the left/south flank. The active front receded slightly from 2005 to 2010 and the portal was a further 1 km up valley. Much of the lower 500 m appeared to have become a stagnant mass of thermokarst seamed with small streams, the greatest loss since the late nineteenth century. However, the active ice was still within 100 m of Calciati's position indicating little meaningful change in over a century and less than 0.2 % for the 53 km main ice stream.

Unresolved questions concern the effects on the terminus of surges of Hispar tributaries, and cycles of aggradation and trenching due to landslide dams on the Hispar River below the snout. Several of Hispar Glacier's tributaries are known to have surged (see Chap. 9). Mason (1930) reported the exceptional advance of the Yengutsa in 1901, which reaches the Hispar valley about 2 km below the Hispar terminus. He was unaware of actual tributaries that surge although there are reports of surges of Garumbar in the late nineteenth century and of Pumari Chissh in 1890–1891. The latter surged again in 1988–1989 (Wake and Searle 1993). In 2007 there was a surge of the huge north flank Khiang Kunyang, whose watershed includes Distaghil Sar (7,885 m), Kunyang Chissh (7856 m) and Trivor (7,720 m). All the surges introduced large bulbs of ice into the main glacier, pushing it aside and causing substantial thickening. The Khiang ice fills the main valley for several kilometres but was still 5–6 km from the terminus in 2012. It should cause a re-advance within a few years. In general, these events mean that neither the observed advances nor wastage of the terminus necessarily reflect or track climate change. Some or all fluctuations may have been triggered by surging or override normal mass balance parameters. This is further complicated by the evident responses of the Hispar ice tongue to sedimentation following landslide damming in the valley below. This may explain the extensive moraines on the north flanks below and well above the height of the present terminus, but seemingly unrelated to it. The separation of the two may have followed from subsequent and ongoing trenching of the Hispar River as it lowers the landslide barriers. It is another mechanism whereby raising and extension, or lowering and retreat of terminal ice may occur that complicate or are unrelated to climate or mass balance responses.

### *12.3.4 Panmah Glacier*

Although sharing watersheds with Biafo and Baltoro, Panmah's fluctuations seem quite unrelated. It has a more limited record with long gaps. Positions can be bracketed fairly well at the time of Adolf Schlagintweit's visit in 1857 and Godwin-Austen's in 1861. The latter's watercolour of the terminus area[3] is detailed enough to compare with later photographs and from his survey station (Hewitt 2007a). The 1861 snout was also hundreds of metres back from 1929 according to observations by an Italian expedition (Savoia-Aosta 1936). However, the meaning of fluctuations is again complicated by the evidence of at least five surging tributaries (Chap. 9). The surges of Chiring in 1995 and Maedan in 2001 provide evidence to reconstruct other surges in the mid-nineteenth century that had been missed (Hewitt 1998a). Some of the highest lateral moraines are definitely markers of surge heights and not trim lines recording mass balance or climate change (Hewitt 2007b). The glacier is also affected by damming of the Dumordo River by the Tsok rock avalanche, about 6 km below the terminus, which led to large-scale aggradation at and below the snout. There has been subsequent, ongoing trenching of the valley fill (Hewitt 1998b, pp 66–71).

### 12.3.5 *Batura Glacier*

The terminus of Batura Glacier changed little in the late nineteenth century or in the twentieth to the mid-1940s, when it stood close to the Hunza River (Goudie et al. 1984, p 449).[4] It then retreated roughly 800 m to 1966. Then there was a small advance of 150 m through the 1970s (BIG 1980). In the early 1980s, it again retreated 100 m, the frontal ice cliff declining in slope and becoming covered with debris and vegetation (Goudie et al. 1984). By 1990 the terminus was 392 m further back than in 1973. Between 1992 and 2007, it moved forward slowly almost 100 m (Shroder and Bishop 2010). Thus, after the 1940s the Batura terminus became almost as erratic as Biafo's. Its terminal lobes also spread into a broad main valley where it has been affected by a series of Holocene landslide blockages up- and downstream (Hewitt 2006b).

### 12.3.6 *Chogo Lungma Glacier*

> I was much struck with ...'Arundu' [= Arandu]... there is so much that is novel and curious...Not the least of these to see agriculture going on close up to a glacier of so large a size.... (Godwin-Austen 1864, p. 47)

[3]In the Royal Geographical Society map library

[4]Sources are Woodthorpe 1885–86 (qu. Mason 1930); Younghusband (1896) in 1889; Knight (1893) in 1891; Price-Wood in 1907 and Etherton 1909 (qu. Mason 1930, 240); Mason (1930) in 1913; Skrine in 1922 (qu. Mason 1930); Visser and Visser-Hooft (1935–1938) in 1925; Survey of Pakistan 1966 and IKP 1980 (qu. Goudie et al. 1984); BIG (1980) in 1978–1979; Shroder and Bishop (2010, pp F220–225).

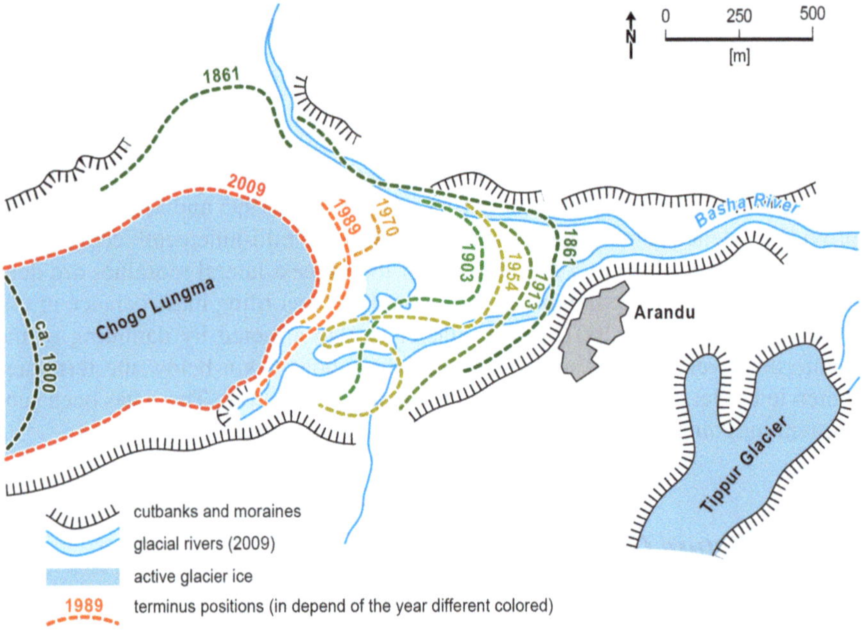


**Fig. 12.4** Variations in the terminus of Chogo Lungma glacier, mainly after Kick (1956, 1962). Changes since 1992 are based on author's field visits and satellite coverage

The timing and direction of modern fluctuations of Chogo Lungma also differ from any described so far (Fig. 12.4). The first Western visitor in 1835 reported it slowly advancing (Vigne 1842, p 2, p 285).[5] However, local people reported its position at the beginning of the nineteenth century had been 2–2.5 km up valley from Arandu village, much further back than at any time since. By 1861 the glacier stood at the margins of Arandu '…its fields touching the ice' (Godwin-Austen 1864, p 47). The same visitor described trees recently uprooted by the advancing glacier and the side valleys along the left/north flank impounded by ice, creating a series of lakes with drowned trees and icebergs in them (see also Mercer 1975, p 382). At that time, or within a few years, the ice reached its highest known level and advance on record.

At a meeting where Workman and Workman (1905) presented their 1902–1903 photographs, Godwin-Austen remarked upon the absence of ice margin lakes and felt that '… the north bank of the glacier all the way up…[had] very much receded… from what it was in 1861' (ibid, p 265). Oestreich (1911–1912, p 29), who was with the Workmans in 1902–1903, thought the glacier was receding. However, in 1913 it was closer to the village again, and advancing to the great alarm of the villagers

[5]According to Godwin-Austen (1864, p 51) '1.5 mile distant' and, according to Kick (1962, p 225), '2 km'.

(Dainelli 1928; De Filippi 1932, p 88). On the basis of a 1902 Workman photograph, Kick (1962, p 225) used stereoscopic reconstruction to show the snout almost as close to the village as in 1861, possibly closer, in spite of having a lower ice surface. There are some uncertainties about the ice and village boundaries in Godwin-Austen's time. Kick proposes the most advanced LIA position may have been in 1913 or thereabouts. Given the far greater ice levels along the glacier, a more advanced position in 1861 or thereabouts seems more likely, but the differences over half a century are minor.

By 1954 Chogo Lungma had retreated 100–150 m from the 1913 position. It suggests a very slow retreat over a relatively small distance and left it more advanced than in 1902 (Kick 1956, p 346). By 1989, however, it had retreated 400–500 m and as much again by 2010. It appears to be the one large glacier now responding to global warming as widely predicted. However, during the late LIA, it hardly behaved as expected and remains more advanced now (2012) than in the early nineteenth century.

### *12.3.7 Minapin Glacier (36° 10′ N; 74° 35′ E)*

The Minapin Glacier falls northwards 4920 m in 16 km from the Rakaposhi Range main crest into the Hunza valley. Its record is widely cited and is among the most detailed but also most puzzling.[6] Local reports given to Hayden (1907) in 1906 described a historical pattern of retreat to the 1880 s but that, 'five generations ago', the terminus was more than a kilometre beyond the 1913 position. This is supported by relatively fresh moraines suggesting a major (early nineteenth century) advance and the furthest known in the LIA. From the 1880 s to 1913, a series of large, rapid advances are reported.

Between 1892 and 1913 the glacier advanced over 1.5 km. In the latter year, according to Mason (1930, p 230) '… this glacier now descends to a lower altitude than any other in the whole Himalaya and Karakoram'.[7] An equally large retreat followed to the 1960s, but only minor changes since then. By 1935 the Vissers (1935–1938) found it had retreated 600–700 m from the 1913 position. A series of other reports tracked further but slower retreat through 1959 (Goudie et al. 1984, p 418). There was a slow 150 m advance in the 1960s and 1970s, and return to slow retreat in the early 1980s. Presently, the terminus seems to be at or close to its 1889 location but more advanced than in much of the late nineteenth century. In 2010 the portal was 100 m forward of its position in 1959–1961 and slightly behind that of 1980 (Goudie et al. 1984, p 420). The right/north part of the tongue was 100 m further back than in 1980. The left part was steeper and more active.

[6]Minapin appears in Oerlemans (2001), and the IPCC 1999 and 2001 reports, each of which shows a large, uninterrupted recession throughout the twentieth century. The data seem mistaken.

[7]The height was roughly 2,150 m. In the early nineteenth century, it may have been below 2,050 m. In either case the elevation span was then over 5000 m.

There are some other unresolved problems. Positions in 1889 and 1892 do not conform to the advances and retreats reported. Distances to the terminus, given from a bridge built after 1889 and present to the 1980s (summarised in Mason 1930 and Goudie et al. 1984), leave 1 km of retreat and advance unaccounted for. Fluctuations in my figure are for retreats and advances reported between visits, not distance-to-bridge estimates. Meanwhile, there have been suggestions Minapin is surge type, and the record could support that (Kotlyakov 1997, Plate 406). It helps explain limited change since the late 1950s during almost 55 years of increased global warming and raises the possibility of a large advance in the near future.

### *12.3.8 Ghulkin Glacier (36° 25′ N; 74° 50′ E)*

This glacier has also received a lot of attention, being a constant threat to the Karakoram Highway and its predecessor and to adjacent village lands. It is 18.5 km long, flows due east and falls 5110 m from Shishpar Pk (7,611 m) to the Hunza River. The terminus is at roughly 2,500 m. Local stories say it blocked the river long ago, but not in reports since 1885. The most advanced stages on record have been *in the* twenty-first *century*. In 1885 it was some 550 m further back than today. In 1913, Mason (1930, p 235) found it '330 yards' from the river, its most advanced recorded position until the present. It had retreated to 1,000 m from the river by 1925, then advanced steadily 500 m to the mid-1960s. After a slight retreat, it again advanced to 1980 when it was 420 m from the river. There were small retreats through the late 1980s to the mid-1990s. Since then it has advanced to reach its most advanced position, if with repeated short retreats and advances every year or two. Related sudden floods and debris flows disrupt the Karakoram Highway and cause damage to nearby villages (Richardson et al. 2009). Ghulkin is the only glacier in the set that is more advanced now than at any time in the last 130 years. In this it is remarkably different from its neighbour Pasu Glacier despite adjacent source zones and similar morphology. The Ghulkin has a much heavier debris cover.

### *12.3.9 Pasu Glacier*

The first observations from 1885 are contradictory; one saying the glacier reached the Hunza River right bank and others that it was either 1.6 km or 'three miles' (4.8 km) away (qu. Mason 1930). Mason, and Goudie et al. (1984), accepted the second figure of Woodthrope. By 1907 the glacier was 2.4 km from the river (Mason 1930), the largest well-established retreat. Between 1907 and 1913, its largest recorded advance occurred and quite rapidly over 1200 m. Mason (1930, p 237) found extensive evidence of recent glacier retreat suggesting the advance had been even greater. From 1913 to 1965 there was a steady retreat and then little change except for wastage of dead ice and an expanding lake at the terminus (Plate 12.4). The absence of significant change in the past 35 years suggests Pasu may also be surge type, although no one seems to

**Plate 12.4** Pasu Glacier terminus area in the early twenty-first century. Great thinning and retreat are evident, but except for decay of stagnant ice and deepening of the terminal lake, very little has changed in 35 years. The view is from the glacier's Great Lateral Moraine as discussed in the text (Hewitt 2004)

have classed it as such. In 2005 active ice was about 2.3 km from the Hunza right bank, not as far a retreat as in 1907. The negligible areas of heavy debris on the surface may be less to do with debris supply than the comparatively steep, crevassed ablation zone. The massive lateral moraines suggest this was not always the case.

## 12.4 Erratic Advances

In all, the picture from the large glaciers reviewed is quite erratic and broadly vindicates Mason's (1930, pp 218–219) view quoted above. On the other hand, his idea of three different regimes and the glaciers he places in them are not supported. A general net reduction applies in most cases since the early twentieth century but not a dramatic one. The view of Goudie et al. (1984) that 'the dominant regime since 1850 has been retreat' was not supported, and it is unclear why they treat advances between 1870 and 1970 as 'secondary trends'. There is no conformity among them. It seems likely they subscribed to the widely held view that the so-called Great Lateral Moraine, a feature of most large Hunza glaciers, is an LIA feature which more recent work calls into question (see below).

The late nineteenth and early twentieth centuries seem chaotic, as do developments since the mid-1990s. To add to the confusion, most of the same glaciers came into near conformity between the 1920s and 1980s, with modest, net retreat and, at most, some minor, short-term advances. Pronounced differences between adjacent glaciers confirm Mason's view, as illustrated by comparing Biafo, Panmah and Baltoro, or Ghulkin, Pasu and Batura. In at least half the cases, the last major advance of the LIA appears in the early twentieth century. This would support Kick's (1989) view of a late LIA culmination. However, indirect evidence and local reports suggest at least as many glaciers had greater advances before the 1850 s indicating a more complex LIA.[8] At some point before the 1930s, the record does place all glaciers well in advance of their present-day positions but, most surprising, no example is presently at its greatest retreat of the past 150 years, if some are close.

None of the types of Karakoram glaciers explored in this book seem to relate closely to, or predict, the fluctuations observed. Alpine types are no more consistent than Turkestan types. No patterns relate to such features as orientation, length and relative steepness, or location within the range. A common assertion is that the extent of debris-covered ice makes the Karakoram distinctive (Kick 1989; Shroder et al. 2000; Scherler et al. 2010). However, while all the glaciers just traced have extensive debris mantles, except Pasu, the story of their fluctuations is quite lacking in common patterns or anything clearly explained by the extent of the debris. No significant debris mantle changes have been observed in the glaciers examined, although large landslides, such as at Bualtar, and surges have caused big changes in some others (Hewitt 2009). In general, heavy debris covers seem to be a highly conservative factor, and one in which the Karakoram differs little from the rest of the Hindu Kush–Himalaya.

Mason (1930) laid great stress on the distinction between 'longitudinal' and 'transverse' glaciers, and Goudie et al. (1984) agree, finding in the Hunza valley, the two types 'demonstrate different fluctuation histories' (p 432). However, there seems no reason why Pasu would be called 'mainly longitudinal' but Ghulkin, its neighbour 'transverse'. Their ice streams are exactly parallel, their basins of nearly identical geometry, and yet they behave very differently. Why is Khurdopin 'transverse' but Virjerab 'longitudinal'? And neither term seems to make sense for the compact basins with two or more major ice streams such as Panmah, Toltar–Baltar, Kukuar or the Rimo system. In any case, broader coverage reveals as much variety among as between 'longitudinal' and 'transverse' types. Examples of each in the ten cases described above are as erratic as the other. Meanwhile, the assignment of classes has focused on lower and main ice streams. It ignores the diversity and ice stream geometries of whole basins. For example, the main Hispar ice stream may be 'longitudinal', but more than half of its ice comes from 'transverse' tributaries. Similarly, the large source basins of Biafo or Siachen are hardly like their lower 'longitudinal' sections.

With 44 confirmed surge types, and some estimates suggesting they comprise as many as one third, they may point to conditions that affect glacier change in general, specifically through unstable dynamics (Chap. 9). The scale and timing of tributary surges seem likely to disturb main glacier mass balance and upset relations to climate change. The record is too short and incomplete to reveal all surge-type tributaries and too sketchy to confirm how many main glaciers are surge type. To reiterate, however, rhythms of advance and retreat in surge-type glaciers have little or no apparent relation

[8] They include Biafo, Baltoro, Chogo Lungma, Pasu, Minapin and Hispar Glaciers.

to climate change; 'mini' instabilities may be even less systematic. There may well be – must be – long-term statistical trends, as climatic influences on thermal regimes and the debris stream as well as mass balance work their way through the set of regional glaciers. If so, they remain to be identified.

## 12.5 The Great Lateral Moraine (GLM)

Present-day Karakoram glaciers have played an important part in efforts to reconstruct regional Quaternary changes. Likewise, deduced sequences of glaciation have influenced work on and ideas about today's ice cover. While Quaternary reconstruction lies outside the scope of this book, there is a need to have a comparative sense of recent and anticipated glacier changes in relation to the longer term. A brief survey of discussions of widely observed features helps identify some salient issues.

Along the lower ablation zones of many Karakoram glaciers are bare cliffs in old ice margin deposits. They rise from the ice edge and culminate above in prominent lateral moraines (Plate 12.5). Essentially, these are the higher, or most well developed of the 'Transitional Landsystems II and III', described earlier: lateral margin and terminal sediment–landform assemblages (Chap. 8). Formerly they were regarded as a single regional phenomenon and called the Great Lateral Moraine (GLM). Meiners (1998, p 55) describes them as '…a very well-marked and well-formed lateral, partly

**Plate 12.5** The Great Lateral Moraine (GLM) at Kukuar Glacier, Chalt valley, Hunza, with 'trim-line' 100–150 m above the ice, on *left* and *right* flanks of the photograph. Kukuar is a surge-type glacier. A much older equally large lateral moraine is seen 400 m above the glacier on the opposite flank (Hewitt 2004)

high moraine, which surrounds the glacier tongues'. A more explicit German term is *Ufermoranen-Dammen; 'embankment moraine dam* [or *barrier]'* (Wiche 1961; Haserodt 1989, p 212). These impressive deposits, around today's termini or within a kilometre or two of them, have been critical in shaping ideas of glacier change (Owen 1988; Shroder 1993; Kamp and Haserodt 2004; Iturrizaga 2006).

The GLM used to be identified strictly with the Little Ice Age (LIA). Some equated it with the '1850 moraines' in the European Alps. According to von Wissmann (1959), '...in High Asia [generally] the moraines... originate in the high stand glaciation at the middle of the past [nineteenth] century...'. Kick (1989) challenged this for the Karakoram and Chogo Lungma in particular. His arguments were similar to those of Mason (1930) who thought that '... the majority of glaciers in the region were in a condition of maximum advance between about 1905 and 1915...'. Meanwhile, on the basis of tree ring data, Haserodt (1984, p 83) found: '...a minimum age of the GLM of Bagrot [Glacier, near Gilgit] of 280 years ... closer to the beginning of the "Little Ice Age" period...'.

Since the mid-1980s, most studies have invoked several glacial expansions and a more complicated history going back much earlier than the LIA, even to the late Pleistocene. What all existing interpretations do agree upon is that the GLM records climate-driven glacier fluctuations. Evidence assembled here calls that view into question. It supports a diversity of origins, in time and process. This, in turn, makes for an awareness of broader, more complex conditions governing glacier change in the recent past.

### *12.5.1 The GLM at Bualtar and Barpu Glaciers*

Moraines from ice levels 200 m or higher surround the lower Bualtar Glacier and cap some tens of cubic kilometres of GLM or related deposits (Plate 12.6). An impressive GLM also surrounds the adjacent, lower Barpu Glacier, which formerly flowed into the Bualtar (Plate 12.7). The GLM of both glaciers is discussed in several influential investigations. Haserodt (1984, p 93) investigated the features at 'Hoppar' and sketched a cross-profile. He placed the highest moraines and ice depositing them in the 'seventeenth, eighteenth or nineteenth C'. In a later paper he describes the features as 'High Stand moraines' of the LIA, defined by '300, 75–100, and 10–25 year vegetation growth' (Haserodt 1989, pp 214–217). These are estimates not age determinations from tree rings, as at Bagrot.

Kalvoda (1992, Plate XXVI/2, p 189) does not use the term GLM but describes the same features. He calls them '...Glacigenous sediments of the lower part of the Bualtar valley-glacier tongue...Huge, in some places up to 160 m high walls of lateral moraines dating from the distinct advance of this glacier...'. However, he adopts a much-expanded time frame, proposing that 'The highest [moraines]... formed during the Hunza phase of glaciation in the upper Pleistocene'.

Kalvoda and Goudie (2002, pp 112–115) also ascribe the 'huge walls of lateral moraines [close to Nagar village]...' to '...the last advance of the valley glacier in

**Plate 12.6** Lower Bualtar Glacier, looking down valley from Shishkin showing GLM remains on both flanks and massive erosion of the slopes descending to the debris-covered ice. Terraced fields of Hopar are above the GLM on the opposite side (Hewitt 1986)

**Plate 12.7** A view down on the Barpu Glacier showing about 15 km of the *right* flank GLM. Gannish Chhissh or Spantik Peak (7,027 m) is in the *right*, *top* background (Hewitt 1987)

**Plate 12.8** Bualtar Glacier margin during passage of thickened ice, protected by rock avalanche debris and with fresh lateral moraine fed by it. Note how the ice rides up over the growing moraine. The GLM and masses of its collapsed materials are above and to the *right* (Hewitt 2010)

the upper Pleistocene' (113). They place the main, 'high stand' Bualtar GLM in the 'Ghulkin I stade'. It has been described as an 'expanded foot', or 'minor valley glaciation', and assigned an age of a 25.7–21.8 ka (Shroder et al. 1993, p 154; Kalvoda 1992; Owen 2006, p 15).

Again, these interpretations assume that the GLM is a record of climate-related glaciations. However, some reservations are immediately in order. Both Bualtar and Barpu were shown to be surge-type glaciers (Chap. 9). What does that imply for their GLMs? Glacial fluctuations and ice-margin deposition in surge-type glaciers relate to cycles unique to each ice mass (Sharp, 1988; Jiskoot 2011). In the late LIA, when most nearby glaciers were advancing, Bualtar retreated. Recently, when most others were retreating, it advanced (Gardner and Hewitt 1989).

There is no reason to assume either glacier only became surge type in recent centuries. As long as they have been so, massive moraine building would be likely to occur only or mainly in surges (Plates 9.2 and 9.5). In fact, they seem to explain some distinctive GLM features described below.

There is also recent evidence of landslides affecting glacier activity. In 1986, massive rock slope failures deposited about 20 million m$^3$ of debris onto Bualtar Glacier (Hewitt 1988). By 2012 the debris sheet, some 4.5 km$^2$ in area, had been transported 10 km down glacier. It suppressed ablation and created an area standing 10–20 m above the surrounding ice. The abundant rock avalanche material was shed at the margins to create prominent ridges of lateral moraine along the ice edge (Plate 12.8). The deposits are glacial moraines in geometry but their composition and timing depend upon rock avalanche materials (Hewitt 2009a). The relevance to the GLM depends, of course, on whether they recur.

**Plate 12.9** View along and down the Bualtar *left* flank GLM, showing massive rotational slumps in GLM materials collapsing towards the glacier. The photo site of Plate 12.8 is below and in the *middle* ground, beside the ice. Note the former road carried down on the nearest slump block (Hewitt 1987)

There are local accounts of a similar landslide onto the glacier in the late nineteenth century, supported by field inspection of the source area. Horizons in the GLM cliff at Hopar have sedimentary characteristics of rock avalanche matrix materials, apparently one or more prehistoric events. There are remnants of three prehistoric rock avalanches that descended onto Barpu Glacier (ibid). The largest, mentioned earlier as a rock slope failure influencing interfluve development, was at Spantik Peak (7,027 m) and travelled 11 km down glacier (Chap. 8). The distinctive crystalline limestone from the peak caps a series of steep, high lateral moraines down the glacier's opposite, right flank for some 16 km. Much of this landslide debris was emplaced by ice riding up over pre-existing moraines, probably in a subsequent surge event (Hewitt 2002, p 368).

On millennial time scales, landslide forcing could well be a source of large moraine-building episodes (Hewitt et al. 2011a). However, in the case of the Bualtar GLM, the landslides and surges, or the glaciations formerly invoked, do not offer a convincing explanation, together or alone. This relates to two other phenomena.

Contrary to an impression conveyed in the literature, the great cliffs in former glacial deposits at Bualtar do not mark an ice contact surface. They are not really 'trimlines' that represent the highest, last *or any* past glacier expansion. They are *erosion* features from an extended period of degrading of earlier GLM deposits (Plate 12.9). Moreover, massive gravitational collapses of GLM materials mean that Bualtar ice and moraines must have been tens of metres higher than today and much closer to the valley centre.

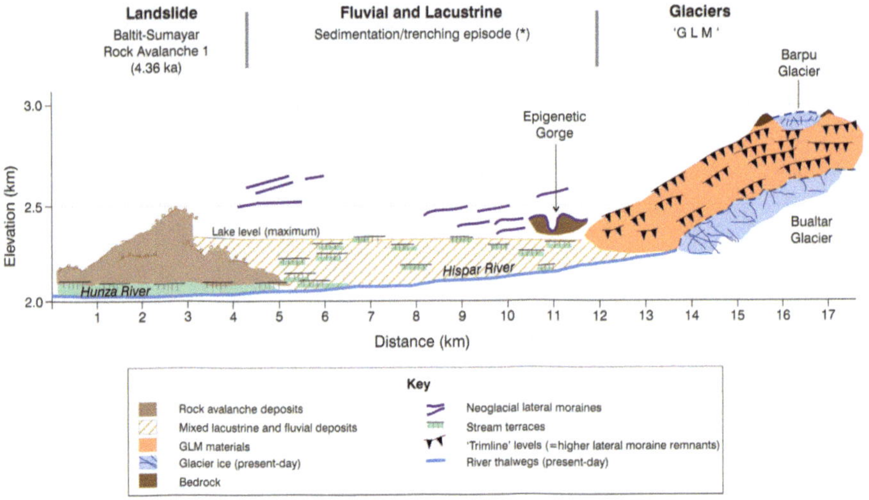


**Fig. 12.5** Schematic longitudinal cross section from Baltit–Sumayar landslide dam to the Bualtar GLM

A second notable feature, given the sheer volume of GLM sediments, is how closely their location and morphology conform to the present ice tongue. They represent ice surface levels up to 250 m higher, but the great changes *vertically* are not matched by comparable advances or lateral expansion of the ice. Consider how, in recent years, some 10–15 m thickening of the lower Bualtar prefigured an advance of 3.5 km. Lateral moraines spread outward some tens of metres and 5–15 m vertically, even though confined within the old GLM (Hewitt 2009). The latter is ten times higher, but no comparably large advances or lateral expansion are evident. It seems to indicate important constraints in GLM development. In this context another recent discovery assumes prime importance.

## 12.5.2 Post-glacial Landslides and Fragmented Drainage

At its junction with the Hunza, the Hispar River was blocked by a large, late-Holocene landslide. A massive rock slope failure from the Ultar Glacier Basin above Baltit descended into and dammed both the Hunza and Hispar Rivers (Hewitt 2001). Thick lacustrine deposits record the lakes impounded. The Baltit–Sumayar barrier is 8–10 km below the Bualtar junction and some 400–450 m high. The dam crest was close to that of the highest GLM remnants beside the present Bualtar terminus (Fig. 12.5). The dam impounded sediment from the whole Hispar Basin, enough to produce heavy and rapid sedimentation to depths of some hundreds of metres in the lake. Initially the Bualtar tongue in its present position would have been floated by the lake but, in time, would have been elevated by sediment build-up in the valley. Subsequently the landslide dam was breached and has been degrading ever since. It is not fully cut through and remains the local base level for the Hispar and Bualtar valleys.

The 'high stand' GLM moraines are at elevations and related in morphology to the landslide dam. It offers an alternative to explain the timing and genesis of the main sediment build-up through control over local base level. Indeed, it is unnecessary to require a massive thickening of the ice. Rather, a build-up of sediment beneath as well as beside the glacier is indicated, much as seen at Ghulkin Glacier today (Owen 1994), in the 'ice-marginal ramps' of Kuhle (1990) and the well-known Hatunraju Glacier in Peru's Cordillera Blanca (Clapperton 1993). Such developments have been identified with debris-covered glaciers and how sedimentation at their margins can create 'giant bounding moraines' or a 'moraine-dammed' condition (Benn et al. 2003). Specifically this occurs with 'decoupled margins' where outwash discharge is not sufficient to carry sediment away (ibid 403). It should be emphasised that this is fairly rare even in the most heavily debris-covered Karakoram glaciers today. The Bualtar Glacier itself, despite being far below its GLM levels, is not 'decoupled' or 'moraine-dammed' but a 'coupled, outwash head type' (ibid). The Baltit–Sumayar landslide and blockage offers an alternative mechanism to create the massive build-up of sediment. In addition to blocking sediment movement where there are very high rates of delivery, over time it would reduce the slope of the glacier. No great change in ice thickness or frontal position would then be required, if any.

If the hypothesis is valid, it is likely that the higher GLM deposits were emplaced while the impoundment was intact and within a few centuries of the landslide occurrence. It would give ample time for several glacier surges and even for one or more landslides to play a part in emplacing the highest parts. The idea of a moraine-dammed condition and surge-driven episodes seems to explain some striking examples of 'breach lobes', sections of ice which overwhelm or burst through the bounding moraines to emplace secondary lobes (ibid 307). Remains of breach lobe moraines, where ice tongues had cut through the highest surviving GLM, occur at five places on the Bualtar left bank. The GLM remnants as a whole suggest what is said to be typical of such cases, a complex history of 'repeated superposition of moraines around the margins' (ibid 381).

Thereafter, the main trend has been trenching and degrading of the valley fill, a response tied to downcutting through the landslide dam. Progressive lowering of the glacier bed and surface does not imply mass balance change, but follows from removal of ice-margin and subglacial deposits as local base level was lowered. It is not to say that there have been no changes in glacier thickness and length, but they are not required to explain these developments and may have been masked by them. It seems, therefore, that the GLM is not purely glacial in origin. Its timing and scope are bound up with episodes of valley fill sedimentation and trenching affecting the Hispar and Hunza valleys. Again, it must be asked whether this is a unique case or has more general implications.

### 12.5.3 Broader Implications

A sample of 40 Karakoram glaciers with well-developed GLM was looked at to see how far conditions or Holocene developments similar to those described at Bualtar are present. In most cases, the timing, heights and volumes of sediment in the GLM are also found to depend, in part at least, on conditions other than climate change and (neo)glaciation (Table 12.3). A third involves surge activity, and massive rock slope failures are known in a quarter. Most importantly, in all cases, the valleys downstream have been blocked by one or more cross-valley landslide barriers (Hewitt, 1998; Hewitt 2006a).

The termini of the Siachen, Kondus, Baltoro, Chogo Lungma, Biafo, Batura, Hispar and Bagrot Glaciers, for example, have well-developed GLM and several other features in common. They have heavy supraglacial moraine over the lower 15 km or more of the glacier tongue. Each terminus rests on and has advanced and retreated over valley fill tens of metres thick, possibly hundreds. Ice levels, terminus behaviour and pro-glacial sediment removal have been partly constrained by valley fill levels, including episodes of downstream aggradation or trenching. Downstream of the termini are stream terraces with upper levels generally lower than the high stand GLM, but continuous with associated glacier margin troughs. Recently it has been shown that, in every case, there are one or more cross-valley landslide barriers a few kilometres downstream of the termini (Hewitt 2001, 2006a, b). These have controlled and still control local base levels, stream thalwegs, sediment movement and storage.

In fact, among several hundred intermediate and large Karakoram glaciers, it is difficult to identify even one whose lower tongue does not terminate where landslide-controlled aggradation and trenching are apparent. In this way, developments in pro-glacial valleys have modified glacial sedimentation, with or without climate-driven glacier change, with or without changes in the ratios of ice and debris delivered to glacier margins. It is an aspect of the Bualtar situation that seems prevalent around the termini of the larger Karakoram valley glaciers, linking them to developments along the upper Indus streams (Hewitt et al., 2011a).

Until recently, the Baltit–Sumayar deposit was interpreted as glacial moraine. It was assigned to the last Pleistocene trunk glacier in the Hunza valley (Derbyshire et al. 1984, p 488). The lacustrine deposits upstream were assigned to glacial lakes, river terraces to trenching of glacial deposits. Age estimates for the last trunk glaciation, the so-called Ghulkin I stade, would date the deposits back to 45,000–50,000 ka. This seems to be the origin of Kalvoda's (1992) views. However, the landslide and the GLM-related episode of disturbance due to it have been bracketed by a well-constrained $^{10}$Be terrestrial cosmogenic nuclide age. Quartz-rich rock was retrieved from boulders exposed on the surface of the landslide where it entered the Hispar valley (Hewitt et al., 2011b). The age determined is 4.36 ka +/− 0.14 ka.

Reasons for rejecting glacial deposition of Baltit–Sumayar, apart from the new age determination, depend mainly on lithology. The deposits consist entirely of granodiorite from the Ultar massif, emplaced over metamorphics that compose the

**Table 12.3** A sample of 40 Karakoram glaciers with well-developed GLM, highlighting conditions or Holocene developments similar to those described at Bualtar

| Glacier | Basin | Surges | RAs on ice | RA dam | Glacier block | GLOFs? |
|---|---|---|---|---|---|---|
| N. Terong | Nubra | ? | X | X | X | X |
| Kondus | Saltoro | ? | X | X | | |
| Charakusa | Hushe | X (T) | | X | X | X |
| Aling | Hushe | X (T) | X | X | X | X |
| Baltoro | Braldu | X (T) | X | X | | |
| Panmah | Braldu | X (T) | | X | X | |
| Biafo | Braldu | X (T) | | X | | X |
| Kutiah | Stak | X | | X | X | X |
| Mani | Phu'gam | ? | X | X | X | |
| Hinarche | Bagrot | ? | | X | | X |
| Virjerab | Shimshal | ? | | | X | X |
| Khurdopin | Shimshal | X | | X | X | X |
| Yazghil | Shimshal | ? | ? | X | X | X |
| Malangutti | Shimshal | | X | X | X | X |
| Koz Yaz | Chapursan | ? | | X | X | |
| Yashkuk Y. | Chapursan | ? | | X | X | X |
| Kuk-i-J. | Chapursan | ? | | X | X | X |
| Murkhun | Hunza | ? | X | X | | X |
| Batura | Hunza | ? (T) | | X | X | X |
| Pasu | Hunza | | X | X | X | |
| Ghulkin | Hunza | | X | X | X | X |
| Hispar | Hunza | X (T) | | X | X | X |
| Garumbar | Hunza | X | | X | X | X |
| Barpu | Hunza | X | X | X | X | X |
| Bualtar | Hunza | X | X | X | X | X |
| Silkiang | Hunza | ? | | X | | |
| Karambar | Hunza | X | | X | X | X |
| Minapin | Hunza | | | X | | |
| Kukuar | Hunza | X | | X | X | |
| Jaglot | Hunza | X | | X | X | |
| Shani | Naltar | | | X | | |
| Kutu N. | Naltar | | | X | | |
| Bhurt | Karambar | | | X | | |
| Karambar | Karambar | X | | X | | X |
| Pehkin | Karambar | | | X | X | |
| Chillinji | Karambar | X | X | X | X | X |
| Chatteboi | Karambar | ? | X | X | | X |
| Karambar P. | Karambar | | X | X | X | X |
| Chiantar | Yarkhun | | | X | | |
| Pechus | Yarkhun | ? | | X | X | X |

Notes: 'RA dam' and 'glacier block' refer to events that interrupted streams and affected local base level below the glacier. 'X' identifies an influence on the GLM. 'T' refers to surges of tributary of glacier named; '?' means only indirect evidence, such as surge-indentifying patterns on the glaciers (Chap. 9). A blank means no evidence for an influence, not that it is definitely absent

*RA* rock avalanche. *GLOF* glacier lake outburst flood

Hunza–Hispar valleys here and continue far upstream of the landslide (Hewitt 2001). A trunk glacier descending the Hunza or Hispar valleys would surely have deposited some if not mainly metamorphic debris here, as in older moraines that do exist above the landslide deposit.

Meanwhile, the glacial view of Bualtar GLM implies 45 ka+of no *net* erosion, which is hard to accept. The present channels of the Hunza, Hispar and Bualtar valleys do not cut deeper than the level of the Baltit–Sumayar deposit, which itself overlies river gravels and fill of unknown but considerable depth. If it is 45 ka old or more, it marks the beginning of an extraordinary 'freeze' in landscape change. If the GLM deposits have 20 ka ages or more, they also involve a surprising slowing of landscape change. Recall that this valley has been termed 'the steepest place on Earth' (Miller 1984). As described earlier (Chap. 1), it has some of the highest known rates of uplift and denudation (Burbank et al. 1996; Zeitler 1985).

The TCN age for the Baltit–Sumayar landslide implies around 4 ka of zero net erosion. Nevertheless, it dramatically reduces the time frame in which massive aggradation above present stream levels occurred, and trenching of several hundred metres that has removed 1/3–1/2 of the GLM and related sediments. The glaciation hypothesis means zero net incision and no incision in bedrock where tectonic reconstructions suggest several hundred metres of uplift, which goes against a well-established view that tectonic uplift and erosion are quite closely matched (Searle 1991; Shroder et al. 1989).

Studies of the glacier margins have focused so much on the complicated interactions of ice dynamics, debris and outwash that other, especially non-glacial, influences have been neglected. This is probably a mistake in the Karakoram, perhaps all of the Himalaya (Fort 1995). There is a variety of transglacial processes crossing and affecting ice-margin developments (Iturrizaga 2006). Short- or long-term shifts in local base levels and large-scale geomorphic instabilities and adjustments affect glacier margin deposition and must be accommodated by it. They can generate changes in ice levels with or without mass balance change. They make the glacier subject to the same local constraints widely identified in fluvial and lacustrine systems. The fact that so many large landslide deposits were misinterpreted as moraines adds to the challenges for existing Quaternary sequences, many of which are based on them (Hewitt 1999, 2001).

Finally, the great influence of debris covers seen today may not be a good indication of what happened in the main Pleistocene glaciations, when the glaciers were much larger. At times of maximum glaciation, the extent of rock walls in Karakoram glacier basins was much less. It would reduce the roles of avalanches, rock slope failures and, hence, debris covers. Debris supply at the ice margins was likely less and finer grained sub- and englacial material more important. Surges may have been less or absent. If they occurred it was likely on a much larger scale, but encountering much less superficial debris. In all, conditions observed today, both on and off the glaciers, must be applied very cautiously to the major glaciations of the past. Valley floor conditions should have changed substantially over time as glaciers retreated and other processes took over, possibly with an unfolding paraglacial influence (Hewitt 2009b).

## 12.6 Concluding Remarks

Glacier inputs and ablation ultimately reflect climate. With due account given to orographic and topoclimatic conditions, climate is the overarching control on the presence and scale of a mountain glacier cover. There must be snowfall and sufficient periods of sub-zero temperatures and limits on solar radiation to sustain ice. The same applies in the Karakoram and, in broad terms, to glacier changes there. Observations prior to the 1920s at latest confirm a greater ice cover during parts of the colder LIA. Earlier human history suggests that the glacier cover in centuries before the LIA was less than today (Kreutzmann 2006). Reductions affected most glaciers through the twentieth century and continue in many, but not all, cases to the present time.

There is evidence of considerable ongoing climate changes in the high Karakoram, but far from a simple matter of temperature 'cause' and melting 'effect'. The descriptive results above challenge a simplistic view in which 'global warming' has the same result everywhere. Chronic irregularities of terminus behaviour seem related to complex adjustments over the extreme elevations and rugged terrain, as well as to a distinctive climate regime. These and other specifics of each basin were shown to complicate mass balance, glacier dynamics and thermal regimes, and the debris stream. Looking for a few standard indicators, especially from reported terminus behaviour, may be a mistake. However, in part this is because the time frames over which climate and glaciers (co)operate, especially with so many and such large glaciers, are not easy to grasp in human terms or in relation to urgent requirements of contemporary societies. Nevertheless, this must be the main concern of a final chapter.

## References


Armstrong RL (2010) The glaciers of the Hindu Kush–Himalayan region: a summary of the science regarding glacier melt/retreat in the Himalayan, Hindu Kush, Karakoram, Pamir, and Tien Shan mountain ranges. ICIMOD, Kathmandu

Associazione Macromicro (2010) Sulle tracce dei ghiacciai 1909–2009 Un secolo di cambiamenti climatici sui ghiacciai del Karakorum. Catalogo, Rome

Auden JB (1935) The snout of the Biafo glacier in Baltistan. Rec Geol Surv India 68:400–413

Bajracharya SR, Mool GP (2009) Glaciers, glacial lakes and glacial lake outburst floods in the Mount Everest region, Nepal. Ann Glaciol 50(53):81–85

Benn DI, Kirkbride MP, Owen LA, Brazier V (2003) Glaciated valley landsystems. In: Evans DJA (ed) Glacial landsystems. Hodder Arnold, London, pp 370–406

Bhambri R, Bolch T, Kawishwar P, Dobhal DP, Srivastava D, Pratap B (2012) Heterogeneity in Glacier response from 1973 to 2011 in the Shyok valley, Karakoram, India The Cryosphere, Discussion Paper, 6, doi: 10.5194/tcd-6-3049-3078, 2012

BIG (Batura Glacier Investigations Group) (1980) The Batura glacier in the Karakoram mountains and its variations. Sci Sin 22(8):958–974

Bishop MP et al (2008) Advancing glaciers and positive mass anomaly in the Karakoram, Himalaya, Pakistan. Paper presented at the American geophysical union fall meeting, San Francisco, 15–19 December

Bolch T, Kulkarni AV, Kääb A, Huggel C, Paul F, Cogley JG, Frey H, Kargel JS, Fujita K, Scheel M, Bajracharya SR, Stoffel M (2012) The state and fate of Himalayan glaciers. Science 336:310–314

Bonardi L (2008) Terra glacialis. Special issue, mountain glaciers and climate changes in the last century. Servizio Glaciologico Lombardo, Milan

Burbank DW, Leland J, Fielding E, Anderson RS, Brozovic N, Reid MR, Duncan C (1996) Bedrock incision, rock uplift and threshold hill slopes in the northwestern Himalayas. Nature 379:505–510

Calciati C, Koncza M (1910) The basin of the Hispar glacier. In: Workman FB, Workman WH (eds) The Call of the snowy Hispar. Constable, London

Clapperton CM (1993) Quaternary geology and geomorphology of South America. Elsevier, Amsterdam

Cogley G (2012) No ice lost in the Karakoram. Nature Geoscience. online March ngeo1456. cogley

Conway WM (1894) Climbing and exploration in the Karakoram-Himalaya, 3 vols. Appleton, New York

Dainelli G (1928) Relazioni scientifiche della Spedizione Italiana De Filippi nell'Himalaia, Caracorum e Turchestan Chinese (1913–1914). Series II, 10 vols (1924–35). Zanichelli, Bologna

Dainelli G (1959) Esploratori e alpinisti nel Caracorum. Unione Tipografico-editrice Torinese, Turin

De Filippi F (1912) La spedizione di S.A.R. il principe Luigi Amedeo di Savoia Duca degli Abruzzi nel Karakorum e nell'Himalaya occidentale (1909). Zanichelli, Bologna

De Filippi F (1932) The Italian expedition to the Himalaya, Karakoram and eastern Turkestan, 1913–14. Edward Arnold, London

Derbyshire E, Jijun L, Perrot FA, Xu S, Waters RS (1984) Quaternary glacial history of the Hunza Valley, Karakoram Mountains, Pakistan. In: Miller KJ (ed) International Karakoram project, vol 2. Cambridge University Press, Cambridge, UK, pp 456–95

di Savoia-Aosta D (1936) La Spedizione Geografica Italiana al Karakoram 1929. S.A. Arti Grafiche Bertarelli, Milano

Fitzharris BB, Lawson WJ, Owens I (1999) Research on glaciers and snow in New Zealand. Prog Phys Geog 23(4):469–500

Fort M (1995) The Himalayan glaciation: myth and reality. J Nepal Geol Soc 11(2 Special Issue):257–272

Fowler HJ, Archer DR (2006) Conflicting signals of climate change in the Upper Indus basin. J Climate 19(17):4276–4429

Gallucio A, Scotti R (2008) Lombard glaciers face the climate change (XX-XXI centuries). In: Bonardi L (ed) Terra glacialis. Special issue, mountain glaciers and climate changes in the last century. Servizio Glaciologico Lombardo, Milan, pp 105–123

Gardelle J, Berthier E, Arnaud Y (2012) Slight mass gain of Karakoram glaciers in the early twenty-first century. Nat Geosci 5:322–325

Gardner JS, Hewitt K (1989) Surge of the Bualtar glacier, Karakoram ranges, Pakistan: a possible landslide trigger. J Glaciol 32:129–140

Godwin-Austen HH (1864) On the glaciers of the Mustagh range (Trans-Indus). Proc Royal Geog Soc 34:19–56

Goudie AS, Jones DKC, Brunsden D (1984) Recent fluctuations in some glaciers of the western Karakoram mountains, Hunza, Pakistan. In: Miller KJ (ed) International Karakoram Project, vol 2. Royal Geographical Society, London, pp 411–455

Haeberli W, Hoelzle M, Suter S (1998) Into the second century of worldwide glacier monitoring: prospects and strategies. UNESCO, Paris

Haritashya UK, Bishop MP, Shroder JF, Bulley HN (2007) Satellite observations of glacier advances and retreat in the western Karakoram. In: AGU Fall Meeting, Abstracts

Haserodt K (1984) Abflussverhalten der Flüsse mit Bezügen zur Sonnenscheindauer und zum Niederschlag zwischen Hindukusch (Chitral) und Hunza-Karakorum (Gilgit, Nordpakistan). Mitteilungen der Geographischen Gesellschaft, München, 96:29–36

Haserodt K (1989) Chitral (Pakistanischer Hindukusch) Strukturen, Wandel und Probleme eines Lebensraumes im Hochgebirge zwischen Gletschern und Wüste. In: Haserodt K (ed) Beiträge und Materialen zur Regionalen Geographie, vol 2. Institut für Geographie der Technischen Universität, Berlin, pp 182–233

Hayden HH (1907) Notes on certain glaciers in northwest Kashmir. Rec Geol Surv India 35:127–137

Hewitt K (1964) A Karakoram ice dam. Indus (Lahore) 5:18–30

Hewitt K (1968) Studies in the geomorphology of the mountain regions of the Upper Indus Basin. 2 vols. Unpublished Ph.D. thesis, University of London

Hewitt K (1969) Glacier surges in the Karakoram Himalaya (Central Asia). Can J Earth Sci 6:1009–1018

Hewitt K (1982) Natural dams and outburst floods of the Karakoram Himalaya. In: Glen J (ed) Hydrological aspects of alpine and high mountain areas. International Hydrological Association. (I.A.H.S.) Publication No. 138, Exeter, UK, pp 259–269

Hewitt K (1998a) Glaciers receive a surge of attention in the Karakoram Himalaya. EOS Trans Am Geophys Union 79(8):104–106

Hewitt K (1998b) Catastrophic landslides and their effects on the Upper Indus streams, Karakoram Himalaya, Northern Pakistan. Geomorphol 26:47–80

Hewitt K (1999) Quaternary moraines vs catastrophic rock avalanches in the Karakoram Himalaya, northern Pakistan. Quat Res 51(3):220–237

Hewitt K (2001) Catastrophic rockslides and the geomorphology of the Hunza and Gilgit basins Karakoram Himalaya. Erdkunde 55:72–94

Hewitt K (2004) Geomorphic hazards in mountain environments. In: Owens P, Slaymaker O (eds) Mountain geomorphology. Hodder Scientific, London, pp 187–218

Hewitt K (2005) The Karakoram anomaly? Glacier expansion and the "elevation effect", Karakoram Himalaya. Mt Res Dev 25:332–340

Hewitt K (2006a) Disturbance regime landscapes: mountain drainage systems interrupted by large rockslides. Prog Phys Geogr 30(3):365–393

Hewitt K (2006b) Rock avalanche dams on the transHimalayan Upper Indus Streams: a survey and assessment of hazards-related characteristics. Ital J Eng Geol Env Special issue I, Rome, pp 61–66

Hewitt K (2007a) Tributary glacier surges: an exceptional concentration at Panmah glacier, Karakoram Himalaya. J Glaciol 53(181):181–188

Hewitt K (2007b) Rediscovering colonized landscapes: the first Europeans at the Mustagh Pass, Karakoram Himalaya, Inner Asia. In: Gervers M, Bulag U, Long G (eds) The exploitation of the landscape of central and inner Asia, Toronto studies in central and inner Asia, 9. Asian Institute, University of Toronto, pp 41–67

Hewitt K (2008) Rock avalanches that travel onto glaciers: disturbance regime landscapes, Karakoram Himalaya, Inner Asia Geomorphology. Int J Pure Appl Geomorphol 103:66–79

Hewitt K (2009a) Rock avalanches that travel onto glaciers: disturbance regime landscapes, Karakoram Himalaya, Inner Asia. In: Crosta G (ed) Geomorphology: Special issue, 103:66–79

Hewitt K (2009b) Paraglacial rock slope failures, disturbance regimes and transitional landscapes, Upper Indus Basin, northern Pakistan. In: Knight J, Harrison S (eds) Periglacial and paraglacial processes and environments, vol 320, Special Publications. The Geological Society, London, pp 235–255

Hewitt K (2010) Gifts and perils of landslides: catastrophic rockslides and related landscape developments are an integral part of human settlement along upper Indus streams. Am Sci 48:410–419

Hewitt K (2011a) Rock Avalanche dams on the trans Himalayan Upper Indus streams: a survey of late quaternary events and hazard-related characteristics. In: Evans SG, Hermanns RL, Strom A, Scarascia-Mugnozza G (eds) Natural and artificial rockslide dams. Springer, Berlin, Heidelberg, pp 177–191, Chap. 6

Hewitt K, John J, Clague JJ, Gosse J (2011b) Rock avalanches and the pace of late Quaternary development of river valleys in the Karakoram Himalaya. Geol Soc Am Bull, doi:10.1130/B30341.1, 16p. 14 June 2011

Hewitt K, Wake CP, Young GJ, David C (1989) Hydrological investigations at Biafo glacier, Karakoram Himalaya: an important source of water for the Indus River. Ann Glaciol 13:103–108

ICIMOD (2007) The melting Himalayas: regional challenges and local impacts of climates change on mountain ecosystems and livelihoods. Technical paper X, International Centre for Integrated Mountain Development, Kathmandu, 33p

ICIMOD (2011) The status of glaciers in the Hindu Kush-Himalayan region. International Centre for Integrated Mountain Development, Kathmandu

Immerzeel WW, van Beek LPH, Bierkens MFP (2010) Climate change will affect the Asian water towers. Science 328:1382–1385

IPCC, Climate Change (2005) The physical science basis. Working group I: contribution to the fourth assessment report of the intergovernmental panel on climate change, Cambridge University Press

Iturrizaga L (2006) Transglacial landforms in the Karakoram (Pakistan): a case study from Shimshal Valley. In: Kreutzmann H (ed) Karakoram in transition: culture, development and ecology in the Hunza Valley. Oxford University Press, Karachi, pp 96–108

Jacob T, Wahr J, Pfeffer WT, Swenson S (2012) Recent contributions of glaciers and ice caps to sea level rise. Nature 482:514–518

Jiskoot H (2011) Glacier surging. In: Singh VP, Singh P, Haritashya UK (eds) Encyclopedia of snow and glaciers. Springer, Dordrecht, pp 415–428

Kääb A, Berthier E, Nuth C, Gardelle J, Arnaud E (2012) Contrasting patterns of early twenty-first century glacier mass change in the Himalayas. Nature 488:495–498

Kalvoda J (1992) Geomorphological record of the Quaternary Orogeny in the Himalaya and the Karakoram, vol 3, Developments in earth surface processes. Elsevier, Amsterdam

Kalvoda J, Goudie AS (2002) Landfrom evolution in the Nagar region, Hispar Mustagh Karakoram. In: Kalvoda J, Goudie AS (eds) Geomorphological variations. Nakladatelstvi, Prague, pp 87–126

Kamp U, Haserodt K (2004) Quaternary glaciations in the high mountains of northern Pakistan. In: Ehlers J, Gibbard PL (eds) Extent and chronology of glaciation, vol 3, South America, Asia, Africa, Australia, Antarctica. Elsevier, Amsterdam, pp 293–311

Kerr T, Owens I (2008) Glaciers and climate change over the last century in the Aoraki/Mt Cook regions, New Zealand. In: Bonardi L (ed) Terra glacialis. Special issue, Mountain glaciers and climate changes in the last century. Servizio Glaciologico Lombardo, Milan, pp 209–22

Kick W (1956) Der Chogo-Lungma-Gletscher im Karakorum. Zeit Glaziol Gletscherk 3(3):335–347

Kick W (1962) Variations of some Central Asian glaciers. In: Proceedings of the general assembly; Symposium on the regimen of existing glaciers. International association science hydrology, vol 55. Helsinki, pp 223–29

Kick W (1989) The decline of the last Little Ice Age in High Asia compared with that in the Alps. In: Oerlemans J (ed) Glacier fluctuations and climate change. Kluwer, Dordrecht, pp 129–142

Knight EF (1893) Where three empires meet: a narrative of recent travel in Kashmire, western Tibet, Gilgit and the adjoining countries. (Reprint 1980) Ferozesons, Lahore

Kotlyakov VM (ed) (1997) Atlas snezhno-ledovykh resursa mira [World atlas of snow and ice resources]. Moscow, Russian Academy of Sciences. Institute of Geography [in Russian]

Kotlyakov VM, Severskiy I (2009) Glaciers of central Asia: current situation, changes, and possible impact on water resources. In: Braun LN, Hagg W, Severskiy IV, Young G (eds) Assessment of snow, glacier and water resources in Asia. IHP/HWRP, Koblenz

Kreutzmann H (2006) Karakoram in transition: culture, development and ecology in the Hunza Valley. Oxford University Press, Karachi

Kuhle M (1990) Ice marginal ramps and alluvial fans in semiarid mountains: convergence and divergence. In: Rachochi AH, Church M (eds) Alluvial fans: a field approach. Wiley, New York, pp 55–68

Kuhle M (2004) The pleistocene glaciation in the Karakoram mountains: reconstruction of past glacier extensions and ice thicknesses. J Mt Sci 1(3):17–298

Li Jijun, Derbyshire E, Shuying X (1984) Glacial and paraglacial sediments of the Hunza Valley, North–west Karakoram, Pakistan: a preliminary analysis. In: Miller KJ (ed) International Karakoram project, vol 1. Cambridge University Press, Cambridge UK, pp 496–535

Liu SY, Ding YJ, Li J, Shangguan DH, Zhang Y (2006) Glaciers in response to recent climate warming in western China. Quat Sci 26:762–771

Mason K (1930) The glaciers of the Karakoram and the neighbourhood. Rec Geol Soc India 63(2):214–278

Mayer C, Lambrecht A, Belo M, Smiraglia C, Diolaiuti G (2006) Glaciological characteristics of the ablation zone of Baltoro glacier, Karakoram, Pakistan. Ann Glaciol 43:123–131

Mayewski PA, Jeschke PA (1979) Himalayan and trans-Himalayan glacier fluctuations since AD 1812. Arct Alp Res 11:267–287

Meiners S (1998) Preliminary results concerning historic to post-glacial stages in the NW-Karakorum (Hispar Muztagh, Batura Muztagh, Rakaposhi range). In: Stellrecht I (ed) Karakoram-Hindukush-Himalaya: dynamics of change, vol 4/1. Rüdgers Köppe Verlag, Köln, pp 49–70

Mercer JH (1975) Glaciers of the Karakoram. In: Field WO (ed) Mountain glaciers of the northern hemisphere, vol 1. Cold Regions Research and Engineering Laboratory (CREEL), Hanover, pp 371–409

Miller KJ (1984) The International Karakoram Project, vol 2. Cambridge University Press, Cambridge

Mott PG (1950) Karakoram survey 1939: a new map. Geogr J 116:513–537

Oerlemans J (2001) Glaciers and climate change. AA Balkema, Lisse, Netherlands

Oestreich K (1911–1912) Der tschogletscher in Baltistan. Z Gletsch 6:1–30

Owen L (1994) Glacial and non-glacial diamictons in the Karakoram mountains and western Himalaya. In: Warren WP, Croots D (eds) The formation and deformation of glacial deposits. Balkema, Rotterdam, pp 9–24

Owen L (2006) Quaternary glaciation. In: Kreutzmann H (ed) Karakoram in transition: culture, development and ecology in the Hunza Valley. Oxford University Press, Karachi, pp 12–23

Paterson WSB (1994) The physics of glaciers, 3rd edn. Pergamon, New York

Pecci M, Smiraglia C (2000) Advance and retreat phases of the Karakorum glaciers during the 20th century: case studies in Braldo Valley (Pakistan). Geogr Fı´s Din Quat 23(1):73–85

Raina VK (2009) Himalayan glaciers: a state-of-art review of glacial studies, glacial retreat and climate change. Ministry of Environment and Forests, New Delhi

Rees HG, Collins DN (2006) Regional differences in response of flow in glacier-fed Himalayan rivers to climatic warming. Hydrol Process 20(10):2157–2169

Richardson SD, Quincey DJ, Nasab N (2009) The 2008 glacier outburst floods from Ghulkin Glacier, Karakoram, Pakistan. Glacier hazards workshop, Universität für Bodenkultur, BOKU, Vienna

Scherler D, Bookhagen B, Strecker MR (2011) Spatially variable response of Himalayan glaciers to climate change affected by debris cover. Nat Geosci. doi: 10.1038/NGEO1068, 4p

Schmidt S, Nüsser M (2009) Fluctuations of Raikot glacier during the last 70 years: a case northern Pakistan. J Glaciol 55:949–959

Schmidt S, Nüsser M (2012) Changes of high altitude glaciers from 1969 to 2010 in the trans-himalayan Kang Yatze Massif, Ladakh, northwest India. Arct Antarct Alp Res 44:107–121

Searle M (1991) Geology and tectonics of the Karakoram mountains. John Wiley, New York

Sella V (1987) Vittorio Sella with the Italian expedition to Karakorum in 1909, catalogue of the exhibition, Embassy of Italy, Pakistan. Fondazione Sella, Biella

Seong YB, Owen A, Bishop MP, Bush A, Glendon P, Copland L, Finkel RC, Kamp U, Shroder JF Jr (2007) Quaternary glacial history of the central Karakoram. Quat Sci Rev 26:3384–3405

Sharp M (1988) Surging glaciers: behaviour and mechanisms. Progr Phys Geogr 12(3):349–370

Shroder JF Jr, Bishop MP (2010) Glaciers of Pakistan. In: Williams RS , Ferrigno JG Jr (eds) 2010 Satellite image atlas of glaciers: Asia. United States Geological Survey, Denver, Professional Paper 1386-F, Washington, DC, pp F201–F257

Shroder JF Jr, Khan, MS, Lawrence RD, Madin IP, Higgins SM (1989) Quaternary glacial chronology and neotectonics in the Himalaya or Northern Pakistan. In: Malinconico L Jr, Lillie RJ (eds) Tectonics of the Western Himalayas. Geological Society of America, Special paper 232, pp 275–294

Shroder JF Jr, Owen LA, Derbyshire E (1993) Quaternary glaciation of the Karakoram and Nanga Parbat Himalaya. In: Shroder JF (ed) Himalaya to the sea: Geology, geomorphology and the quaternary. Routledge, London, pp 132–158

Shroder JF Jr, Bishop MP, Sloan V, Copland L (2000) Debris-covered glaciers and rock glaciers in the Nanga Parbat Himalaya, Pakistan. Geogr Ann 82A:17–31

Sigurdsson O (2008) Glaciers of Iceland: peculiarities and variations during the last century. In: Bonardi L (ed) Terra glacialis Special issue mountain glaciers and climate changes in the last century. Servizio Glaciologico Lombardo, Milan, pp 53–62

Smiraglia C, Mayer C, Mihalcea C, Diolaiuti G, Belo M, Vassena, G (2008) Himalayan-Karakoram glaciers: results and problems in the study of recent variations of major non-polar glaciers. In: Bonardi L (ed) Terra glacialis. Special issue, Mountain glaciers and climate changes in the last, pp 149–164

Survey of Pakistan (1966) Atlas of Pakistan, Rawalpindi

Tsevetkov DG, Osipova GB, Xie Z, Wang Z, Ageta Y, Baast P (1998) Glaciers of Asia: In: UNESCO (ed) Into the second century of worldwide glacier monitoring: prospects and strategies. UN Educational, Scientific and Cultural Organization, Paris, pp 177–196

UNESCO (1998) Into the second century of worldwide glacier monitoring: prospects and strategies. UN Educational Scientific and Cultural Organization, Paris

Vigne GT (1842) Travels in Kashmir, Ladak, Iskardo, the countries adjoining the mountain-course of the Indus and the Himalaya, north of the Punjab, 2 vols (Reprint 1987) Indus, Karachi

Visser PhC, Visser-Hooft J (1935–1938) Wissenschaftliche Ergebnisse der niederländischen Expeditionen in den Karakorum und die angrenzenden Gebiete in den Jahren 1922, 1925 und 1929–30. E.J. Brill, Leiden

von Wissmann H (1959) Die heutige Vergletscherung und Schneegrenze in Hochasien mit Hinweisen auf die Vergletscherung der letzten Eiszeit. Akademie der Wissenschaften und der Literatur in Mainz. Abhandlungen der mathematisch-naturwissenschaftlichen Klasse 14:1103–1431

Wake CP, Searle MP (1993) Rapid advance of Pumarikish glacier, Hispar glacier basin, Karakoram Himalaya, correspondence. J Glaciol 39(131):204–206

Wiche E (1961) Klimamorphologische Unterschungen im westlichen Karakoram. Tagungsbericht u.wiss. Abh.Dt. Geographentag, Berlin 1959. 190–203, Wiesbaden

Williams RS, Jr, Ferrigno JG (eds) 2010 Satellite image atlas of glaciers: Asia, United States Geological Survey, Denver, Professional Paper 1386-F, Washington, DC

Workman FB, Workman WH (1905) From Srinagar to the sources of the Chogo Lungma glaciers. Geogr J 25:245–268

Yafeng S, Liu S, Shangguan D, Li D, Ye B (2006) Peculiar phenomena regarding climatic and glacial variations on the Tibetan plateau. Ann Glaciol 43:106–111

Younghusband FE (1896) The heart of a continent: a narrative of travels in Manchuria, across the Gobi desert, through the Himalayas, and Chitral, 1884–1894. John Murray, London; reprinted (1994) Book Faith, New Delhi

Zeitler PK (1985) Cooling history of the NW Himalaya, Pakistan. Tectonics 4:127–151

Zemp M, Hoelzle M, Haeberli W (2009) Six decades of glacier mass balance observations: a review of the worldwide monitoring network. Ann Glaciol 50:101–111

# Chapter 13
# Glaciers in Human Life

**Abstract** The following three main concerns are addressed: (1) mountain peoples closely involved with glaciers, (2) down-country relations in Pakistan, China, India and Afghanistan, especially for water resources and risks, and (3) trans-boundary questions. In the latter, glacierised areas are involved in resource and geostrategic agendas, cultural and religious questions between countries and, not least, recurring and ongoing armed conflicts. There has been a tendency, beginning in the earliest modern work, to detach glaciers and glacial science from human contexts. Much has been investigated in both areas but their interrelations largely ignored, in particular how scientific glacier knowledge might engage with local knowledge for mutual benefit and to help address mountain land concerns. This bears on problems of some urgency in adapting to glacier and other climate-related changes. Lately, there has been the so-called 'melting Himalayas' issue, possible water crises due to loss of ice through global climate warming. It is a justifiable concern if, to date, not as evident in Karakoram glaciers or clear in its implications as for some other High Asian mountains. A brief survey is given of how mountain communities relate to and view the glaciers, mainly their agricultural and pastoral activities, and how modernization is affecting them. A case study of the Hopar villages, Barpu and Bualtar Glaciers, illustrates something of the complexities and the risks from glacier hazards. Interest in the water resources of the Indus and Yarkand Basin is overwhelmingly about supplies for the larger populations, cities and industries of the lowlands. The two rivers, specifically their main stems, may well have the world's highest ratios of glacier meltwaters to numbers of inhabitants dependent on them. How glaciers relate to other cryosphere elements in terms of water resources, notably to snowfall and permafrost, remains largely unknown. How they relate to rapid socio-economic changes and large-scale water resource projects is even murkier. Many developments are necessarily poorly informed by knowledge of the glaciers because monitoring and research are so limited. Their present state and potentially adverse impacts of regional temperatures and other changes are equally uncertain. Meanwhile, the glacial waters of the Indus flow through four countries and, in the Yarkand Basin, China has issues with the indigenous cultures of Tibet and Xinjiang Province. Future prospects seem likely to depend

K. Hewitt, *Glaciers of the Karakoram Himalaya: Glacial Environments, Processes, Hazards and Resources*, Advances in Asian Human-Environmental Research,
DOI 10.1007/978-94-007-6311-1_13, 

more on relations between Pakistan, India, Afghanistan and China than on glacier change. The Indus Waters Treaty is generally considered a model for resolving differences and avoiding conflict over trans-boundary water resources. Nevertheless, for decades the upper Indus Basin has been subject to armed intervention and recurring wars. Relations between the countries concerned and the Treaty are under considerable stress these days. Of special concern is how conflict has blocked and disrupted scientific work on the glaciers and continues to do so, not least the much-needed improvements in monitoring the cryosphere and responding to glacier change.

**Keywords** Mountain communities • Transhumance • Water resources • Ecological services • Amenities • Hazards • Irrigation systems • Indus Waters Treaty • Conflict

## 13.1 Introduction

To begin with, certain matters need emphasis. Firstly, the Karakoram is an *inhabited and historic* mountain range. The valleys of the upper Indus streams have been settled for more than a thousand years. For much longer still, paths to, and some over, the glaciers have linked the region to the civilizations of the Indus Basin and Inner Asia (Plate 13.1). Some highly valued places record this (Hedin 1917–1922; Dani 1989).

**Plate 13.1** Baltit Fort, dated back about one thousand years. In the valley below are graffiti and inscriptions dating back another thousand years at least and from a range of civilisations (Neelis 2006). In the background are peaks and glacier watersheds of the Rakaposhi range. The fort sits on and is partly constructed from debris of the Ultar Glacier moraines and the Baltit-Sumayar rock avalanche, described in Section 12.5 (photo Hewitt 2006)

Secondly, one can hardly overestimate the challenges of the national and transnational water supplies to which the glaciers contribute. The lowlands are largely desert and semi-desert areas of water deficit. Their dense populations and economic survival depend on elaborate hydraulic systems that manage and allocate water from the mountains. Herein lies the main reason for the 'global' attention the glaciers have been receiving. And it is commonly asserted in influential media and political speeches that water resources in India and Pakistan are at or near crisis conditions (Wirsing and Jasparro 2006). To be sure, the more pressing developments and concerns have nothing to do with glaciers or, indeed, global climate change. Nevertheless, these pose major medium-term threats for existing water supplies, of greater floods and droughts, above all how societies will respond to reports of 'disappearing glaciers'.

Thirdly, however, a special responsibility here is to draw attention to the lives and needs of people in the region closer to the glaciers. In the mountains are some millions of inhabitants, including those most directly dependent upon snow and ice resources. They too have some acute socio-economic problems. There are threats to environmental sustainability including, but not only, from climate and glacier change. Rapidly growing towns and regional centres add to problems as well as the fate of thousands of villages and some of the largest surviving mountain agro-pastoral economies. The diversity of relations to glaciers among these societies is only hinted at by that of the glacial environments identified thus far. Of special concern are political and economic conditions that link highland and lowland developments, village and metropolis.

Each of the topics raised has its special features and claims to importance. How they interact creates some of the most pressing and intractable questions. Inevitably, perhaps, violent clashes get the most attention. The sheer scale and urgency of water demands in surrounding lowlands loom largest for those in authority. Typically, however, modern glacier science is resolutely detached from human contexts – as can be said of this book so far! In high mountains the problem is often reinforced by stereotypes that romanticise them as wilderness or colonial images of the Himalayas as among 'the last frontiers', 'The Third Pole', even 'empty lands' (Hewitt 2007). It is not that such notions lack their critics (Debarbieux and Gillet 2000; Price 2006; Hewitt and Mehta 2012) or to deny some sensitive investigations of Karakoram societies which avoid them (Stellrecht 1998; Kreutzmann 2006). Unfortunately, even the latter have little to say specifically about human relations to glaciers. Only rarely do the many studies and travel writings identify links to, or local knowledge of, the glaciers not arising solely from visitors' preoccupations. There are a few exceptional investigations of people and glaciers in other regions (Ladurie 1971; Tobias 1986; Cruikshank 2005; Orlove et al. 2008). They show what might be done in the Karakoram, and what appears below is partly inspired by them.

Recent and urgent concern over glacier losses does underscore their human significance. More difficult is to get beyond broad generalisations about 'the water towers of mankind', 'the melting Himalayas' or 'the disappearing glaciers' (Jianchu et al. 2007; Viviroli and Weingartner 2006). As the previous chapter showed, this has been compounded by poorly informed and exaggerated assumptions about how, or how fast and uniformly, the glaciers are changing. Similarly, while a considerable effort identifies glacier hazards, it has much less to say about the condition and responses of people most at risk (Said 1998; Derbyshire and Fort 2006). Danger is more often attributed to altitude and remoteness, alpine and steepland conditions, climate change or extremes of weather, if not 'backwardness'. Equal attention needs to be given to the long-term adaptations of mountain dwellers and their emerging vulnerabilities in a changing socio-economic environment. All sorts of modernization plans are proposed for them. Critiques of how modern developments have undermined rather than improved their safety are equally compelling (Ives 1997). Efforts are under way to address many of these questions but, at the time of writing, most of the work remains to be done for peoples of the Karakoram.

The original plan for this book was to give as much space to the human aspects of the glaciers as the geophysical. However, the 70-year absence of a modern review of the latter created unusual demands. The limited scope of materials specifically on people's relations to glaciers thwarted a larger effort in this regard. For the moment, a sketch of the more salient issues and some case studies must suffice towards a sense of the scope of human concerns and practices in relation to glaciers.

## 13.2 Inhabited Mountains

Impressions of a 'mountain fastness' notwithstanding, few if any glacier basins in the Karakoram lack long-entrenched uses and rights, histories, resources, symbols and attachments for local communities. A host of place names identify uses, old stories and events and the different linguistic groups living there now or in the past (Hughes 1984; Kreutzmann 2006; Hewitt 2008). Of special interest are mountain pastoralists and their seasonal migrations in pursuit of pasture, known as *vertical transhumance*. It reflects greater humidity as well as warmer summer conditions high up in the Karakoram (Miehe and Miehe 1998) and brings the most widespread and frequent interactions with glaciers (Ehlers and Kreutzmann 2000; Clemens and Nüsser 2008; Kreutzmann 2012). Extensive pastures surround the glaciers up to 4,500 m or more. They have been in use for centuries and the vegetation is greatly modified by seasonal grazing (Saunders 1983; Nüsser 1998; Janjua 1998). Along and overlooking the glaciers are countless temporary summer settlements. Bio-resources are extracted beside the glaciers, especially in the well-watered valley side troughs, including wood for fuel, charcoal and construction, herbs and medicinal plants, and wild game.

Snowfields and small ice masses are critical for water supply of innumerable farms and villages and for the few Karakoram towns. In most places, local technologies cannot cope with the volumes and large fluctuations of glacial rivers. Crops depend on meltwater from snowfields and small ice masses high up (Plate 13.2). The waters are led across and down the mountainsides by elaborate networks of channels to the terraced fields around settlements (Whiteman 1985; Butz 1989). Some crops may be grown at temporary summer settlements as high as 3,500 m. A few permanent villages and croplands are found beside glaciers as high as 3,000 m. The majority are at lower and drier elevations, although, as noted, almost

**Plate 13.2** Impressive terraced field systems at 3,000 m and higher in Naltar valley, near Gilgit. Although the valley is more humid than most in the Central Karakoram, permanently settled areas and the fields depend upon irrigation water brought from snowfields and small glaciers up above (Hewitt 2004)

all have summer activities in high valleys and along the glaciers. Modernization has brought many small-scale hydroelectric projects also tapping water from smaller snow- or ice-fed streams.

Traditional farming and herding continue to be widespread, but menfolk increasingly work for cash outside their village. This leaves much of the traditional economy in women's hands. Men who act as guides, porters, drivers and cooks in trekking and mountaineering expeditions are among the more numerous visitors to the glaciers (Emerson 1986; MacDonald and Butz 1998; Butz 2006). Some, who join the army or supply its needs, may go to frontier posts high up, including the battleground of the Siachen Glacier (see below).

In the Yarkand Karakoram there are no settlements, only temporary visitors on expeditions and military assignments. In the trans-Himalayan upper Indus, there are few towns, but those that exist have been growing fast, notably since completion of the Karakoram Highway (KKH) in the 1970s. This links the lowlands of Pakistan with China and has encouraged a phenomenal amount of road building such that almost every community now has four-wheel drive access at least. Nevertheless, most people still reside in villages where material life revolves around traditional practices and a close interdependence between animal raising and crop growing. They include quite complicated seasonal and land use patterns. Traditional divisions of labour, especially by gender, are involved in the management of soil, irrigation water and manuring of fields, planting, weeding and harvesting, orchards and pasture lands. Village activities commonly spread into the valleys of one or more glaciers. There are different protocols affecting who goes to the glaciers and high pastures or when they go (Azhar-Hewitt 2011).

Variations in practices and divisions of labour across the region partly reflect environmental differences, mainly the history of distinctive cultural practices in more than a dozen distinct ethnolinguistic groups. Communities are spread over more than 2,000 m elevation, in valleys with greater or lesser dryness, more and less reliable and distant water sources and exposure to or shading from sunlight. Nearly all residents are of the Islamic faith but from at least four distinct branches. These encourage somewhat different approaches to land and life, differing social contacts within and outside the region, sometimes violent conflict (Afridi 1988; Hashmatullah 1987). A case study will give some sense of the range of interactions with the cryosphere.

## 13.3 Hopar Villages and Barpu–Bualtar Glaciers: A Case Study

The Hopar villages are in the uppermost areas of Nagar in the Hunza valley and are among the higher permanent settlements above and beside present-day glaciers (Plate 13.3). Village land consists of sediments deposited beside glaciers and by streams and slope processes that descended to them. People's lives are

**Plate 13.3** Hopar villages sit above Bualtar Glacier, whose debris-covered ice is visible to the right and below the cliff (Hewitt 1987)

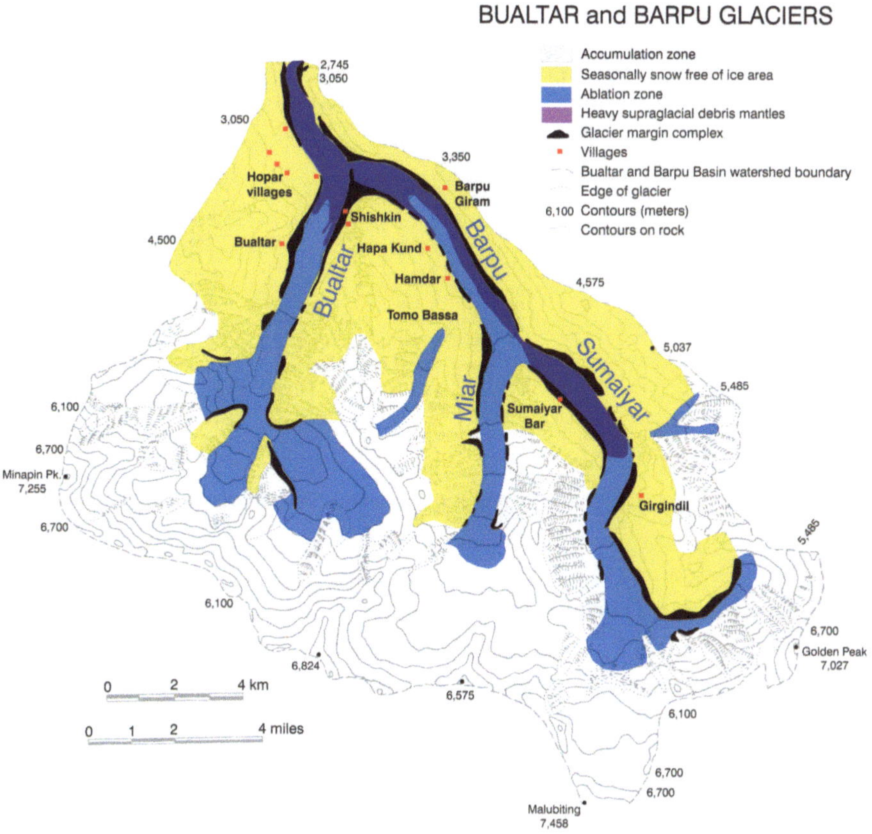


**Fig. 13.1** Map of Barpu–Bualtar Glaciers and main sites of human activity

closely intertwined with two main glaciers. Bualtar passes immediately below the cliff at the edge of the villages. Barpu Glacier terminates on the opposite side of the valley. At elevations around 3,000 m, the climate is semiarid and relatively hot in summer. Meltwater from higher up is essential for irrigating fields, driving flour mills, domestic and other uses. The glaciers are too difficult or unreliable to tap so that snowfields and minor ice masses in a small tributary valley are the main sources. Barpu's outlet stream is now tapped for hydroelectric power.

Encounters between villagers and the glaciers occur on virtually every day of the year. Children scurry across the ice to Shishkin, an abandoned settlement on the far side of Bualtar Glacier, in search of kindling for home fires or grass for animals. Sometimes they or their mothers take small animals to graze there. Throughout the year hunters have business further up the glacier and shepherds go to check on the yaks, kept out of the villages even in winter. Most encounters are, however, in the summer months when large herds are taken to pastures and temporary settlements along them (Figs. 13.1 and 13.2).

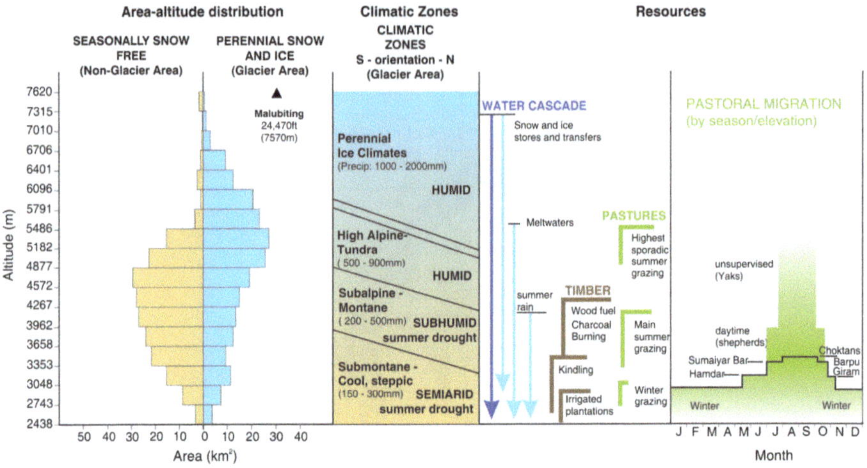


**Fig. 13.2** Patterns and verticality of human activities in relation to Barpu–Bualtar Basins related to Fig 13.1

## *13.3.1 Glacier Transhumance*

The search for pasture and taking of the animals away from the croplands begins as soon as the snow melts from village fields and when planting begins at the vernal equinox. However, most animals start their great summer journey up the glaciers on a particular day, the third week of June at or close to the summer solstice. The exact date varies according to the lunar calendar. Then, great numbers of goats, sheep and cattle are assembled at the head of the cliff overlooking Bualtar Glacier. They must descend the path to its surface and cross the ice to reach grazing lands along the Bualtar right (north) side and the Barpu Glacier.

Much skill is required where routes cross the glaciers, getting animals safely on and off the ice as well as in travelling over it and avoiding crevasses. The numbers of routes, camps and summer settlements indicate the scope of links to glacier behaviour (Figs. 13.1 and 13.2). Pasture usage and care follow from well-established rights that divide them among the five villages. Specific summer pastures and summer settlements or *chiragha*'s are as follows[1]:

(i) Boroshal's pastures are at *Hunoona, Bualtar, Hamdar*
(ii) Ghoshoshal's: *Hapakund*
(iii) Hakalshal's: *Supputhar, Daltar, Shakashal, Miar*
(iv) Holshal's: *Shaltar, Shishkin, Sumaiyar bar, Murukmaidan*
(v) Rathul's: *Rush Phari, Dhazanchi, Czuming*

In addition, common use is allowed of pastures around *Choktans, Girgindil, Malangushi Hari, Yakazina* and *Shuja Basa*. Until the 1970s these belonged to the

[1] Thanks to Dr. Farida Azhar-Hewitt (pers. comm.) for these materials from her field notes.

**Plate 13.4** Shimshal's summer high pasture at Showari (4,735 m) looking east towards Shaksgam valley. Here women care for the animals (Hewitt 1999)

former ruler of Nagar, the Mir. For the most part just yaks are seen there, and shepherds only visit intermittently. The same applies to *Borosh* and *Shabera* up the Bualtar. However, local reports reinforce indications on the ground that transhumance has declined since former times. Modern developments, mostly those that attract men away from the villages, seem more important than results of overgrazing or climate change reported elsewhere in the Himalaya. At places like *Shaltar* and *Hamdar*, many fine terraced fields have not been cultivated for several generations.

For the people of Hopar pastoral activities are almost entirely men's work. Only a few of the lower *chiraghas* like *Hapa Kund* are visited by women to help with crops of potatoes and animal fodder. They take care of the weeding among other specific duties. Otherwise, between 5 and 15 men from each village spend the summer at the pastures. These days, the shepherds tend to be youths and older men. Meanwhile, with the animals away, village activities focus on crops and mainly women's work weeding them and taking the greenery that results for the small milk animals that stay behind. Except for these, or to help with ploughing and threshing cereals in late summer, the herds do not return to the villages until late October after crops are harvested.

In other parts of the mountains, arrangements can be different. In Hushe and Shimshal valleys, for example, women go to the high pastures and, with occasional male visitors, remain there alone through the summers (Plate 13.4). In other cases, such as the upper Braldu and villages like Askole, whole families go to the charagahs. In Naltar, and around Nanga Parbat, the high pastures are mainly occupied by Gujar people, migrant groups who otherwise do not control the land and leave the mountains in winter (Nüsser 1998).

**Plate 13.5** The 1987 surge of the Bualtar Glacier. Shepherds stand at a place where they had regularly taken their animals across the ice to high pasture, no longer possible (Hewitt 1987)

### *13.3.2 Glacial Hazards and Responses at Hopar*

In one sense, the Hopar villages, even the whole larger community of Nagar, are gifts of the glaciers, part of the evolution of the Hunza and lower Hispar valleys (Hewitt 2010). This has only come about through the skills and hard work of generations of farmers and herders. The Hopar villagers are particularly well adapted to the resources and amenities provided by the glaciers. They also confront a range of dangers related to them.

Some glacier hazards are direct such as crevasses or paths suddenly closed by ice movement and erosion. Bualtar and the Sumaiyar tributary of Barpu are surge type. When a surge occurs, serious dangers arise such as landslides and glacier outburst floods. Routes to pastures and other resource lands can be closed. Other hazards such as animals lost, injured by falling or from rock falls and avalanches, are not specifically glacial but involve life around glaciers.

A recent lethal disaster occurred in the 1970s when an avalanche dammed the stream that provides irrigation water to Hopar, impounding a lake high above the villages. Just as a group of men were trying to release it, the dam burst. Some were immediately drowned; others died as the flood destroyed parts of Ratul village, flour mills along the stream and many acres of crop land and orchard. Eventually some relief funds were given to repair the damage and open up new land among moraines further down the valley.

Catastrophic landslides fell onto Bualtar Glacier in 1986 (Hewitt 1988). A surge of the glacier in 1987 brought increased ice-margin erosion, deposition, recurring lakes and outburst floods (Plate 13.5). These have compounded long-term loss of land.

**Plate 13.6** View along the cliff edge at Hopar and over the lower Bualtar. Large earth slumps are evident and, down the cliff, remnants of former fields, irrigation channels and the old road (Hewitt 2005)

### *13.3.3 Landslides and Land Loss*

The great build-up of deposits around the lower Bualtar and Barpu Glaciers provided relatively attractive sites for settlement and cultivation. For a century or more, however, ice levels have declined, exposing near-vertical walls in the old lateral deposits at the edge of Hopar village land. Along fronts of more than 20 km, a series of large, slow landslides have been eating into village lands. They have removed some 4–5 $km^2$ in the past three or four decades. The direct impacts include land loss and deep fissures that cut irrigation systems and paths. Down the cliff, 50–100 below the villages, are sections of broken road, stone terraces, fields and irrigation channels carried away by the landslides (Plate 13.6). There are recurring difficulties of access to and across the glacier. According to weather and season countless smaller stone and earth falls occur.

In the 1980s when our team of Canadians and Pakistani engineers were monitoring the Bualtar and Barpu Glaciers, the Hopar village elders asked us to investigate the landslides and land loss. The hazards of glacier and landslide seemed immediately evident. They arose in societies that seemed closely adapted to constrained spaces at the glacier margins, to the steep slopes above permanent and temporary summer settlements. Important as these facts are, however, some more critical underlying factors were hidden from us.

**Plate 13.7** View of the Hopar settlements from Shishkin side, showing how they occupy moraines and terraces above the debris-covered Bualtar Glacier. Debris from the 1987 landslides covers the ice in the middle-ground, having been transported 7 k down the glacier. The Great Lateral Moraine discussed in Chap. 12 overlooks the ice in the foreground and margins of Hopar settlements (Hewitt 2002)

### *13.3.4 Hidden Hazards and Invisible Distress*

When the glaciers were building the high moraines at Hopar and Shishkin opposite, obviously they were much higher than today. They are widely believed to mark the glacier maximum of the Little Ice Age and part of the 'Great Lateral Moraine' described in Chap. 12 (Kick 1989; Meiners 1998). Yet the glacier seems never to have been much wider or more advanced than at present in spite of being 150–250 m higher! Could it thicken so much, advance so little, yet remain as narrow? In recent years, some 10–15 m thickening led to an advance of 3.5 km and outward spread of some tens of metres. For old moraines ten times higher, no comparable advances or lateral expansion is evident (Plate 13.7).

An alternative that fits the facts more convincingly was described in Chap. 12 (12. 5).

Meanwhile, it turned out that the worst impacts are not directly nor, mainly, in the physical environment. They relate to two other contradictory findings – that there is no lack of land that could be reclaimed but is not and, also, that away from the main settlements even more land had been abandoned recently than lost to the landslides. Even so, the rate of population growth in the villages is notoriously high. Only gradually did the real disaster emerge, and how its roots lie in the gendered division of traditional labour and pressures of modernization.

A key discovery came not from the work on glacier and geo-hazards, but with the women of Hopar by Azhar-Hewitt (1989). She noted the absence or extreme rarity of women living into the 40s – that many women had unusual levels of stress, hypertension and recurrent problems associated with child-bearing and excessive menstrual bleeding. She concluded the women were severely overworked and not getting adequate nourishment or medical care. This was in a quite different area from the landslides and land losses, but not independent of them, a hidden disaster magnified by land loss. Their worst impacts played out in 'women's work'.

The loss of land was occurring mainly where women spend their time and wrecking facilities on which they are most dependent – places closest to the villages and the spaces where women must work in producing food for family and village, wash clothes and care for small animals. They had to work longer hours and go further afield, which meant more walking and carrying. Subsistence production is women's main role in the villages and worth a lot more than appears.

The technical conclusion was that the landslides could be monitored and land losses predicted *but not prevented*. It meant that the only way to offset the loss to the traditional economy was to reclaim other available land. However, that is men's work, along with building the essential water channel network. Men are unwilling or unable to help unless cash and tools were made available.

In fact, this had been behind the elders' request in the first place! They wanted a report by outside 'experts' to get official recognition and convince government or an NGO to provide funds and equipment. It is always harder with so-called creeping disasters which, like many droughts, lack sudden spectacular damage and loss of life, even when having larger net impacts (Smith and Petley 2009).

The economy of each household, their hopes for their children and their expectations are far from being dependent on women and the subsistence economy alone, though it is most directly impacted by the glaciers. Many men have gruelling jobs down country, on the roads and in the army or carrying heavy loads for mountaineering groups. Neither they nor their women are ready to give up the things that only money can buy. Households with too few or no men to bring in cash suffer some of the worst difficulties; such are the promises and pressures of the outside world, and not unfamiliar ones. In part, women's work means cash earnings need not go into subsistence, or not all of them, but can go into household and other improvements, pay for children's school books and clothing as well as helping families through the ups and downs of the modern wage economy. In the end it has been the building of a clinic and hospital and better access to health care and medicines for women or the migration of families to growing towns that have reduced the risks of land loss.

In sum, the worst impacts of the landslides were being borne primarily by village women but hidden as a tertiary hazard within the traditional gender division of labour and pressures of modernization. What the example shows is that, indeed, glaciers and their changes can have huge impacts on life and livelihoods, but in ways that play out according to the particular social context and other pressures on the communities at risk. This can be said of the range of hazards for mountain peoples (Deckens 2007a, b; Hewitt 2010; Hewitt et al. 2008; Azhar-Hewitt and Hewitt 2012). Much the same can be anticipated of the dangers from climate change and what is required to find effective ways to adapt.

## 13.4 Glaciers and Yarkand River Waters

There are no settlements in the Karakoram valleys of the Yarkand River. Most human issues arise where the waters enter the southern Tarim Basin in Xinjiang Uyghur Autonomous Region, western China where the river is important for agriculture, municipalities and developing industries (Yang and An 1991). In areas that can be reached by glacial flood waters, specifically glacier lake outburst floods, there are some 1.8 million people (Chap. 10). Also 38 million ha of irrigated land, major irrigation canals and six hydro-power plants are at risk from inundation. A railway causeway crosses the flood plain where the river enters the lowlands.

Recently, the use of and plans for Yarkand waters have expanded sharply and will continue to do so. Demands increase with rapid urbanisation and expanding cities like Kashgar, with engineering schemes for agricultural and industrial purposes, in all of which waters coming from the Karakoram glaciers are assuming a major role. Two large reservoirs are planned in the gorges just before the river reaches the open basin.

## 13.5 The Indus Waters

In total the Indus Basin covers about 1 million km$^2$. About half lies in the mountainous headwaters which supply almost 75 % of all river flows and are snow covered for some part of the year at least. The perennial snow and ice cover is a small fraction of the whole, probably around 22,000 km$^2$, of which about 70 % is in the Karakoram. The area covered by glacier ice itself is barely half of this. Nevertheless, as shown earlier, precipitation is highest in these areas, and annual ice ablation can be many times larger than precipitation, giving them a significance out of all proportion to the area involved.

The largest part of the basin is in Pakistan, with about two-thirds of some 240 million inhabitants and in India with about one-third. A further nine million people live in the Kabul River basin in Afghanistan, dependent on meltwaters from the Hindu Kush and Hindu Raj. In the trans-Himalayan Indus valleys are about 1.3 million people and more than 75 % in Gilgit–Baltistan, including most of the Karakoram and areas disputed by India and Pakistan. A small part of the high mountain area is in China, without permanent inhabitants.

There are major issues and concerns involving all the countries within and around the Indus Basin. Some are returned to below. Nevertheless, the greater portion of trans-Himalayan Indus waters, and specifically those from Karakoram glaciers, flows to Pakistan. Certain developments have made them an increasingly vital resource for that country. Its exceptional and growing dependence on snow and ice arises from the Indus Waters Treaty, signed by India and Pakistan in 1960 (Michel 1967). The Treaty gives the eastern Punjab tributaries largely to India, the Sutlej, Beas and Ravi which are predominantly rain fed. Seasonal snowmelt is also

**Table 13.1** Comparative magnitudes of various aspects of water supply and use of the Indus streams in Pakistan (MAF = million acre-feet)

| Water supply/use | Orders of magnitude MAF | Comments |
|---|---|---|
| Western Indus Rivers | 142 | |
| Diverted (twenty-first century) | 104 | 67 MAF at Independence in 1947 |
| Groundwater recharge | 55 | |
| Rainwater use | 50 | |
| Groundwater extracted | 39 | |
| Outflow to Arabian Sea | 35 | In some recent years less than 2 MAF |
| Tarbela Dam storage | 11.6 | Net useable storage 9.7 MAF |
| System losses | 8.6 | Mainly canal seepage, evaporation |

significant but glaciers only minimally. These rivers have been central in the historical development of Punjab Province, Pakistan's most populous and powerful economic and political unit, often called its 'breadbasket'. Replacement works for the Punjab especially have been one of the largest tasks, using waters from the western Indus streams. And it is for these that snow and ice are the major sources and, for the Indus main stem, glacier waters (Table 13.1). Thus, an important consequence of the Treaty is to replace waters derived largely from rainfall and secondarily from snowmelt, with largely snowmelt and glacier-fed rivers. As already shown, the latter have distinctive seasonal rhythms, higher elevation source areas and differing reliability and climate relations, compared to rain-fed streams. Historically, they have had much less monitoring and forecasting experience compared to the monsoonal rivers.

For farmers and public utilities, the timing and seasonal demand can be more crucial than total volumes of water. Most agricultural, municipal and industrial uses are locked into relatively inflexible timetables through the year. This partly explains the massive increase in groundwater extraction in recent decades, notably by tube wells in the Punjab. Well water is more readily matched with demand than meltwaters. As it happens, however, this has also become the most overstressed resource in the whole basin.

The scope, history and direction of water development exert an influence on water resources (Meadows and Meadows 1999). Existing uses guide and constrain future ones. The required changes in mindsets and practices from a once-prevailing dependence on monsoonal rains and well water to snow- and ice-fed streams put unusual pressures on managers and public expectations. Demands are in a state of flux if influenced much more by factors other than climate change or mere glaciers.

The situation at the start of the new millennium is that, according to a World Bank (2012) study:

> Pakistan has harnessed the Indus River to bring 35.7 million acres under irrigation… land in otherwise desert conditions…. Today, the country has the world's largest irrigation system, one of great technical, institutional and social complexity… [and it] accounts for 1/4 of the country's GDP, 2/3 of employment and about 80 % of exports.

The total length of canals alone is almost 570,000 km. Yet nothing in or behind these figures is static (World Bank 2005). Agricultural employment involves 60–70 % of the work force but has been dropping steadily and changing in character, while cities expand more rapidly than rural populations. In the countryside, major changes affect the size and distribution of incomes and the status of labour, with differing impacts on landless and landowning groups, big and small landowners. There is increasing mechanisation and expansion of commercial and export crops over subsistence agriculture (ibid, p. 23). There are issues of increasing migrant labour throughout the countryside and including the northern mountains. Most go to mushrooming cities like Karachi, a megacity with an estimated 21 million population, or Lahore, the second largest with almost 6.5 million. All such changes alter patterns and pressures of water demand.

Despite, or perhaps because of, the billions spent on water development, reports suggest that per capita water *availability* today is only 1/5th of what it was in 1950 and falling (ibid, p. ix). To date, little of this reduction can be attributed to climate change, although that is likely to become an ever-larger, more threatening factor. Equally alarming is that fresh water reaching the lower Indus in Sindh Province has declined to barely 6 % of what it was in the late nineteenth century. In some years it is close to zero, almost entirely due to engineering works for extraction and diversion of flows (IUCN n.d.). For some decades this has had extremely adverse effects on water supply and the ecology of the lower Indus riparian lands, on the fishermen, farmers and towns of Sindh Province (Hewitt 1977).

Meanwhile, since completion of Tarbela Dam in 1974, the largest project to date on the main Indus, 'losses' of potentially useable water in the systems downstream have almost doubled. Most are due to artificial conditions including seepage from canals, evaporation from surface waters or from high water tables and waterlogged land, and due to water theft. They are equal to more than one-quarter of the storage capacity of hydraulic works. As noted, groundwater extraction, mainly through tube wells of which there are now almost 650,000, has been one of the fastest growing developments. They were funded in part by efforts to control and reduce waterlogging as water tables rose with canal seepage and increased irrigation. Well water now accounts for about 60 % of farm use in the Punjab, probably at, if not above, sustainable capacity (ibid, p. 15).

There is widespread evidence of deteriorating water quality in streams, groundwater and canals. Some waters are increasingly saline, and salt continues to build up in places, although salinisation is no longer perceived as the great threat it once was (FAO 2012). Agricultural wastes including pesticides and fertiliser are widespread threats, but pollution from cities with inadequate sewage treatment and unmanaged industrial wastes is the most threatening and affects the more heavily populated areas (World Bank 2005).

These remarks offer no more than a glimpse of the issues and complexities that arise for water development generally. Harnessing of waters from the mountains, which must be the focus here, raises its own unique challenges.

## 13.6 Significance of Snow and Ice Contributions

For the whole Indus system, snowfall is a much larger source of runoff than the glaciers. Rainfall–runoff is much larger again. On the other hand, glaciers are possibly the largest untapped source of fresh water, 'untapped' mainly because of timing, most glacier yields coinciding with the summer monsoon. Historically, it has meant glacier waters, or their equivalent in down-country flows, largely went to the sea, but with an unknown, surely important, part going to groundwater storage. With Tarbela Dam, storing late summer waters for winter use, a larger contribution from the glaciers has been realised, to some extent from the barrages lower down the Indus main stem. Even so, the outflows to the Arabian Sea mainly occur during peak glacier yields. In most years they may well be larger than the effective glacier contribution, although no one has yet managed to separate out the actual contributions week-by-week or year-by-year.

As already shown, the most substantial glacier contributions come from sub-basins with watersheds rising above 6,500 m and singling out the Karakoram where many rise above 7,000 m. Basins with a large snowmelt contribution have substantial areas above 3,500 m but not the highest mountains. Summer monsoonal rains dominate flows from areas below about 3,500 m, overwhelmingly so in the Front Ranges of the southern Himalayan flank and in the considerable areas of rain-fed agriculture in the Plains. Winter and spring rains are important because of timing more than the total contributions and, of course, coincide with snowfall and snowpack storage in the mountains. Monsoon rainfall is probably the most important source of recharge of the groundwater aquifer, followed by seepage from the irrigation system. It would be helpful to know how much snow and glacier meltwaters contribute, or could contribute, given the widespread speculation about how climate and cryosphere change will affect the aquifer, but this remains uncertain.

The Karakoram and other main ranges have the larger or only glacierised areas and glacial meltwaters because of high elevation watersheds, However, as also shown, the actual ice surface and meltwater sources are largely in a lower elevation band between 3,800 and 5,000 m (Chap. 2). What happens in these elevations is critical to annual water yields, but caution is needed over a tendency to treat all areas between 4,000 and 6,000 m as the same, which sometimes happens in general hydrological climate models (Rees and Collins 2006; Armstrong 2011). Yes, 80 % of the glacier cover occurs here. However, *there are no glaciers* in *90 %* of the upper Indus Basin at this elevation range. Far more of the trans-Himalayan basin consists of high-arid areas towards the Tibetan Plateau and drier, lesser ranges. Sub-basins draining to the Indus from the north flank of the Greater Himalaya are forested (or deforested) and with small or no glaciers. Each of these has a very different hydrological regime from the glacierised basins (Fowler and Archer 2006). Each responds differently to climate change.

For the main stem Indus and flows to Tarbela Dam, my own work suggests glacier basins may provide as much as 40 % of average annual flows, mainly from the Karakoram. The snowmelt contribution coming from the Karakoram and much vaster areas as well is normally larger, probably around 50 %. Rainfall makes up the balance of about 10 %, mainly from the Northwest Himalayan watersheds. There are large variations in amounts and relative contributions from year to year. For regional water resources, more critical concerns involve timing in relation to demands outlined above, rather than total amounts, and extreme high or low events bringing major floods or drought.

It has been shown how area–altitude relations are critical for peak yields from glaciers and their timing. At their lower elevations heavy debris covers suppress ablation throughout the year, adding to the importance of what happens in mid- to upper ablation zone areas. In these, for ice ablation to start, seasonal snowfalls and the cold season freeze–thaw carapace must first be removed to expose glacier ice with thin or no debris covers. There are few data, but experience at Biafo suggests these are not trivial aspects, especially in year-to-year variations and responses to climate change.

The different summer and winter climatic regimes in the mountains introduce large seasonal variations and the way they combine with high mountain influences (Chap. 1). The 'sub-Mediterranean' regime of the upper Indus Basin is mainly associated with winter precipitation and summer drought. The 'sub-Monsoonal' regime brings winter drought and summer precipitation. Relating these to water supply is not quite the same as to glacier mass balance. The melting of winter snowfall is important in annual runoff, whereas summer snowfall itself mainly goes to the glaciers, while there is usually reduced stream flow and little or no precipitation in the more arid valleys. Ablation yields are strongly affected by the strength of the monsoon and summer storminess. A strong monsoon or strong winter Westerlies can complement each other. Heavy winter precipitation (snow cover) tends to reduce ablation, while low winter snowfall encourages more ablation, which compensates for less snowmelt. These also prefigure the worst case conditions, the rarer instances of strong winter Westerlies *and* strong summer monsoon which can result either in 'glacial flood' conditions if melting is proportionate to precipitation or in 'glacial drought' if snow stays on the ground and goes mainly into glacier storage. Similarly if winter snowfall is low but unusual conditions severely limit summer ablation, another rare case, then 'glacial drought' results. An increase in such occurrences is one possible outcome of global climate change, greatly complicating how to respond. In many ways these are the 'nightmare scenarios' for water planning, more so than 'disappearing glaciers'. They do create further uncertainties about what the latter would mean if and when it happens.

Large, unresolved questions concern different components of water yields from glacier basins. A common misconception is that it is all and only about ice ablation. This is the primary or only interest of glacier mass balance studies and the key to strictly glacier water yields. However, even if the UIB gauging stations were close to or at glacier termini, it would still be necessary to identify the range of sources, including:

(i) Ice ablation
(ii) Melting of seasonal snowfall on the glacier
(iii) Melting of seasonal snowfall in ice-free basin areas
(iv) Melting of the seasonal freeze–thaw 'carapace' that develops on these subtropical glacier surfaces (Chap. 8)

(v) Rainfall within the basin, usually tiny but not always so and likely to grow with warming
(vi) Rock glaciers, dead ice and (degrading?) permafrost within glacier basins

Contributions from Karakoram glacier basins probably follow the order of the list, but with differing responses to changing conditions. Ice ablation tends to be much the larger factor from heavily glacierised areas and basins. However, such measures as could be found suggest the other sources may involve as much as 25 % of average yields from glacier basins, with unknown but possibly large differences from year to year (Chap. 6). Away from glacier basins and as more and more of the trans-Himalayan is considered, then, progressively, seasonal snowmelt overhauls average ice ablation. This leads to an overall picture in which the amounts of components other than ice ablation, and their fluctuations, equal or exceed estimates of what climate and glacier changes will involve, say, over several decades at least. Without being able to disentangle the various components and their variability within total yields, it is difficult to say what part of water supply is changing in response to climate.

In discharge hydrographs from glaciers, it is seasonal conditions in ablation zones, notably high flows, which stand out and seem most likely to decide how climate change plays out. Nevertheless, they also vary with:

(a) Length of ablation season, which varies vertically with the migration of above-zero temperatures
(b) Ablation season weather and dustiness
(c) Area of glacier ice exposed to ablation which increases with elevation and, for how long, which tends to decline with elevation, according to seasonal snowfall and the extent of the freeze–thaw 'carapace'
(d) The distribution of debris mantles and the spectrum of thick versus thin covers

In heavily glacierised areas ice exposed to ablation is a decisive factor. In this sense, it is the 'golden egg' of glacier hydrology and explains why 80–90 % of water yields from glacier basins occur in two to three months of the year. However, as shown earlier, how the 'egg' is pealed, so to speak, brings many variations, something in which the other factors are critical. Unlike rainfall–runoff or snowfall–snowmelt relations, the roles of annual precipitation and temperatures are buffered by intervening conditions in glacier basins and can even seem unrelated. Each ablation season unfolds according to how the various factors work together.

In the upper Indus system, Archer (2003) demonstrates differing runoff regimes related, in part at least, to the relative importance of glaciers, snowfall and rainfall. Based on weather station and river discharge data, he showed that flows of the Shyok, Shigar and Hunza Basins are poorly correlated with winter precipitation, but strongly related to seasonal temperatures. These rivers drain the high Karakoram and yields from glacier basins are large. By contrast, the Gilgit, Astor and Kunhar Rivers, and the Ladak Indus above the Shyok confluence, have runoff well correlated with winter precipitation. Their mountains are generally not as high as the main Karakoram, and they have extensive winter snowfall and much smaller glacier covers. It suggests snowmelt contributions predominate.

Conditions on small or minor glaciers and areas where they prevail are likely to be different again. Seasonal snowfall, directly or as redistributed by wind and avalanches, probably plays a much greater role. With relatively lower elevations, snowfall and rainfall assume a much greater significance in water supply and climate responses, their hydrology differing accordingly. These are also basins with high concentrations of rock glaciers (Chap. 11).

## 13.7 Glacier Geography, National and Trans-boundary Issues

Most waters from glaciers throughout the Hindukush–Karakoram–Himalaya (HKH) originate within the boundaries of one country but flow into and across one or more others. The larger populations and extent of dependence on glacier waters are in surrounding lowlands, not necessarily in the same country as the glacier headwaters. This applies to the three great Himalayan rivers, the Indus, Ganga and Tsangpo-Brahmaputra. In each, watershed areas where the glaciers are embrace several countries. The rivers draining them cross several countries. The Indus glaciers and meltwaters originate within the boundaries of *four* countries and drain to the lowlands of *three*.

The largest quantities and relative shares of glacier meltwater involve the main stems of the Indus and Tsangpo-Brahmaputra. Both derive mainly from *trans-Himalayan* glacier systems. In the Indus system, the larger fraction of glaciers are drained through the NW Himalayan or Nanga Parbat 'Syntaxis'. The Tsangpo-Brahmaputra drains through the East Himalayan or Namche Barwa 'Syntaxis'. These two comprise some of the most geologically active, steep and high relief terrain on Earth, nonetheless being closely scrutinised for further large hydroelectric projects.

The main glacier areas and mass in the upper Indus Basin are in the Karakoram Mountains, most of these controlled by Pakistan but disputed and claimed by India. Other significant glacier areas are under Indian control, mainly in the eastern Karakoram, Ladakh and Zanskar Ranges, including territories disputed by Pakistan. China controls a small area of the Aksai Chin with some glaciers also disputed by India. Finally, meltwaters from the Hindu Kush and Hindu Raj glaciers draining to the Kabul River start in the northwestern Pakistan, flow through Afghanistan, then return to Pakistan through the densely populated and strategically located Peshawar basin.

## 13.8 Conflict or Cooperation?

Superlatives have seemed warranted by the Karakoram and its glaciers on various levels, from the dimensions of the mountains and the ice cover to the numbers of people who depend on its glacier-fed rivers. None of these are more compelling than the past extent and potential for future armed conflict. The mountain range is invested by tens of thousands of troops and divided among Pakistan, India and China who have fought a number of wars in the region. There is an ongoing conflict on glaciers themselves, the 'Siachen War', as well as various uneasy stand-offs along the Cease Fire Line. The latter was brokered by the United Nations following the first war between Pakistan and India in 1947–1948. The UN 'blue hats' have had one of their longest international duties policing the Line of Control. Even so, further wars broke out in 1971 and 1999 and countless skirmishes since Indian forces occupied Siachen Glacier in 1984. Pakistan followed suit, occupying ice immediately to the west, mainly on the Saltoro, Kondus and Baltoro Glaciers. Upwards of 250 military camps and bases are reported in the Karakoram, related to the Siachen conflict alone, many on or beside the glaciers (Noorilhudah 2012). Estimates vary enormously but seem not less than USD $300 million per year for India and Pakistan close to $200 million – $0.5 billion in round figures. Possibly it is as high as a $1billion, for a conflict with, fortunately, almost no fighting, which would be much more costly. Deaths, however, are on the order of 8,000, more than 80 % said to be from accidents, avalanches, altitude sickness, frostbite and other non-warfare problems (Noorilhudah 2012, pp. 24–25). One of the worst events was an April 7, 2012, avalanche and landslide that destroyed a Pakistan army camp at Gayari, just below Saltoro glacier. It killed 129 soldiers and 11 civilians.

Turning to the western Karakoram, its people have been constantly affected by the wars in Afghanistan, valleys closed by occupying forces, the flows of arms, fighters and refugees they have generated, many using glacier passes. Similarly, the Kashmir dispute has flared up into repeated armed actions. There has been spillover from unrest in Tibet and Xinjiang Province and other Inner Asian conflicts. A large fraction of deaths in all these events are of soldiers recruited from the mountain areas, irreplaceable losses of able-bodied men and fathers for small communities, apart from the grief caused. This is hardly different from the rest of the Greater Himalayan Region. Few places are unaffected by war, simmering unrest, states of emergency, disenfranchisement of their people and developments enforced by armed intervention. A major problem is that the main responsible decision makers come from outside the mountains.

In any event, few parts have been more affected by conflict and violence than the Greater Karakoram Region and none where the glaciers are as involved. Nor should it be implied these are only recent or post-Independence developments. The Karakoram was situated at the core of what Kipling called 'The Great Game' of the nineteenth century empire, between Britain, Russia and China (Hopkirk 1992). The notion continues to resurface in the supposedly post-imperial world (Karrar 2009; Walberg 2011).

This is not the place to go into the nature or history of these conflicts or to assess the merits or defects of various claims and actions. One must underscore how massively they affect people of the region and pose ongoing threats to the Karakoram environment, the possibilities for sound and safe development and harmony between the mountains and lowlands. A small but crucial part is how often glacier research and monitoring have been frustrated or prevented. Some reasons for hope or sane dialogue may follow from how the 'Line of Control' has remained largely intact and respected by India and Pakistan for so many decades.

One could emphasise how, elsewhere in the world, glaciers or scientific research and cooperation involving them provide one of the few triumphs of diplomacy and restraint over conflict, specifically in the Arctic and Antarctica. There are major differences from High Asia, not least in the numbers of inhabitants, the political actors involved and differing roles of Cold War and nuclear deterrence. Nevertheless, the ongoing international cooperation in polar regions and favourable outcomes for scientific discovery there seem worth emphasising rather than sabre-rattling.

## 13.9 Concluding Remarks

The larger and most pressing issues relating to the Karakoram glaciers no doubt concern water supplies in the Indus and Yarkand lowlands. They involve much the greater numbers of people, by far the most productive economic sectors, politically influential centres and responsible institutions. Conversely, treating the high mountains and glaciers merely as 'water towers' and focusing only on the quickest ways of tapping them for down-country needs seem likely to endanger both. The conditions that govern the availability and variability of water yields have to be respected, and how these relate to the health and changing state of the glaciers. There is a special responsibility of states and the international community for the well-being of people who live in the mountains.

One can only conclude by stressing the need for attention to the mountain environments and peoples as the context for assessing the glaciers and responding to changes affecting them. They involve many developments unrelated to the cryosphere or global climate change. Nevertheless, the various problems are converging to create the 'perfect crisis', one which has been magnified by unfounded assumptions about, and lack of established knowledge of, the state and future of the Karakoram glaciers.

## References

Afridi BG (1988) Baltistan: a history. Emjay Books, Peshawar

Archer DR (2003) Contrasting hydrological regimes in the upper Indus Basin. J Hydrol 274:198–210

Armstrong RL (2011) The glaciers of the Hindu Kush-Himalayan region: a summary of the science regarding glacier melt/retreat in the Himalayan, Hindu Kush, Karakoram, Pamir, and Tien Shan mountain ranges. ICIMOD, Kathmandu

Azhar-Hewitt F (1989) Women's work, women's place: the gendered life-world of a high mountain community in northern Pakistan. Mt Res Dev 9(4):335–352

Azhar-Hewitt F (2011) The other side of silence: the lives of women in the Karakoram Mountains. IUniverse, Bloomington

Azhar-Hewitt F, Hewitt K (2012) Technocractic approaches and community contexts; viewpoints of those most at risk from environmental disasters in mountain areas, northern Pakistan. In: Gaillard JC, Kelman I (eds) Climate change modeling for local adaptation in the Hindu Kush-Himalayan Region. Community, environment and disaster risk management review, pp 57–77

Butz DAO (1989) The agricultural use of meltwater in Hopar settlement, Pakistan. Ann Glaciol 13:35–39

Butz D (2006) Tourism and portering: Labour relations in Shimshal, Gojal Hunza. In: Kreutzmann H (ed) Karakoram in transition: culture, development and ecology in the Hunza Valley. Oxford University Press, Karachi, pp 394–403

Clemens J, Nuesser M (2008) Animal husbandry and utilization of alpine pastures in the Nanga Parbat region of Northern Pakistan: Comparison of Raikot and Rupal valleys. In: Israr-ud-Din (ed) Proceedings of the third international Hindu Kush cultural conference, Karachi. Oxford University Press, Oxford, pp 71–81

Cruikshank J (2005) Do glaciers listen? Local knowledge, colonial encounters, and social imagination. University of British Columbia Press, Vancouver

Dani AH (1989) History of the Northern Areas of Pakistan. National Institute of Cultural and Historical Research, Islamabad

Debarbieux B, Gillet F (2000) Mountain regions: a research subject. In: Proceedings of the international mountain research workshop, June, Autrams, University of Geneva, Switzerland

Deckens J (2007a) 'The snake and the river don't run straight': local knowledge on disaster preparedness in the Eastern Terai of Nepal. ICIMOD, Kathmandu

Deckens J (2007b) Local knowledge for disaster preparedness: a literature review. ICIMOD, Kathmandu

Derbyshire E, Fort M (2006) Geomorphology and mountain hazards of the Hunza Valley. In: Kreutzmann H (ed) Karakoram in transition: culture, development and ecology in the Hunza Valley. Oxford University Press, Karachi, pp 73–95

Ehlers E, Kreutzmann H (eds) (2000) High mountain pastoralism in northern Pakistan. Erdkundkisches Wissen, 132, Bonn

Emerson P (1986) An inhabited wilderness. In: Tobias M (ed) Mountain people. University of Oklahoma Press, Norman, pp 88–96

Fowler HJ, Archer DR (2006) Conflicting signals of climate change in the Upper Indus Basin. J Clim 19(17):4276–4429

Hashmatullah K (1987) History of Baltistan, 1939, Lok Virsa translation, Islamabad

Hedin S (1917–1922) Southern tibet: discoveries in former times compared with my own researches in 1906–1908, vol. 7, History and exploration in the Karakoram Mountains. Lithographic Institute, Swedish Army, Stockholm

Hewitt K (1977) Desertification, development and the "Admirals" of Manchar lake, Pakistan. Econ Geogr 53(6):358–363

Hewitt K (1988) Catastrophic landslide deposits in the Karakoram Himalaya. Science 242:64–67

Hewitt K (2001) Catastrophic rockslides and the geomorphology of the Hunza and Gilgit Basins, Karakoram Himalaya. Erdk 55:72–94

Hewitt K (2004) Geomorphic hazards in mountain environments. In: Owens P, Slaymaker O (eds) Mountain geomorphology. Hodder Scientific, London, pp 187–218, Chap. 9

Hewitt K (2005) The Karakoram anomaly? Glacier expansion and the "elevation effect", Karakoram Himalaya. Mt Res Dev 25:332–340

Hewitt K (2007) Rediscovering colonized landscapes: the first Europeans at the Mustagh Pass, Karakoram Himalaya, Inner Asia. In: Gervers M, Bulag U, Long G (eds) The exploitation of the landscape of Central and Inner Asia, Toronto Studies in Central and Inner Asia, 9. Asian Institute, University of Toronto, pp 41–67

Hewitt K, Clague JJ, Orwin J (2008) Legacies of catastrophic rock slope failures in mountain landscapes. Earth Sci Rev 87:1–38

Hewitt K (2009) Rock avalanches that travel onto glaciers: disturbance regime landscapes, Karakoram Himalaya, Inner Asia. Geomorphology 103:66–79

Hewitt K (2010) Gifts and perils of landslides: catastrophic rockslides and related landscape developments are an integral part of human settlement along upper Indus streams. Am Sci 48:410–419

Hewitt K, Mehta M (2012) Rethinking risk and disasters in mountain areas. Rev geogra alp, Special issue 100–1, Montagne, marginalite et catastrophes, http//rga.revues.org/1633
Hopkirk P (1992) The great game: the struggle for empire in central Asia. Kodansha International, New York
Hughes R (1984) Yasin valley: the analysis of geomorphology and building types. In: Miller KJ (ed) International Karakoram Project, 2. Cambrdige University Press, London, pp 253–288
IUCN (n.d.) The lower Indus River: balancing development and maintenance of wetland ecosystems and dependent livelihood. Water Nat Initiative
Ives JD (1997) Comparative inequalities: mountain communities and mountain families. In: Messerli B, Ives JD (eds) Mountains of the world: a global priority. Parthenon Publishing, New York, pp 61–84
Janjua ZJ (1998) Tradition and change in the Darel and Tangir Valleys. In: Stellrecht X (ed) Karakorum-Hindu Kush-Himalaya: dynamics of change, vol 4/1 Rudiger Koppe Verlag, Koln, pp 415–427
Jianchu X, Ericson M, Hewitt K, Shrestha A, Vaidya R (2007) The melting Himalayas: Regional challenges and local impacts of climates change on mountain Ecosystems and livelihoods. Technical Paper X, ICIMOD (International Centre for Integrated Mountain Development), Kathmandu, Nepal
Karrar HH (2009) The new silk road diplomacy: China's central Asian foreign policy since the Cold War. University of British Columbia Press, Vancouver
Kick W (1989) The decline of the last Little Ice Age in High Asia compared with that in the Alps. In: Oerlemans J (ed) Glacier fluctuations and climate change. Kluwer, Dordrecht, pp 129–142
Kreutzmann H (ed) (2006) Karakoram in transition: culture, development and ecology in the Hunza Valley. Oxford University Press, Oxford
Kreutzmann H (ed) (2012) Pastoral practices in High Asia. Agency of 'development' affected by modernisation, resettlement and transformation. Springer, Dordrecht
Ladurie EL (1971) Times of feast and famine: a history of climate since the year 1000. Doubleday, New York
MacDonald K, Butz D (1998) Investigating portering relations as a focus for transcultural interaction in the Karakoram region of northern Pakistan. Mt Res Dev 18:333–343
Meadows A, Meadows PS (eds) (1999) The Indus River: biodiversity, resources, humankind. Oxford University Press, Oxford
Meiners S (1998) Preliminary results concerning historic to Post-glacial stages in the NW-Karakorum (Hispar Muztagh, Batura Muztagh, Rakaposhi Range). In: Stellrecht I (ed), vol 4/1. pp 49–70
Michel A (1967) The Indus Rivers: a study of the effects of Partition. Yale University Press, New Haven
Miehe S, Miehe G (1998) Vegetation patterns as indicators of climatic humidity in the western Karakoram. In: Stellrecht I (ed), vol 4/1. pp 101–126
Neelis J (2006) Hunza-Haldeikish revisited: epigraphical evidence for trans-regional history. In: Kreutzmann H (ed) Karakoram in transition: culture, development and ecology in the Hunza Valley. Oxford University Press, Karachi, pp 159–170
Noorilhudah (2012) Cover story: into the valley of death. Newsline, Karachi, May 18–23. http://www.newslinemagazine.com/2012/05/into-the-valley-of-death-the-siachen-conflict
Nüsser, M. (1998): Nanga Parbat (NW-Himalaya): Naturräumliche Ressourcenausstattung und humanökologische Gefügemuster der Landnutzung. Bonner Geogr. Abh. 97.
Orlove B, Wiegandt E, Luckman BH (eds) (2008) Darkening peaks: Glacier retreat, science and society. University of California Press, Berkeley
Price M (ed) (2006) Global change in mountain regions. Sapiens Publishing, Duncow
Rees HG, Collins DN (2006) Regional differences in response of flow in glacier-fed Himalayan rivers to climatic warming. Hydrol Process 20:2157–2169
Said M (1998) Natural hazards of Shigar Valley, Northern Areas of Pakistan. In: Stellrecht I (ed), vol 4/1. pp 251–268

Saunders F (1983) Karakoram villages: an agrarian study of 22 villages in the Hunza, Ishkoman and Yasin Valleys of Gilgit District. UN Food and Agricultural Organization, Pakistan

Smith and Petley (2009) Environmental hazards: assessing risk and reducing disaster, 5th edn. Routledge, London

Stellrecht I (ed) (1998) Karakorum-Hindu Kush-Himalaya: dynamics of change. Rudiger Koppe Verlag, Koln, pp 319–337

Tobias M (ed) (1986) Mountain people. University of Oklahoma Press, Norman

Viviroli D, Weingartner R (2006) Mountains as "water towers" for humanity: state of knowledge and significance in a changing world. In: Price M (ed) Global change in mountain regions. Sapiens Publishing, Duncow, pp 16–18

Walberg E (2011) Postmodern imperialism: geopolitics and the great games. Clarity Press, Atlanta

Whiteman FTS (1985) Mountain Oases: a technical report of agricultural studies in the Hunza, Ishkoman and Yasin valleys of Gilgit District. UN Food and Agricultural Organization, Integrated rural development/Department of Agriculture, PAK/80/009, Gilgit

Wirsing RG, Jasparro C (2006) Spotlight on Indus River diplomacy: India, Pakistan, and the Baglihar Dam dispute. Asia-Pacific Center for Security Studies, http://www.apcss.org/Publications/APSSS/IndusRiverDiplomacy.pdf

World Bank (2005) Country water resources assistance strategy: water economy: running dry. Report No. 34081-PK, Agriculture and Rural Development Unit, South Asia region, Pakistan

Yang HA, An RZ (1991) Yarkand River Basin (interior drainage area of Tarim Basin), Glacier Inventory of China, vol 3. Science Press, Beijing

# Index



K. Hewitt, *Glaciers of the Karakoram Himalaya: Glacial Environments, Processes, Hazards and Resources*, Advances in Asian Human-Environmental Research,
DOI 10.1007/978-94-007-6311-1, © Springer Science+Business Media Dordrecht 2014

MIX
Papier aus verantwortungsvollen Quellen
Paper from responsible sources
FSC® C105338

If you have any concerns about our products,
you can contact us on
**ProductSafety@springernature.com**

In case Publisher is established outside the EU,
the EU authorized representative is:
**Springer Nature Customer Service Center GmbH**
**Europaplatz 3, 69115 Heidelberg, Germany**

Printed by Libri Plureos GmbH
in Hamburg, Germany